SOTHIS PRESS

Sothis Press aims to completely reimagine the public domain book with an emphasis on universal access to captivatingly designed print editions. We believe that technology will enable every single book ever published to be perpetually accessible in beautiful print form. Sothis is beauty in ubiquity.

www.sothis.press
1 Waterhouse Square, London, ECIN 2ST

Hydraulic Elevators

Their Design, Construction, Operation, Care and Management

William Baxter

SOTHIS PRESS

ISBN 978-1-0368-0184-7

This edition is published by
Sothis Press © 2023. The company
is registered in England and Wales.
The Company Number is 14900389.

HYDRAULIC
ELEVATORS

THEIR DESIGN, CONSTRUCTION, OPERATION
CARE AND MANAGEMENT

BY

WILLIAM BAXTER, JR.

Author of "Practical Talks on Electricity,"
"Switchboards," Etc.

1910

McGRAW-HILL BOOK COMPANY

239 WEST 39th STREET, NEW YORK

6 BOUVERIE STREET, LONDON, E. C.

CONTENTS

CHAPTER		PAGE
I.	Fundamental Principles of Different Types	1
II.	Counterbalancing—The High-Pressure System	11
III.	Plunger Hydraulic Elevators—Plunger Elevators	18
IV.	Horizontal Cylinder Types—"Pushing" and "Pulling" Types	21
V.	Counterbalancing the Lifting Ropes of Elevator Cars	25
VI.	General Arrangement of Hydraulic Elevator Systems	30
VII.	Vertical Hydraulic Elevators—Simple Low-Pressure Vertical Type Otis Elevator with Hand-Rope Control	33
VIII.	Low-Pressure Vertical Elevators, Lever Control	45
IX.	Pilot-Valve Control—Regulating Valves—Speed Regulating Valves	52
X.	Safety Devices	60
XI.	Good Features of Magnetic Valve Control; Obstacles that Prevent its General Use—Descriptions of Control Systems	66
XII.	Magnetic Controller for Battery Current	70
XIII.	Double-Power Hydraulic Elevator System	79
XIV.	Otis, Running-Rope System	85
XV.	Principal Causes of Disordered Mechanism and How Prevented or Removed—Directions for Packing Pistons or Valves	89
XVI.	Automatic Devices Used for Stopping Cars at Top and Bottom Landings; Their Care and Their Value as Safety Appliances	100
XVII.	Effect of Stretching of the Ropes	106
XVIII.	Construction of Traveling-Sheave Frames, Foundations and Supports, to Avoid Piston and Cylinder Friction	112
XIX.	Why the Piston is Weighted	120
XX.	The Electrical Features of Vertical Cylinder Elevators, with Magnet Control; Operation and Care of Pilot Valves and Connecting Mechanisms	123
XXI.	Horizontal Hydraulic Elevators	131
XXII.	Description of the "Pulling" Type of Horizontal Elevator, Showing the Operating Principle of the Whittier Machine	142
XXIII.	The Marse-Williams; Construction and Operation of the Valves	147
XXIV.	Crane Horizontal "Pushing" Machine: How the Stop-Motion Gets Out of Adjustment; the Cylinders and Other Parts	155
XXV.	Crane Horizontal Machine; Description of Automatic Stop Valve; Packing the Main Piston and Cylinders	159
XXVI.	Construction Details of the Whittier Machine with Special Reference to the Stop-Valve Mechanism	165
XXVIII.	Overhead Sheaves and Bearings; Kind of Lubricant Best Suited for Them; Discharge Pipe Relief Valves; the Use of Strainers; When Required	171
XXVIII.	Discussion Relative to the Details of the Morse and Williams "Pulling" Machines	176

Contents

Chapter		Page
XXIX.	High-Pressure Hydraulic Elevators	187
XXX.	Operation of the Main and Pilot Valves of the Otis Vertical Elevator—the Electrical Control Sometimes Employed	198
XXXI.	Construction and Operation of the Accumulators and the Automatic Valves Used With Them	205
XXXII.	Automatic Stop-Valves With Otis High-Pressure Machines	212
XXXIII.	Adjustment and Care of Automatic Stop-Valves and Mechanism of High-Pressure Elevators; How to Pack the Different Parts; Kinds of Packing Used	221
XXXIV.	Stop-Valves Used With Accumulators; General Arrangement of Apparatus in High-Pressure Systems; the Pumps and Their Operation	228
XXXV.	Plunger Elevators	236
XXXVI.	Operation of Main and Pilot Control Valves; Comparison of Different Control and Automatic Stop-Valves	254
XXXVII.	Rack-and-Pinion Valves for Otis Freight Elevators; Type of Valve Used for High Speed; Notes on the Care of the Plunger and Ropes	263
XXXVIII.	Replacing Worn Shoes; How to Disconnect the Plunger; Care Observed in Handling Parts	272
XXXIX.	The "Standard" Plunger Elevator; Principal Differences Between This and the Otis Types; Valve, Construction and Operation	279
XL.	Operation of the Pilot and Main Valves; How the Adjustments are Made	286
XLI.	Construction and Operation; Details of the Highest Type of Passenger Elevator Made by the Standard Plunger Elevator Company	300
XLII.	Practical Instructions in the Care and Management of the "Standard" Plunger Elevator; Essential Features to Look Out For	307
XLIII.	Hand Rope Control for Freight Elevators; Pump and Connections Used with "Safe Lifters"; Locking Device for Plunger Elevators; Hand-Rope Control	313

CHAPTER I

FUNDAMENTAL PRINCIPLES OF DIFFERENT TYPES

The principal mechanism of a hydraulic elevator is very simple, although it does not appear to be so from casual inspection of a first-class installation in a large office building. This is due mainly to the fact that the apparatus is all of large dimensions and so disposed that only a portion of it can be seen from any one point. It is also due to the fact that a number of small parts are added, the function of which is to contribute to the perfection and safety of operation.

The principle of operation of the hydraulic form of elevator is as simple as the construction of the mechanism, and consists in utilizing the force developed by the pressure of water in a hydraulic cylinder to lift a weight, which generally is the elevator car, but in some cases is not. The water under pressure that supplies the elevator cylinder may be derived from street mains, from an open tank placed on an elevation (for example, the roof of the building), or from a pressure tank placed in any convenient location.

The simplest form of hydraulic elevator is illustrated diagrammatically in Fig. 1. The car is suspended from one or more ropes that pass over a sheave *A*, located at the top of the elevator well, and run down to the upper end of a piston rod *R*. This rod carries at its lower end a piston *P* that fits watertight in the lifting cylinder *C*. Water under pressure is admitted to the cylinder above the piston, through the pipe *I*, and the piston is thereby forced downward, hoisting the car. To make the apparatus complete, elementally, all that is required is a valve in the pipe *I* which, when turned in one direction, will connect the cylinder with the pressure tank, or other source of supply, and when turned in the other direction will stop off the supply from the pressure tank and connect the upper end of the cylinder with a discharge tank. When moved to the central position, it must close the outlet from the cylinder so that the water contained therein cannot escape.

It is evident that when the water is permitted to flow out of the upper end of the cylinder, if the elevator car weighs more than the piston P, the piston will run up and the car will run down. If, when the piston is in any position in the cylinder, from the uppermost to the lowest, the flow of water out of the cylinder is stopped, the motion of the piston will be arrested and the car will be stopped.

The load that can be lifted when the piston *P* is forced down will depend upon the pressure of the water; the higher the pressure the greater the permissible load. If it is desired that the elevator be able to lift loads varying anywhere between 100 and 2000 pounds, then the pressure of the water admitted to the cylinder must be sufficient to lift the maximum load. Whether the load lifted is the maximum or the minimum, the same amount of water will be required to lift the car

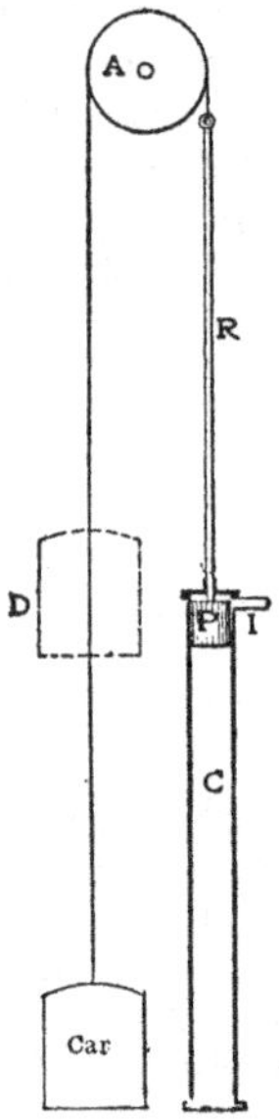

FIG. 1

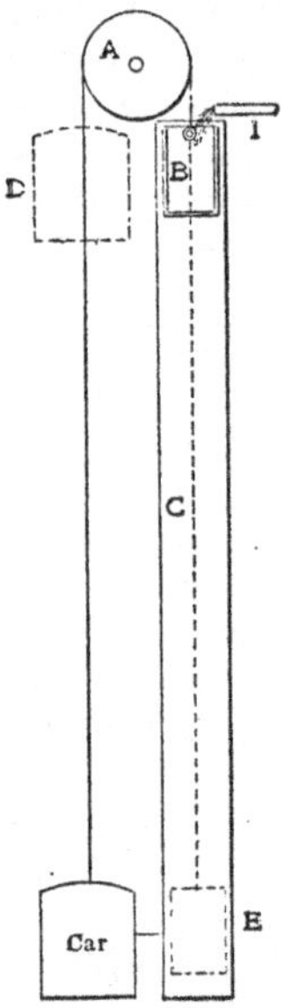

FIG. 2

from the lower position to the upper position *D*, hence the same amount of power will be required if the pressure of the water remains the same. In practice elevators are supplied with water at constant pressure, in nearly every type used; and in these, as shown in the foregoing, the energy absorbed to make a trip with the empty car is equal to that required to lift the maximum load. This is an inherent defect in hydraulic elevators that many inventors have tried to overcome, but no one has succeeded in doing so in a satisfactory manner up to the present time. There is a system in use which partially accomplishes the result, and others have been devised that wholly accomplish it, but these latter possess defects in other directions that render them undesirable.

The elevator represented in Fig. 1 is what is called a "one-to-one" geared machine, being so called from the fact that the car moves through the same distance that the piston does. This class is only susceptible of application in a few cases. It is not suited to the operation of elevators that run to the top of the building, as can be clearly seen from the diagram, the uppermost position, *D*, of the car being less than one-half the distance up to the overhead sheave *A*.

THE WATER-BALANCE ELEVATOR.

A simple way of arranging a one-to-one gear machine so that the elevator may run to the top of the building is illustrated in Fig. 2. This is what is known as a "water-balance" elevator, and is a type that was used quite extensively in some of the large western cities some years ago, and so far as I know may still be in use, but is no longer manufactured. In this elevator the car is suspended from wire ropes that pass over the overhead sheave *A* and carry at their other ends an iron bucket *B*. This bucket weighs less than the empty car, but is of such size that when filled with water it will overbalance the car and its maximum load. If the car is at the bottom of the building and the bucket at the top, as in the illustration, and the operator desires to run the car up, he pulls on a hand rope that opens a valve in the pipe *I* and allows water to flow into the bucket. When a sufficient amount of water has run into the bucket to overbalance the car and its load, the latter will begin to move upward, and then the supply of water is cut off. The velocity of the elevator car is controlled by a brake that grips the guides that the car runs on and is controlled by the operator in the car; by means of the same brake the car is stopped. When the elevator is at the top of the building and the bucket at the bottom, the operator to make the down trip pulls a rope that opens a valve in the bottom of the bucket and lets out water until the bucket becomes lighter than the car.

With this arrangement it is evident that the amount of water used for each trip of the car is not the same, but depends upon the load; the same is true for the down trip, hence the energy used is nearly proportional to the weight lifted and lowered. This statement is intended to apply to the total work done; that is, to an all-day run. It is not true, however, with reference to a single trip, either up or down. On any up trip the amount of water that must be let into the bucket depends not only upon the load, but also upon the amount of water remaining in the bucket from the previous trip. On the down trip the amount of water that must be let out of the bucket depends on the amount it contains as well as upon the load, and is small if the load is great, and large if the load is light, so that it is inversely proportional to the load instead of directly proportional. For an all-day run, however, the statement as made is substantially correct.

The velocity at which the elevator car can be run is very high, and if the operator is sufficiently expert the smoothness of motion is all that can be desired, and stops can be made without any perceptible jar; but if the brake fails to work, the car can shoot up to the top of the building or down the bottom with a velocity great enough to cause disaster.

One of these water-balance elevators was installed in the Western Union building at the corner of Broadway and Dey street, New York, about twenty-five years ago, but was removed some years thereafter. So far as I know it ran without any mishaps, but the element of danger involved is great, and on that account it never was looked upon with favor by eastern elevator experts.

TWO-TO-ONE ELEVATOR.

In order that an elevator may be run to the top of the building by what we may call the regular type of hydraulic elevator, it is necessary that the gear be greater than one-to-one; that is, the car must travel a greater distance than the piston in the lifting cylinder. Fig. 3 is a diagrammatic representation of an elevator geared two-to-one. In this arrangement the lifting ropes running up from the car pass over the overhead sheave *A* and then under a sheave *B* that is supported in a frame attached to the upper end of the piston rod. The ends of the lifting ropes are anchored to a strong support *G*. If the piston moves down through a distance of ten feet, the car will be drawn up through a distance of twenty feet, and when the car reaches the position *D*, the piston will be at the bottom of the cylinder.

This elevator can be made to operate in either of two ways, one by admitting water on top of the piston and the other by admitting water below the piston. If it is to be operated by admitting water on top of the piston, the latter must weigh somewhat less than twice the weight of the empty car, so that the latter may be able, without any load, to descend and draw up the piston. If the elevator is to be operated by admitting water under the piston, then the latter must weigh something more than twice as much as the car when fully loaded, so that it may be able to lift the maximum load. The deficiency in the first case and the excess of weight in the second must be enough to overcome all the friction of the apparatus and impart to the elevator the required velocity. The amount of this unbalanced weight, as it is called, will vary from

300 to 500 pounds, according to the size and speed of the car, and as the gear is two-to-one, the piston will be underbalanced in one case and overbalanced in the other, from 600 to 1000 pounds.

THREE-TO-ONE TYPE.

Elevators geared two-to-one are used only in low buildings, say sixty feet high or less. For higher runs the gear is made anywhere from three-to-one up to eight-to-one. A three-to-one elevator is shown in

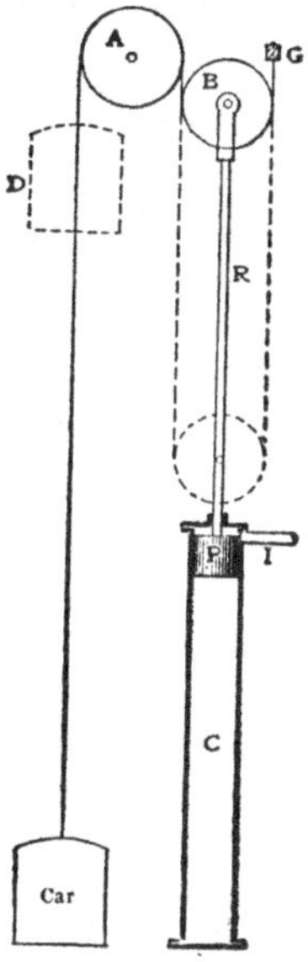

FIG. 3

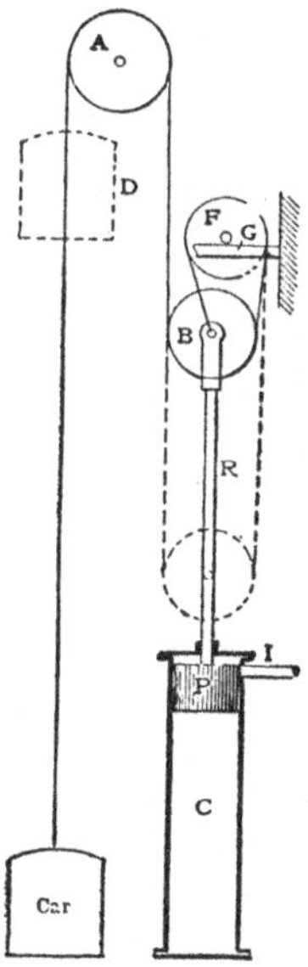

FIG. 4

Fig. 4; here the lifting ropes, instead of being anchored after passing under the sheave attached to the upper end of the piston rod, pass over a stationary sheave *F,* which is firmly secured to a support *G,* and then run down and are secured to the frame that carries the sheave *B.* In this case, the piston *P* travels only one-third the distance traveled by the car, the latter running up to the position *D,* while the piston runs down to the lower end of the cylinder *C.*

This type of elevator, like Fig. 3, can be arranged so that water can be admitted under the piston to push it up, but as it requires three pounds in the piston to balance one pound in the car, the weight would be very great, and so far as I know this arrangement has never been used. In fact, even in machines geared two-to-one it has been used in only one design, which was brought out about ten years ago and installed in the American Tract Society building in New York City and in several buildings in western cities. Its construction differed from Fig. 3 in that a plunger was used in place of a piston and rod, this change being made to simplify the apparatus and also to provide an easy way of obtaining the weight necessary to pull up the fully loaded car. The design was known as the upright-plunger elevator. It is not manufactured at the present time.

DETERMINING THE GEAR.

For reasons that will be explained hereafter, vertical elevator cylinders are seldom made with a stroke greater than 30 or 32 feet; therefore, if the hight of a building is such that a three-to-one gear would require a cylinder of more than 32 feet stroke, the gear is increased to four, six or eight-to-one. The gear is in every case equal to the rise of the car divided by the stroke of the piston, hence a four-to-one gear can be used for a car travel of not more than 125 feet, a six-to-one for 180 and eight-to-one for 240 feet.

It sometimes becomes desirable to determine the gear of an elevator already installed in a building, and this cannot be done readily by dividing the run of the car by the stroke of the piston unless these figures are known, because it is difficult to make the measurements. It is not necessary to go to this trouble, however, as the gear can be determined from an inspection of the lifting ropes running up from the sheave frame attached to the top of the piston rod. Referring to Fig. 1 it will be seen that only one rope runs up from the piston rod, and this is a one-to-one gear. In Fig. 3 one rope runs up from the left side and one from the right side of the sheave *B*, and this is a two-to-one gear. In Fig. 4 there are three ropes running up from the piston-rod sheave, and this is a three-to-one gear. In Fig. 5 there are four ropes running up from the sheaves *B* and *H*, and this is a four-to-one gear. In all higher gears this same relation holds good, so that to find the gear of any elevator all that is necessary is to count the number of ropes, or sets of ropes, that run up from the sheaves attached to the piston rod.

In Fig. 5 it will be noticed that the sheave *H* is placed above the sheave *B*. This arrangement is used in vertical rigging so as to reduce the dimensions of the well in which the sheaves travel. If the gear is greater than four-to-one, so as to require two or more stationary sheaves, these are also usually arranged one above the other. With this construction it is necessary that the lower of the traveling sheaves be of larger diameter than the one above it, and that the upper stationary sheave be the larger. This is necessary in order to prevent the ropes from rubbing against each other, as can be easily understood from inspection of Fig. 5.

In Fig. 4 it will be noticed that the centers of the sheaves *B* and *F* are not in line vertically, *F* being offset to the right. This arrangement is required so as not to have the ropes pull the piston rod to one side.

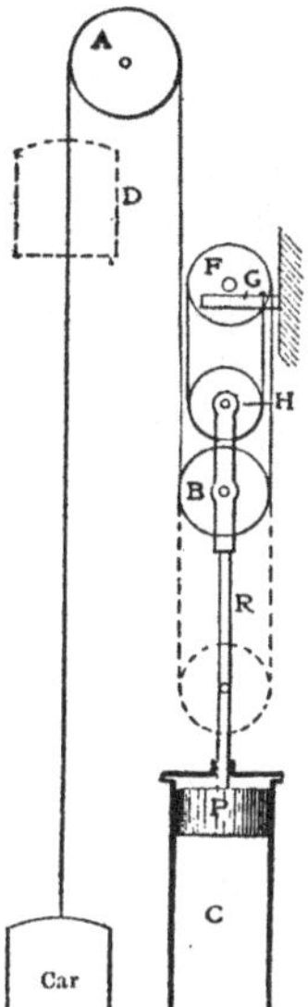

FIG. 5

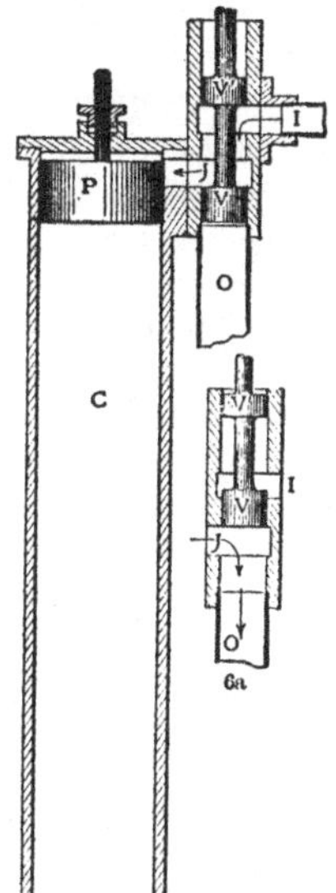

FIG. 6

In Fig. 5 all the sheave centers are on the same vertical line, which arrangement is practical because all the ropes run vertically. With any odd gear, the lowest of the stationary sheaves would have to be offset, as in Fig. 4, and on that account its diameter would be much less than that of the other sheaves. For this reason the only odd gear used is three-to-one, gears above four-to-one being even, such as six-to-one and eight-to-one. In actual elevators the frame that holds the traveling sheaves is provided with shoes to run on guides, so that with an odd gear, if all the sheaves were placed in line, the side thrust would not produce any serious results other than to increase the friction and wear at the upper end of the guides.

SIMPLE BALANCED VALVE.

In the "regular" types so far discussed, water under pressure is admitted to the cylinder to move the piston, and it remains to show how the flow of water is controlled so as to cause the piston to move upward or downward or to stop at any desired point. All this is accomplished by a single valve of very simple construction, which can be readily understood by reference to Fig. 6. In this diagram the valve is shown as consisting of double pistons, *V V'*, mounted on a valve-rod that projects upward from the valve chamber. The actual form of the valve differs considerably in the elevators of different manufacturers, but the principle of operation of all is substantially the same.

Means are provided whereby the operator in the car can move the valve in either direction or bring it to the central position, so as to cause the elevator to ascend, descend or stop. If the valve be depressed to the position in which it is shown in Fig. 6, water from the pressure tank will flow in through the pipe *I* to the valve chamber and passing out through the port *J* will enter the cylinder and force the piston down. If the valve be depressed so as to open the ports wide, as shown, the maximum amount of water will flow into the cylinder, and the piston will be driven down with the maximum velocity. If the valve is depressed only enough to uncover a small portion of the ports, only a small volume of water will pass through, and as a result the piston will travel down at a slower speed. Thus the velocity of the piston, and, therefore, that of the car, can be controlled by varying the position of the valve.

If, when the piston is moving down, the valve be raised so that *V* covers the port *J*, the flow of water into or out of the cylinder will be prevented and the piston will be held stationary in the cylinder. As the water cannot be compressed, it makes no difference where the piston may be, whether near the top or bottom; its motion will be arrested whenever the port *J* is covered. As it is possible for the operator to move the valve almost instantaneously over the port *J*, unless means are provided to prevent such sudden movement, the elevator is liable to be stopped too abruptly. To avoid sudden stops various devices are used, all of which will be fully described hereafter.

If, when the piston has been forced down to the bottom of the cylinder, the valve is raised so that *V* passes above the port *J*, as shown at 6a, then the flow of water from the pressure tank will be prevented, and the water in the cylinder will pass out through the port *J* to the discharge pipe *O*, as indicated by the arrows, and the piston *P* will move up as fast as the water escapes. To vary the velocity of the piston on its upward motion, all that is necessary is to vary the position of the valve so that the part *V* may uncover as much of the port *J* as may be required to develop the desired car speed. The function of the upper piston *V'* is merely to balance the water pressure on the valve.

The arrangement of cylinder and valve illustrated in Fig. 6 was used exclusively in the early days of hydraulic elevators, but is not used at the present time except in high-pressure machines.

It is obvious that when the piston P is at the top of the cylinder the pressure acting on it is not as great as when it is at the bottom, for in the latter position the weight of the water above the piston is added to the pressure. This would not make a material difference with a short cylinder, neither would it count for much if the pressure of the water used to operate the elevator were 600 or 700 pounds. Suppose, however, that the water were under a pressure of only 50 pounds per square inch and that the stroke of the piston were 30 feet; then the weight of water above the piston when the latter is in the lowest position will amount to practically 13 pounds per square inch of piston surface, so that the actual pressure per square inch acting to force the piston down will be 63 pounds, while with the piston at the top of the cylinder it will be only 50 pounds.

EQUAL PRESSURE ARRANGEMENT.

Early in the history of hydraulic elevators the inequality in pressures was overcome by the simple construction illustrated in Fig. 7. The pipe *I* connecting with the pressure tank delivers water to the "circulating pipe" *K*, which connects the upper end of the cylinder with the valve chamber in the manner clearly shown. If the piston *P* be at the bottom of the cylinder and the valve be depressed to the position shown in the sketch, the water above the piston will descend through the circulating pipe *K* to the valve chamber and through the port *J* to the lower end of the cylinder, so that as the piston moves upward the water is transferred from the upper end to the lower end of the cylinder. (The piston is drawn upward by the weight of the car and load.) With the valve in this position both ends of the cylinder are in direct communication with the pressure tank; consequently all the space in the cylinder, the circulating pipe and the valve chamber will be filled with water. Now suppose that the piston is at the top of the cylinder, and that the valve is raised until the lower part *V* is above the port *J;* then the water under the cylinder will flow out, and water from the pressure tank will fill the upper end of the cylinder. The pressure now acting on the piston will be the pressure of the water on top plus the suction of the water under the piston; the latter will decrease as fast as the piston runs down, and the weight of water above the piston will increase at the same rate, so that the net downward effort on the piston remains constant throughout the stroke.

The circulating pipe cannot be used with cylinders longer than about 32 feet, for the same reason that pumps cannot lift water to any greater hight than this; that is, that the pressure of the atmosphere will not hold water up against the under side of the piston to any greater hight. It is for this reason that cylinders are seldom made for piston strokes greater than about 30 feet. In some cases, for one reason or another,

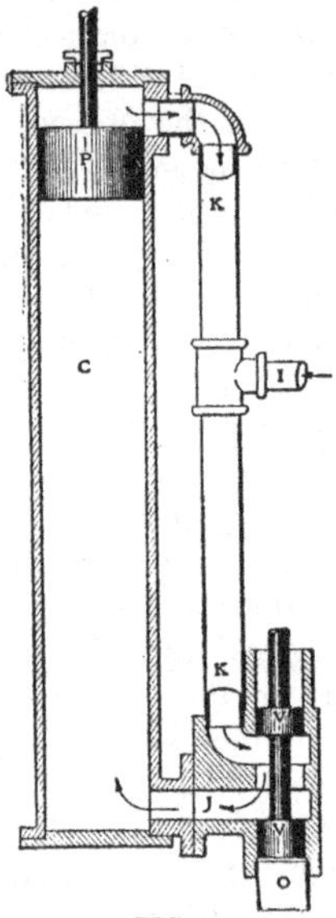

FIG. 7

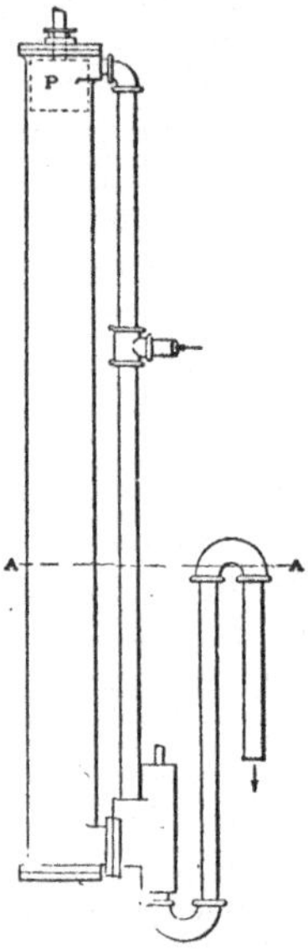

FIG. 8

it becomes necessary to use a cylinder of greater length, but when this is done the discharge pipe is provided with a goose neck, as shown in Fig. 8, the top bend of which rises above the line *A-A*, the distance from this line to the under side of the piston when it is at the top of the cylinder being about 30 feet.

At the present time many elevators are operated with water under a pressure of about 750 pounds per square inch, and with these the circulating pipe is not used, as the inequality in pressure due to the weight of the water is not sufficient to make it necessary. The difference between 50 and 63 pounds is a great deal, relatively, but the difference between 750 and 763 is less than 2 per cent.

CHAPTER II

COUNTERBALANCING—THE HIGH-PRESSURE SYSTEM

Looking at Fig. 9, it can easily be seen that the hydraulic cylinder acts only to lift the car and that the latter is lowered by its own weight combined with that of the load. This being the case, it is evident that the empty car must be heavy enough to pull up the piston, the piston-rods, the traveling sheave and frame and, in addition, to overcome all the friction of the moving parts. As a matter of fact, the weight of the car is more than enough to do this work in almost every case, so that if the car could be made lighter, power could be saved, because there would be less dead load for the hydraulic cylinder to lift on the upward trips. It is not practicable in most cases to reduce the weight of the car, but the same result is accomplished by adding a counterbalance weight on the side of the lifting cylinder, as shown at *W* in Fig. 9.

When the counterbalance is located in the frame of the traveling sheave, as shown in this diagram, its weight must be much greater than that of the proportion of the car weight it is intended to balance. The difference between the two weights will depend upon the gear of the elevator. If the gear is two-to-one, as in Fig. 9, the counterweight will have to weigh twice as much as the portion of the car weight that it balances. If the gear were three-to-one, the counterweight would weigh three times as much, and for a four-to-one gear, four times as much as the portion of the car it balances. From this it can be seen that the weight of the counterbalance could be greatly reduced if it were connected directly with the car by means of ropes running over an overhead sheave, as shown in Fig. 10, and this construction is commonly used, although the whole of the counterbalancing is very rarely done in this way.

A counterweight connected as shown at *w* in Fig. 10 is called an independent counterbalance, and it possesses some advantages as well as some disadvantages. One advantage of an independent counterbalance is that it saves weight, particularly in high-gear machines, as each pound in the counterweight will balance one pound of car, while in the counterweight *W*, the number of pounds required to balance one pound of car is equal to the gear of the machine; in Fig. 10 it would be four pounds.

Another advantage of the independent counterbalance is that the ropes that connect the weight with the car carry a portion of the weight of the latter equal to the weight of the counterbalance, and thus take off this amount of strain from the main lifting ropes, and thereby render the elevator that much safer.

The principal objection to the independent counterbalance is that it interferes with making quick stops of the elevator. This objection is not very great in the case of slow-running cars, but increases rapidly

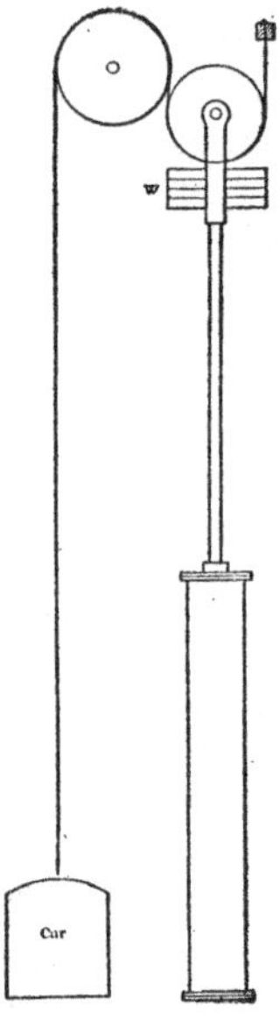

FIG. 9

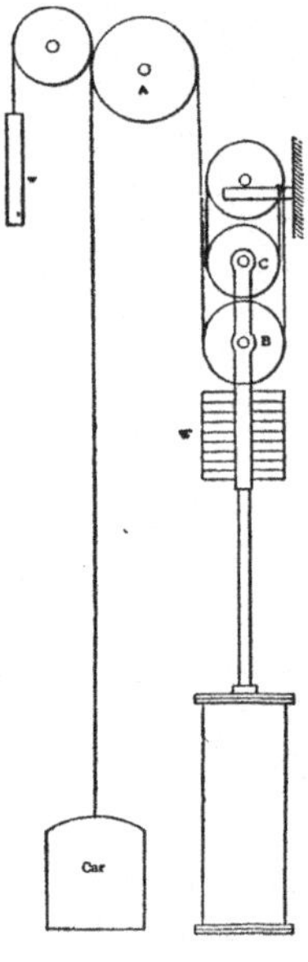

FIG. 10

as the car speed increases. Looking at Fig. 10 it can be seen that if the car were stopped suddenly when running up, the momentum of the weight *w* would tend to keep it running down, and thus draw up the car and slacken the ropes passing over the sheave *A*. It can also be seen that if the weight *w* were nearly as heavy as the car, the tendency to continue on the downward run would be much greater than if its weight were only a small portion of that of the car. The energy stored in the moving parts increases as the square of the velocity, so that if the car speed be doubled, the energy stored in the counterweight will be quadrupled; hence, the higher the car speed, the greater the objection to putting a large portion of the weight in the independent counterbalance.

Any weight used as a counterbalance, whether located at *w* or at *W*, will oppose the stopping of the car on the upward trip, but unless the weight is equal to a large portion of that of the car, this opposition will amount to nothing in practice, because the distance in which a car can be stopped without discomfort to the passengers is much greater than that through which it would be carried by the momentum of a light counterbalance. The object of the counterbalance is to save power by reducing the weight that the lifting cylinder has to raise, and this object can be more fully realized by placing the weight at *W* than by placing it at *w*, because in this way the weight of the empty car can be more nearly balanced. Why this is so can be made clear by a simple calculation.

Suppose that the car in Fig. 10 weighs 1500 pounds, and that if all but 500 pounds of this weight be balanced, this remainder will be sufficient to run the car down at the desired velocity. Now if all this counterweight be put in an independent weight, as at *w*, we will require 1000 pounds, and if it be put in the sheave frame, as at *W*, we will require 4000 pounds. Suppose that the car speed is four feet per second; then, since the energy stored in the moving mass is proportional to the square of the velocity, *w* would be equal to $1000 \times 4 \times 4 = 16,000$, while with the counterweight at *W* the velocity would be reduced to one-fourth, or one foot per second, so that the stored energy would be $4000 \times 1 \times 1 = 4000$. This force is exerted upon the ropes that pass under the traveling sheaves *B* and *C*, so that the force acting to lift the car is proportional to 1000, as compared with 16,000 in the other case.

As the force tending to lift the car, due to the momentum of the counterweight, is sixteen times as great with the weight in *w* as with it in *W*, it follows that with the latter counterbalance a shorter stop can be made than with the former. The distance required to make a stop with the weight in *w* might be greater than that desired, and the only way to reduce it would be by making *w* lighter. It is evident that with the counterbalance at *W* the unbalanced weight of the car can be considerably less than with the weight in *w*, and that the difference increases with an increase in car speed, and with higher gear ratios. From this it will be evident that when the gear of the elevator is high, the momentum of the counterweight at *W* has very little effect upon the quick stopping of the car, and on that account the weight of the latter can be more nearly balanced than it can with the independent weight at *w*. In practice the total counterbalance is put partly in *w* and partly in *W*, the object of so doing being to reduce the amount of iron required, and also to add to the safety of the elevator system by providing additional ropes to hold the car, and by taking some of the strain from the car-lifting ropes. Placing the counterbalance weight, or a part of it, at *W* does not increase the strain on the lifting ropes, as might be supposed on first considering the subject, as this strain is determined wholly by the weight of the car and its load.

HIGH-PRESSURE HYDRAULIC ELEVATORS.

The elevators so far considered are of the type known as low pressure, and are actuated with water pressures ranging between about 40 and 175 pounds per square inch. In the early days of hydraulic elevators the pressure was obtained by pumping water into a tank located upon the roof of the building, and as a column of water of one square inch cross-section weighs about 0.434 of a pound per foot of hight, the pressure seldom exceeded 50 pounds, and in many cases was as low as 30 pounds per square inch. With the advent of high buildings greater car speed was demanded, and this necessitated increasing the power of the elevator machine. This increase could be obtained in two ways, one by making the cylinder of larger diameter, and the other by increasing the pressure. As space in the buildings was valuable, the latter plan was adopted, and in order to obtain the increased pressure, a closed pressure tank was substituted for the open roof tank. The pressure tank had been used years before the time of high-power elevators, in buildings that were so low that the pressure obtained from an open tank on the roof was not high enough to keep the diameter of the cylinder within reasonable bounds; but from this time all elevators installed in high buildings were operated with water drawn from pressure tanks.

The general introduction of the pressure tank resulted in an increase in the working pressure to from 75 to 100 pounds. By this increase the size of the cylinder, the tanks and the piping can be considerably reduced, with a corresponding saving in space and material. With the continued increase in the hight of buildings the pressure has been increased, so that at the present time it is common practice to use pressures up to 180 pounds and even a little higher.

With the view to reducing the size of the cylinders and piping still farther, elevator designers decided to make another increase in pressure, but this time a clear jump was made from 180 to 700 and 800 pounds per square inch, and this higher pressure is now carried on all "high-pressure" systems. The reason why a jump was made instead of increasing the pressure gradually is that the type of apparatus used for the lower pressures is not adapted, in many of its details, to much higher pressures, and since changes in these details were required it was considered advisable to increase the pressure to the highest point that had been found to work with entire success in other fields using similar apparatus, so as to derive the greatest possible benefit obtainable from the use of high pressure.

The differences between high-pressure and low-pressure elevator machines can be pointed out by the aid of Fig. 11, which represents, elementarily, a high-pressure machine geared two-to-one. A plunger replaces the piston and rods of the low-pressure machine, and the water acts to force it out of the cylinder instead of drawing it in. The advantage of this construction is that only one packing is required, and that is located in the end of the cylinder, where it is easily reached. In the low-pressure cylinder, the piston must be provided with a packing, as well as the piston-rods, but as the diameter of the cylinder is generally from 10 to 20 inches, there is ample room for the bolts that tighten up the packing gland, and these can be readily got at when the piston is at the end of the cylinder.

The plunger of a high-pressure machine is finished off true and smooth, but the cylinder is only finished at the end, this part being a casting provided with a pipe connection at the side and a stuffing-box at the outer end. This end casting is bored to fit the plunger. The body of the cylinder is made of extra strong steel piping, the interior diameter of which is considerably larger than the outer diameter of the plunger. As many lengths of pipe are used for the cylinder as may be required, and these are generally connected together by flange couplings.

It will be noticed in Fig. 11 that the cylinder is inverted and that the plunger forces the sheave *B* downward. On this account this type of machine is known as the inverted plunger. Upright-plunger machines geared two-to-one have been made, but higher geared machines would be impractical; whereas the inverted machine can be made of any gear desired, and in reality is seldom made less than four-to-one. In a two-to-one inverted-plunger machine the plunger must weigh considerably less than double the weight of the empty car, otherwise the latter would not be able to run down and lift the plunger up into the cylinder.

If the cylinder were mounted with the packed end up, under the sheave *B,* the weight of the plunger would have to pull up the car when loaded to its maximum capacity, hence the plunger would have to weigh considerably more than twice as much as the car with its maximum load. From this it is evident that for the inverted-plunger machine the weight of the plunger is very much less than for the upright plunger. For any gear greater than two-to-one the upright plunger would not be desirable on account of the great weight of the plunger necessary to lift the car. In the low-pressure system the water is admitted on top of the piston and forces it down, thereby lifting the car, but in the upright plunger the water is admitted under the plunger and forces it up, thus permitting the car to run down.

Inverted-plunger machines are generally geared from four-to-one to eight-to-one. A two-to-one machine cannot be used for very high lifts because the plunger could not very well be made sufficiently light. Aside from this difficulty, however, it is cheaper to increase the gear and thus reduce the length of the plunger and cylinder. The general arrangement of an inverted-plunger elevator geared six-to-one is shown in Fig. 12.

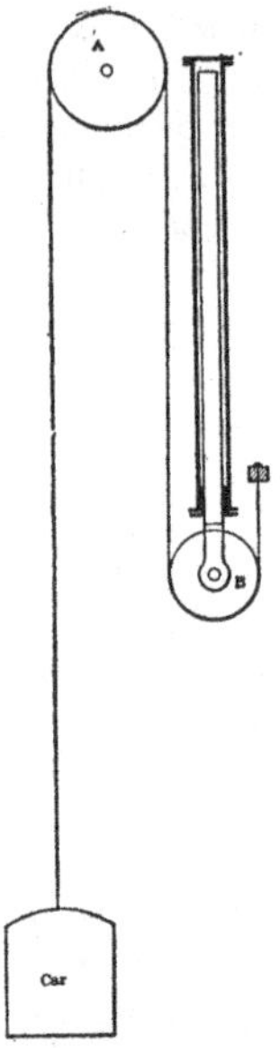

FIG. 11

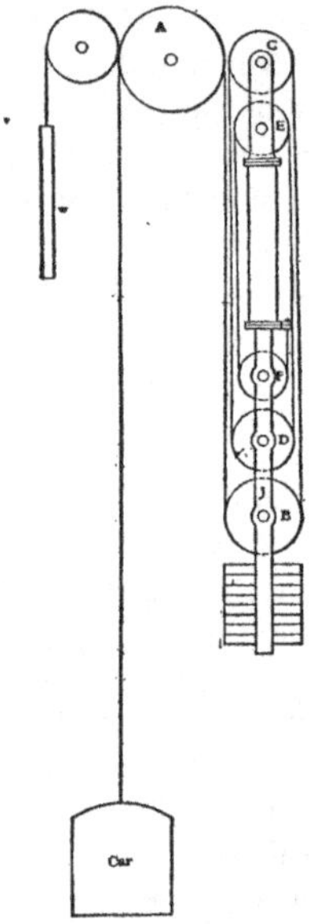

FIG. 12

As will be noticed, the counterbalance weight is partly at *w* and partly at *W*. For medium hight buildings, little weight has to be placed at *W*, as the plunger, sheaves and frame are nearly heavy enough, especially as in such buildings the gear is likely to be not more than four-to-one.

The advantages of the high-pressure system are that considerably less space is required for the machinery and piping, and that the efficiency is higher than that of the low-pressure system. The higher efficiency is due to the fact that the loss of energy in forcing the water through the pipes can be greatly reduced. The energy required to force water through a pipe is made manifest by a reduction, or drop, in the pressure between the point where the water enters and the point where it passes out. The higher the velocity of the water, the greater the drop in pressure, the increase in drop being about as the square of the velocity.

With a low-pressure system operating under about 50 pounds, it might be found to be difficult to keep the loss of pressure between the pressure tank and the lifting cylinder below 20 pounds if the connecting piping were long. This would mean a loss of energy of 40 per cent. If the pressure be doubled, one-half the amount of water will be required, assuming the piston and valve friction to remain the same. As one-half the amount of water is used, the cross-section of the pipes can be reduced to one-half without changing the velocity of the water, and consequently without increasing the loss of pressure above 20 pounds. From this it is evident that the pipe loss can be reduced from 40 per cent. to 20 per cent. by increasing the pressure from 50 to 100 pounds, and at the same time the pipes can be reduced to one-half the cross-section.

If the pressure be raised to 800 pounds, then the plunger area can be reduced to one-eighth, and as one-eighth of the water will be required. the cross-section of the pipes can be reduced to one-eighth without increasing the velocity of the water, and, therefore, without increasing the drop in pressure, still assuming constant friction. By increasing the pressure to 800 pounds, and reducing the pipes to one-eighth of the cross-section, the loss of pressure in the pipes is not increased beyond the original 20 pounds, and this is now only 2½ per cent. of the working pressure.

The foregoing is a better showing than could be made in practice because the loss in the pipes and in the piston friction will increase with the pressure and the reduction in diameter, so that the size of the cylinder and pipes cannot be reduced as much as these calculations show, but the sizes would not have to be increased more than one-half, or, roughly, the cross-sections of the plunger and pipes of the 800-pound system would not be more than one-fifth of those of the 100-pound system.

In low-pressure systems the piping varies between about 3½ inches and 7 inches in diameter, and such piping requires considerable space, especially if there are many bends. In the high-pressure system the piping is seldom over 2½ inches in diameter. The pipes required for low-pressure systems are kept down in size somewhat by proportioning them so that when the car runs at the maximum velocity it will only lift about five-eighths of the maximum load, and with the full load the velocity is about one-half of the maximum. With high-pressure elevators the pipes can be made so as to maintain the maximum car speed with nearly the maximum load, and still be of small size, because the drop of pressure in the pipes is so small a percentage of the total pressure.

CHAPTER III

PLUNGER HYDRAULIC ELEVATORS

PLUNGER ELEVATORS

Fig. 13 illustrates in its simplest form what is known as the plunger elevator. Elevators of this type have been made in this country for many years for short rises, but within the last seven or eight years they have come into use for all classes of service and for buildings of any hight. In Europe plunger elevators have been used for many years, but generally for moderate rises—in fact, what we call high rises in this country are practically unknown in Europe, where buildings over one hundred feet high are exceptional. At the present time there are a large number of plunger elevators in operation in New York that rise over 200 feet, and in one building they rise about 300 feet. Fig. 13 illustrates what is commonly called a sidewalk elevator, which is arranged to run the platform *A* up from the level of the basement floor *B* to the sidewalk *D*. All this diagram requires to complete it is an operating valve in the pipe *I* and suitable guides for the platform to keep the plunger *P* in line with the cylinder *C*.

The construction of this type of elevator is very simple. The cylinder is made of one or more lengths of steel pipe, the lower end being closed by a suitable cap and the upper end being finished off with a casting bored to take the plunger, and provided with a stuffing-box and an inlet at the side for connection with the water-supply pipe. The cylinder is set in a hole in the ground, and is made of a length a few feet greater than the rise of the elevator car. The plunger is made of steel pipe and is turned true and smooth to fit the bore of the top casting of the cylinder. When the rise is greater than one length of the piping used, two or more lengths are joined by means of internal sleeves made long enough to give the joints as much strength as the body of the pipe.

The simple arrangement of Fig. 13 is very satisfactory for short rises, but for any considerable hight it is necessary to counterbalance the car in order to reduce the pressure required to operate the elevator. The plungers are as a rule made of six-inch pipe, which when turned weighs about 16 pounds per foot for standard pipe, and about 25 pounds per foot for extra strong, which is used for very heavy duty. If the rise of the car is 200 feet, and the plunger is made of standard pipe, it will weigh about 3500 pounds, with the end castings. If the car is of equal weight, which is not very far from the average, the total weight will be

7000 pounds, which is a considerable load to lift, especially if the speed is 400 or 500 feet per minute. If a counterbalance of 5000 pounds is provided the power required to operate the elevator will be greatly reduced.

The counterbalance can be arranged in either of two ways, one of which is illustrated in Fig. 14 and the other in Fig. 15. In the first arrangement the gear between the car and counterweight is one-to-one, and in the second it is two-to-one. Each arrangement has its advantages

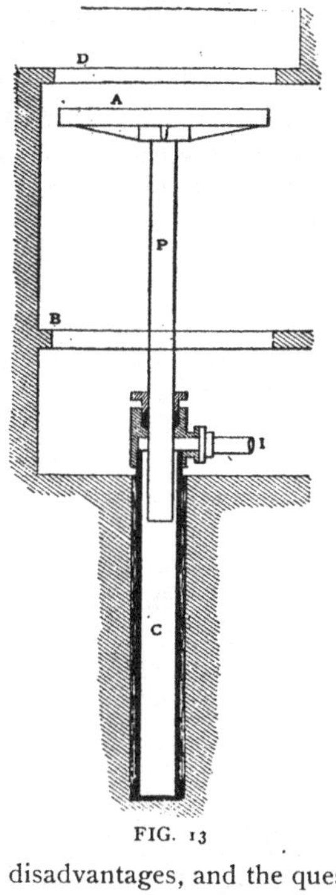

FIG. 13

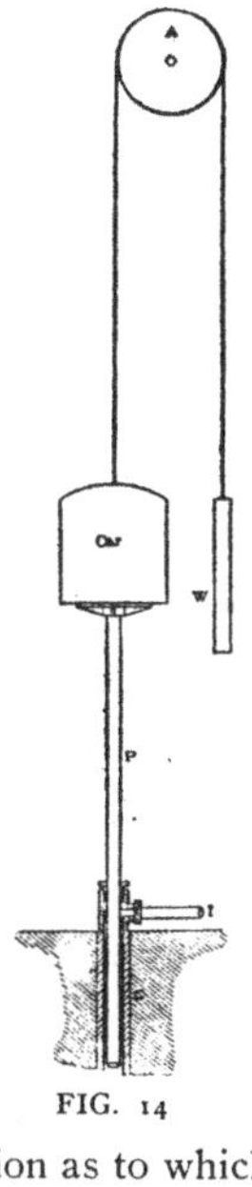

FIG. 14

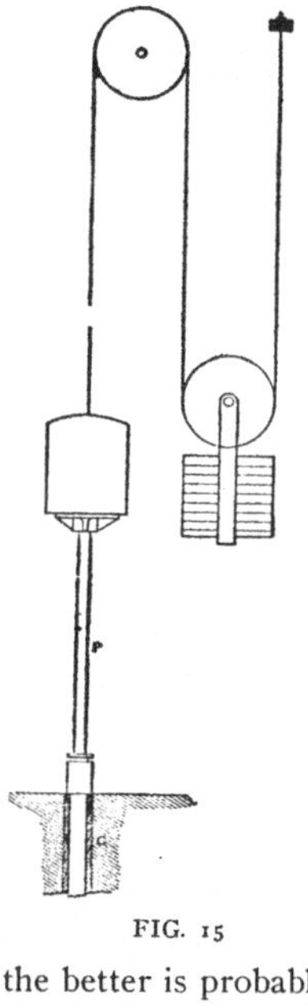

FIG. 15

and disadvantages, and the question as to which is the better is probably a matter of individual opinion. The plan in Fig. 14 is simple and requires less weight than the two-to-one gear of Fig. 15; but to offset this there is the fact that to be able to make a stop on the up trip within a certain distance, the unbalanced weight of the car must be greater. The reason for this difference is the same as that given in the discussion of Fig. 10, Chapter II.

The counterbalance serves not only to reduce the amount of power required to operate the elevator, but also to reduce the compression stress that the plunger is subjected to, and thus to prevent the buckling of the latter when stopped suddenly on a downward stroke. Taking the figures for the rise and the weight of car and counterweight previously given, it will be evident that if the car is empty it will fall short 1500 pounds of balancing the counterweight, therefore the latter actually holds up 1500 pounds of the plunger. This means that if the car is empty, the upper end of the plunger is subjected to tension and not to compression. At a point about 90 feet from the top there will be no stress in the plunger, and below this point a compression stress will appear which will increase at the rate of about 16 pounds per foot.

If a load of 1500 pounds is in the car, then the point where there is no stress in the plunger will be at the upper end. With a load of 2500 pounds in the car, which is about the maximum carried in passenger elevators for the average office building, the compression stress at the top of the plunger will be 1000 pounds, and will increase below this point at the rate of about 16 pounds per foot, becoming about 2600 at the center point and 4200 pounds at the bottom of the plunger. This is the reason why the plunger, although only 6½ inches in diameter, will not buckle under the load even if made 300 feet long. Another fact that accounts for the rigidity of the plunger is that only the portion that is in the air, above the top of the cylinder, is liable to buckle, and the compression stress in this decreases as the car rises, owing to the fact that the passengers leave the car at the various floors as the elevator ascends and when the top of the building is reached the car is nearly if not entirely empty.

The compression stress that the plunger is subjected to could be reduced by increasing the counterbalance, but if this were done the distance required to make a stop would be increased. The unbalanced weight of the car has to be determined with reference to the distance in which stops must be made, and also with reference to the speed of the car. If the stopping distance remains the same, the counterweight must be reduced as the speed is increased, or the weight may remain unchanged if the stopping distance is increased.

The stopping distance is controlled by setting the valve-operating mechanism so that the flow of water cannot be stopped entirely in less than a certain time; while the operator may make a slower stop, he cannot make a quicker one. The effect of making too quick a stop on the up trip would be to lift the plunger off the water in the cylinder a few inches; the vacuum produced would assist in bringing the car to a state of rest. On the down trip the effect of too quick a stop would be to put an excessive compression stress in the plunger.

CHAPTER IV

HORIZONTAL CYLINDER MACHINES—"PUSHING" AND "PULLING" TYPES

Horizontal-cylinder elevators are divided into "pushing" and "pulling" machines. A "pushing" machine is one in which the piston pushes the traveling sheaves away from the cylinder, and a "pulling" machine is one in which the piston pulls the traveling sheaves toward the cylinder. In Fig. 16 a pulling machine is illustrated. In this type of elevator, when the car is at the bottom of the well the traveling sheaves are at *D*, and the piston is at the front end of the cylinder *G*. When the car ascends to the top of the building, the traveling sheaves and cables reach the position shown by the broken lines. It will be evident that the pressure water enters the cylinder between the piston and the cylinder-head and forces the piston to the right, and through the piston-rod the traveling sheaves *D* are pulled toward the cylinder. The weight of these sheaves is taken by the roller *E* and guides *F*. The mode of operation of a pushing machine is shown in Fig. 17. Here it will be obvious that the water entering the cylinder between the piston and the cylinder-head forces the piston to the right and, through the rod, the traveling sheaves *D* are pushed away from the cylinder and reach the position indicated by the dotted lines when the elevator car reaches the top of the building.

Which of these two designs is the better, it is difficult to say. The pushing machine obviates the use of a stuffing-box around the piston-rod, but this is offset by the fact that the diameter of the sheaves must be greater than that of the outside of the cylinder, in order that the lifting ropes may clear the latter; this can be readily seen from Fig. 17. In both designs, the stationary sheaves *B* have to be placed at the side of the elevator well, in order that the lifting ropes may run up in the space between the wall of the well and the side of the car. On this account the cylinder of the pushing machine is located nearer the well than that of the other type. In some buildings the pumps, tanks, etc., are located near the elevator well, but in others they are some distance away, so that in some cases it is advantageous to have the cylinder near the elevator, while in others it is not, as it is always desirable to have it as near to the pumping system as possible, to reduce the length of the pipe connections. The pulling machine is not as well adapted to high pressures as the pushing machine, on account of the packing around the piston-rod, and on that account the latter type is the only one that has been used in high-pressure systems.

Horizontal cylinders are not provided with circulating pipes because none is required, both ends of the cylinder being on the same level. The pressure water passes in and out of one end of the cylinder only, and no water reaches the other end except what leaks by the piston.

Horizontal-cylinder elevator systems are counterbalanced with an

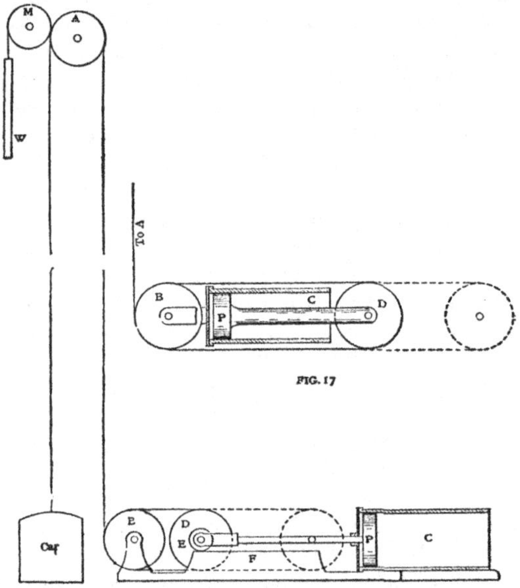

FIG. 17

FIG. 16

independent weight, as shown at *w* in Fig. 16. As can be readily seen, no counterbalancing can be done in the traveling-sheave frame, because this moves in a horizontal direction. Horizontal machines are geared from six-to-one to twelve-to-one. The sheaves by means of which the gearing is effected are not located one behind the other, as in the vertical machines, but are placed side by side on axles that are at right angles to the axis of the piston-rod, as indicated in Figs. 16 and 17. The number of lifting ropes used is two, and each one of these passes around a set of sheaves on one side of the center line of the piston-rod, so that the pull of the ropes may not act to twist the sheave axle out of position. When it is desired to obtain a gear higher than twelve-to-one, the general practice is to use a machine of one-half the desired gear, and double this by using a traveling sheave in the elevator well in the manner illustrated in Fig. 18. With this arrangement, if the gear at the cylinder

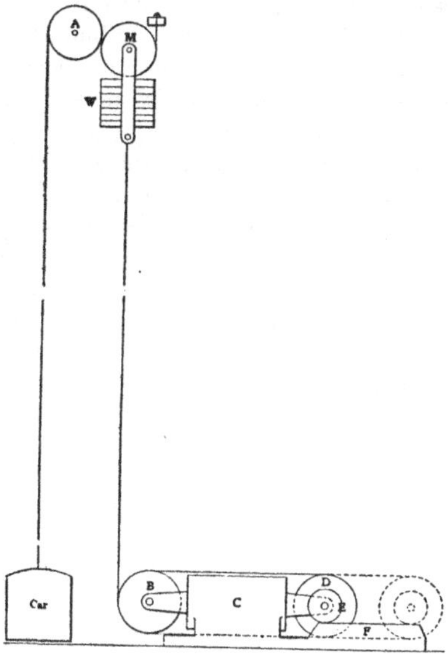

FIG. 18

is eight-to-one, the gear between the car and the piston will be sixteen-to-one. With this construction the counterweight can be placed in the sheave frame, as shown at *W*, and the advantage of this method of counterbalancing realized.

As to a comparison of the vertical and horizontal machines, it is difficult to say which is the better, because the difference between them is very slight. In the horizontal machine the lower side of the piston and cylinder are subjected to a wear from which the vertical cylinder is free, but this objection is of less importance in elevator than in steam-engine cylinders, as the latter are worn more in a lifetime. In the vertical-cylinder machine the piston friction is greater because the stroke is longer, but to offset this in the horizontal machine the sheave-and-rope friction is greater on account of the higher gear. In practice the type of machine used is in most cases determined by the dimensions of the building. If the building covers a good deal of ground, and the space in the basement is not as valuable as elsewhere, the horizontal machine is used. If the building is high and stands on a small lot, the vertical type is preferable, and if the floor space is very small, it may be the only type that can be used on account of the length of the horizontal machine being greater than the greatest dimension of the building site.

When horizontal machines are installed in buildings where the space is contracted, it is common practice to stack the machines one on top of the other, making what are called double- and triple-deckers. When vertical cylinders are used, they are sometimes located in the elevator well at one side of the space in which the elevator travels, but this arrangement is not desirable because the noise made by the water passing in and out of the cylinder is objectionable; moreover, if the cylinder springs a leak, which is not a rare occurrence, the passengers in the car are liable to receive an unwelcome shower bath. The best practice is to locate vertical cylinders in a separate well so located that the lifting ropes may be run directly to the elevator cars over sheaves at the top of the building.

CHAPTER V

COUNTERBALANCING THE LIFTING ROPES OF ELEVATOR CARS

When an elevator car is at the bottom of the well, the weight that has to be lifted comprises that of the car, its load and the ropes hanging in the elevator well above the car. When the car is at the top of the building, only the car and its load have to be lifted. This being the case, it is evident that if the counterbalance is such that when the car is at the top of the building all its weight except 500 pounds is balanced, then when the car is at the bottom floor the unbalanced weight will be 500 pounds plus the weight of the lifting ropes hanging above it. If these ropes were light it would not be worth while to consider their weight, but as they are not light, the power required to operate the elevator can be considerably reduced by providing means to compensate for the varying weight of ropes that must be lifted when the car is at different points in the elevator well. Why this is true, and how the weight of the ropes is compensated for, can be made clear by the aid of Fig. 19.

In Fig. 19 let the top position of the car be 200 feet above the lower position. If the elevator is connected in the way shown it may have four or six lifting ropes, which as a rule are 5/8 of an inch in diameter. These ropes weigh about 0.7 of a pound per foot, which for four ropes makes 2.8, and for six ropes 4.2 pounds; for 200 feet, therefore, the weight will be 560 pounds for four ropes or 840 pounds for six ropes. If, when the car is at the top, its unbalanced weight is 500 pounds, in the bottom position it will be $500 + 560 = 1060$ pounds with four ropes, or $500 + 840 = 1340$ pounds with six lifting ropes. If the maximum load the car is designed to lift is 2000 pounds, the average will be about 1000, so that the work the elevator machine has to do when the car is approaching the topmost position is to lift 1500 pounds, but when it starts from the bottom position the load that must be lifted is 2060 pounds if there are four lifting ropes, or 2340 pounds if there are six ropes. It will be evident that if the weight of the ropes could be eliminated, about 27 per cent. of the power could be saved if there were four ropes, or about 36 per cent. if there were six ropes. As a matter of fact, the saving would be more, for, as may be seen from Fig. 19, when the car reaches the upper position, the traveling sheave *M* will be in the lower position *N*, and the length of the ropes from *G* to *H* will be added to the weight of the counterbalance; therefore, this weight would have to be deducted from the counterbalance in order not to have less than 500 pounds of unbalanced weight when the car is in the top position.

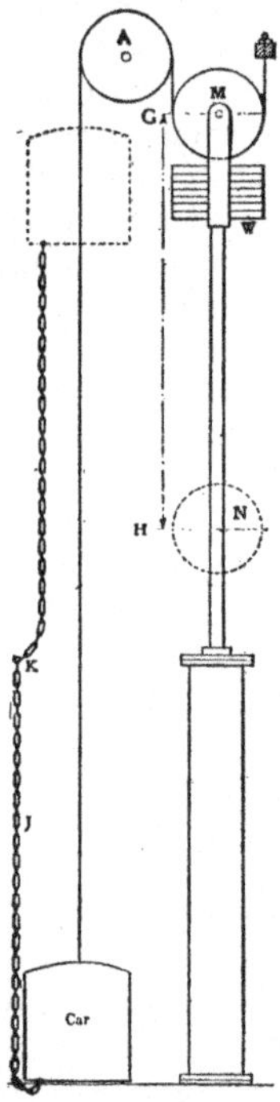

FIG. 19

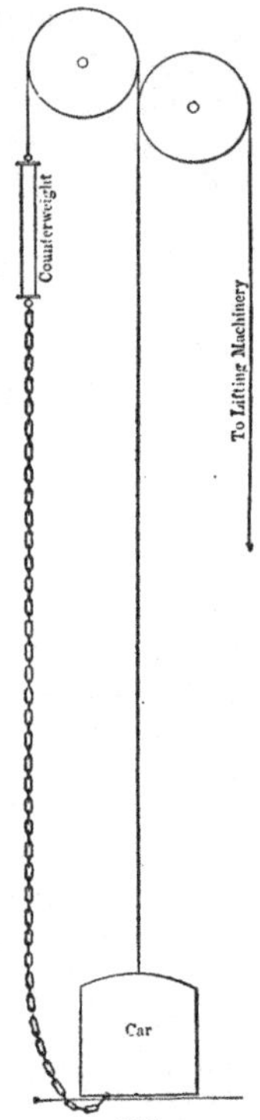

FIG. 20

The weight of the ropes cannot be eliminated, but the same result can be obtained by using a chain *J* secured to the side of the elevator well at a point *K* a trifle above the middle point, the other end being fastened to the underside of the car. When the car is at the bottom, the chain hangs from the point *K*, but as the car ascends, part of the weight of the chain hangs on the car, and when the car reaches the top position all of the weight of the chain will hang from it. Now to make this chain compensate for the varying weight of the lifting ropes all that is necessary is to make it weigh as much as the 200 feet of ropes that hang in the elevator well when the car is at the bottom plus the length of ropes from *G* to *H* when the car is at the top. If the elevator is geared two-to-one, as in Fig. 19, the distance from *G* to *H* will be 100 feet, or one-half the rise of the car. For a three-to-one gear, *G—H* will be equal to one-third the rise of car, and so on for any other gear.

When the compensating chain is secured to the bottom of the car, its weight acts in opposition to that of the counterbalance, and the latter must be increased in weight sufficiently to compensate for the weight of the chain. The chain can be attached to the underside of the counterbalance in some cases, and then its weight acts with the counterbalance so that the latter can be reduced in a corresponding amount.

The best way to connect the compensating chain is to attach one end to the underside of the car, and the other to the counterweight, as in Fig. 20. With this arrangement the weight of chain can be reduced to one-half, because when the car is at the bottom of the well the chain hangs on the counterweight, and when the car is at the top, the chain hangs from the car. In practice, the chain is connected with the car only in almost every case, because there is not sufficient space in the path of the counterbalance for it to hang freely without danger of being caught.

In a horizontal elevator with an independent counterbalance, as shown in Fig. 16, the counterbalance ropes add their whole weight to the counterbalance when the car is at the top of the building, and their whole weight to the car when the latter is at the bottom; hence, the weight of chain required to compensate for the counterbalance ropes is double the weight of these ropes, and to compensate for the lifting ropes a weight equal to that of those ropes must be added.

In practice, usually two compensating chains are used, and although they function perfectly and never give trouble, they are somewhat unsightly, and on that account a method of compensating for the varying weight of the ropes by means of a column of water has been used in some cases. This arrangement is illustrated in Fig. 21. When the piston descends in the cylinder *C* the water under the piston is forced up into the stand-pipe *B*, and this exerts a back pressure that increases as the piston travels downward. When the piston is at the top of the cylinder, the car is at the bottom of the building and, therefore, the weight of the ropes has to be lifted. At this time, however, the water in the stand-pipe *B* is at the level *E*, so that there is no back pressure, and as a result the lifting force is equal to the full pressure of the water admitted above the piston. When the car is at the top of the building the piston is at the bottom of the cylinder, and the water in the stand-pipe is forced up to the level *D*, so that there is a back pressure acting on the underside of the piston equal to the head *E*—*D*.

To make this arrangement compensate fully for the varying weight of the ropes all that is necessary is to make the internal diameter of the stand-pipe *B* such that when the piston reaches the lowest position in

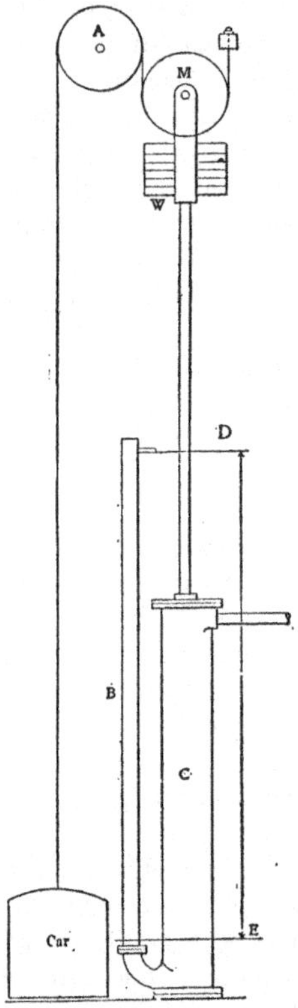

FIG. 21

the cylinder the weight of the water in *B* will be just sufficient to offset the weight of the ropes. It will be noticed that this back pressure pushes upward against the piston and thus holds up a certain portion of the counterweight, so that a weight can be used that would be sufficient to overbalance the car entirely when the latter is at the top of the building if it were not for the assistance given by the back pressure of the water in the stand-pipe. It might be supposed that to obtain the exact weight of water required it would be necessary to make the stand-pipe of cast-iron sections and bore them accurately to the proper size, but in practice it has been found wherever this arrangement has been installed that ordinary steam pipes can be used that are near enough to the proper diameter to compensate for the weight of the ropes within less than four or five per cent. The water supply under the cylinder is kept up to the proper amount by what leaks past the piston, and in fact an overflow is provided at the upper level in the stand-pipe to carry off the excess. This arrangement, of course, cannot be used with a circulating pipe, and it may also be mentioned, that the lower level *E* cannot be more than about 32 feet below the underside of the piston when the latter is in the highest position.

CHAPTER VI

GENERAL ARRANGEMENT OF HYDRAULIC ELEVATOR SYSTEMS

There are several ways in which hydraulic elevator systems are arranged, and these are shown in the sketches from Fig. 22 to Fig. 24. The first one is what is called the gravity system, and is the type first used. A tank is placed on the roof of the building, and the water is pumped into this to obtain the necessary head to actuate the piston in the lifting cylinder *C*. Water passes from this tank through the pipe *A* to the upper end of the cylinder, and forces the piston downward and the car upward. On the return stroke of the piston the water circulates through pipe *B* from the upper to the lower end of the cylinder. On the next down stroke, the upper end of the cylinder is again filled from the pipe *A*, and the water under the piston passes out through the pipe *D* to the discharge tank, whence it is raised to the roof tank by the pump. The pump is stopped and started automatically by a float in the roof tank, which closes the starting valve of the pump when the water reaches the high level, and opens it when the water is drained to the low level.

The system shown in Fig. 23 is known as the pressure-tank system, and differs from Fig. 22 only in having a pressure tank in place of the roof tank. With the gravity system the pressure is generally between 25 and 40 pounds per square inch, according to the hight of the building. With the pressure-tank system the range is between 50 and 200 pounds. In the pressure-tank system the pump is controlled by a pressure regulator that is connected with the pressure tank.

Fig. 24 is a system that has been installed in a number of large buildings by the Otis Elevator Company, and is known as the double-power system. There are two pressure tanks, one carrying about double the pressure of the other. The main pump delivers water into the low-pressure tank, and a booster pump draws from this tank and delivers into the high-pressure one. The elevator cylinder operating valve is made so that when opened part of the way it draws water from the low-pressure pipe only, and when opened further it draws from both the low-pressure and high-pressure pipes. If the load in the car is light, the desired speed can be obtained with the low-pressure water alone, but if the load is so heavy that the low pressure will not give the desired speed, the valve is opened further, and then high-pressure water runs in to help the low pressure and gives the proper velocity of car. A check-valve is included in the low-pressure pipe to prevent the high-pressure water from running back into the low-pressure tank; but unless the load

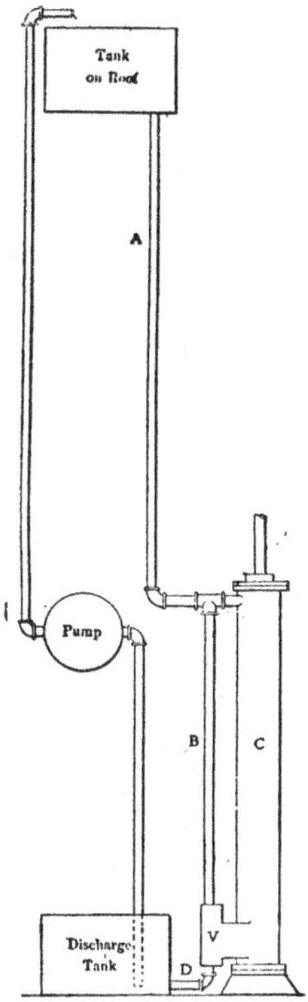

FIG. 22

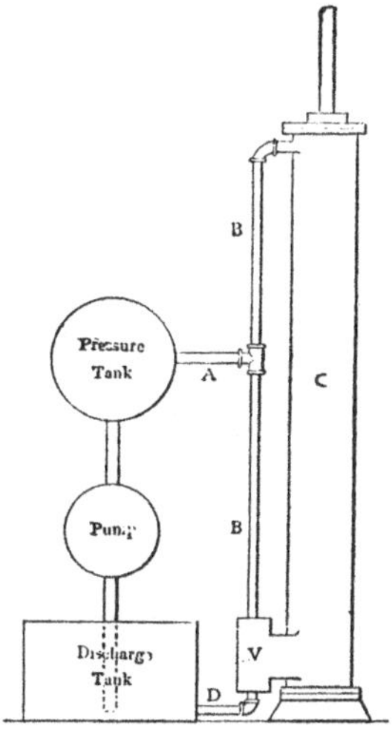

FIG. 23

is very nearly up to the maximum, water will be drawn from both tanks. This system was devised in order to make the power used by the elevator more nearly proportional to the loads lifted. When the low-pressure water is used, the power consumption is about one-half that when the high pressure is used. For loads greater than one-half the maximum, the power required is more than one-half, but less than the whole amount, unless the load is very near to the maximum, in which case all the water used will be drawn from the high-pressure tank.

The three systems illustrated in Figs. 22 to 24 are of the low-pressure type, and can be used to operate one or any number of elevators. The

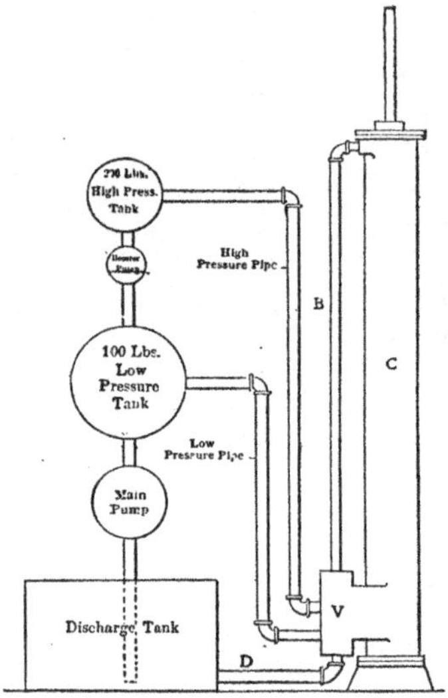

FIG. 24

high-pressure system described in Chapter II is arranged in a manner similar to Fig. 22 and is fully explained in Chapters XXIX to XXXIV.

CHAPTER VII

VERTICAL HYDRAULIC ELEVATORS

SIMPLE LOW-PRESSURE VERTICAL TYPE OTIS ELEVATOR WITH HAND-ROPE CONTROL

The fundamental principles of the different types of hydraulic elevators having been discussed in the preceding chapters, the constructional features of each type will next be considered. This chapter is devoted to the simplest form of vertical low-pressure elevator, such as is installed in buildings from five to seven stories high. An elevator of this class is very clearly represented in Fig. 25, which shows an Otis machine geared two-to-one. Looking at this illustration, it will be seen that the general arrangement is the same as that of diagram Fig. 3, and from this fact it might be inferred that the elevator is not counterbalanced. This, however, is not the case; a counterbalance is used, but it is placed within the cylinder, resting on top of the piston. This construction is very common with two-to-one machines, and even with higher gears. Generally a portion of the counterbalance is placed on top of the piston, so that in such machines the counterbalance weight is divided into three parts, one being within the cylinder, one in the traveling sheave frame, and one constituting the independent counterbalance.

It will be noticed in Fig. 25 that there are two piston-rods *R*. This construction was adopted in the early days of hydraulic elevators partially to increase the safety of the apparatus, but principally to prevent the traveling sheave *B* from twisting around. The ropes tend to hold the sheave from twisting, but they will not prevent slight movements, while the double piston-rods will. Now and for several years past, however, the frame of the traveling sheave has been made in the form of a cross-head running in stationary guides, thus effectually preventing any side movement of the sheave. With this construction the main benefit of the double piston-rods is additional safety; while it is possible for one rod to break or become loose, it is practically impossible for both to give way at the same time.

The arrangement of the cylinder *C*, the circulating pipe *K* and the valve *V*, in Fig. 25, is the same as in the diagram Fig. 7, even the inlet *I*

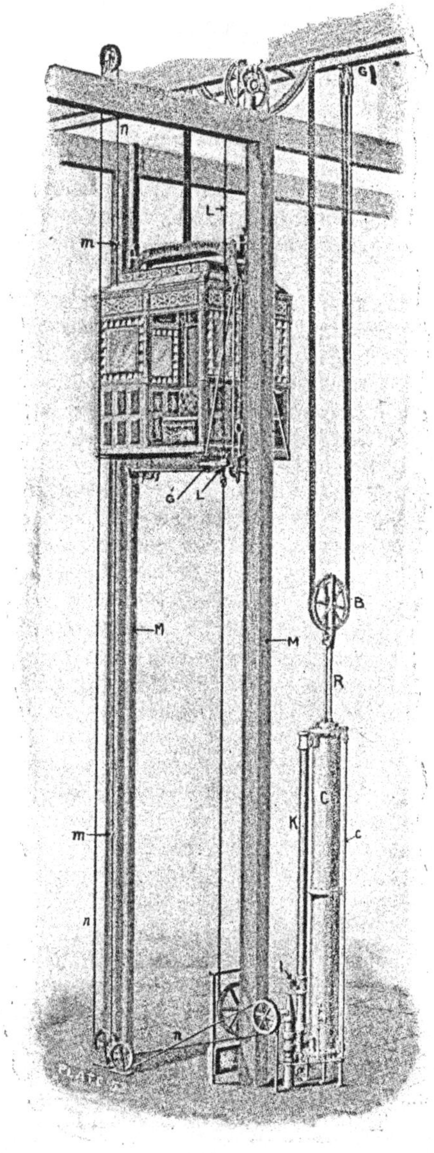

FIG. 25
OTIS VERTICAL HYDRAULIC ELEVATOR FOR MODERATE SPEEDS

being similarly situated. The small pipe *c* is for the purpose of carrying off the drip from the upper side of the top cylinder head, ordinarily, and also for the purpose of draining the water from the upper end of the cylinder, in cases where it is necessary to run the piston to the top of the cylinder to renew or adjust the packing. Some cylinders are arranged to be packed from the upper end and others from the lower end, the latter design being the one generally used in modern machines. As will be noticed, the pipe *c* connects at the bottom of the cylinder with other pipes that connect with the valve chest and the lower end of the cylinder. All these pipes are either to carry off the drip or to draw water from the various parts of the cylinder and valve chest when desired. Globe valves are placed in the drainage pipes so as to keep them closed normally.

The elevator is operated by the movement of the hand rope *n*, which passes around a sheave, at the side of the valve chest, which moves the valve through a rack and pinion gear, thence under two small sheaves at the bottom of the elevator well, and from there upward to the top of the well over another small sheave. One side of the hand rope passes through the elevator car, and by pulling this side up the operator causes the car to descend, while by pulling it down he causes the car to ascend. It will be noticed that near the top and bottom of the well balls *m* and *m'* are placed upon the hand rope. These balls are made of such a size that they cannot pass through the openings in the floor and roof of the car through which the rope passes; therefore, when the car, running upward, strikes the upper ball *m*, the latter goes up with the car and pulls upward the hand rope, thereby moving the control valve back to the stop position. Should the car fail to stop, the valve would be carried beyond the stop position and would connect both ends of the cylinder so as to cause the car to run down. This reversal of the motion of the car cannot occur when everything is in proper adjustment, for under such conditions, when the valve is completely closed, the car will stop. If, however, the car should run away by any mishap, it might run beyond the normal limit of travel, and then the valve would be slightly opened in the opposite direction, just enough to develop a retarding force sufficient to stop the car. The action when the car approaches the bottom floor is the same as when it approaches the top; that is, the lower stop ball *m'* is struck and carried down with the car, thereby closing the operating valve. The balls *m m'* are known as automatic top and bottom limit stops, and constitute one of the most valuable safety devices with which elevators are provided, although this fact is not generally realized as fully as it should be. Most men appear to regard them as convenient devices used to stop the car if the operator fails to do so, and to think that if they were not used, the car would simply strike the bumpers a rather hard blow. This would be the case with very slow running cars, but at high speeds, if the automatic limit stops were not used, serious results would be produced if the operator neglected to stop the car in time.

It can be readily seen that if the capacity of the lifting cylinder is sufficient, the car can be run at a much higher speed than is desirable if the valve is opened to its full capacity. To obviate this difficulty, stops are provided so that the operator when pulling on the hand rope

FIG. 26

cannot open the valve beyond the amount necessary to give the maximum car speed. These stops, which are usually mounted at some convenient point in the elevator well, are set above and below the stop balls *m m'*, so as to limit the distance through which the latter can be moved. In some cases additional stop balls are used, on account of its not being convenient to place stops to act directly upon *m* and *m'*. The positions of these stops that limit the amount of opening of the valve are determined experimentally when the elevator is installed.

To keep the car from striking the sides of the elevator well it is run between guides, shown at *M M* in Fig. 25. In the construction here illustrated the guides are made of hard wood. The car is guided at the top by shoes that fit freely against the guides and are provided with means for adjusting them so as to be neither too tight nor too loose. At the bottom the car is guided by jaws formed in a safety device that was formerly known as a safety plank, but at the present time is generally spoken of simply as a "safety." It received the name of "safety plank" from the fact that it is made of a massive hard-wood plank, varying from 4 inches thick and 11 inches wide in the smaller

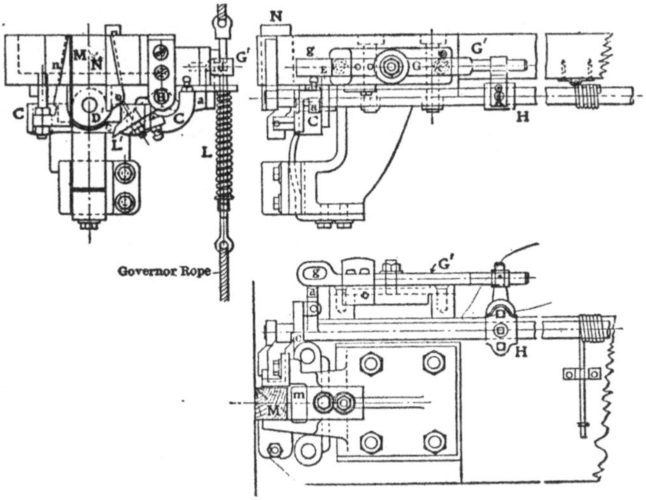

FIG. 27

sizes, to 5 inches thick and 15 inches wide in the larger ones. The jaws of this safety are reinforced with massive iron castings, and on one side are provided with a wedge that can be adjusted in position by means of screws, and on the opposite side with another wedge that can be forced between the guide and the jaw to stop the car if one of the lifting ropes breaks, or the car attains an excessive velocity from any cause. This safety is not very clearly shown in Fig. 25. By the aid of Fig. 26, and the drawing Fig. 27, which shows one end of the safety plank, its construction and operation can be fully understood. In the latter figure the governor rope rod *L* is shown only in the end elevation.

Looking at Fig. 26 it will be seen that the two lifting ropes that run down to either side of the car are connected with the ends of a rocking lever *C*. This lever *C*, as shown in Fig. 27, is pivoted at *D'*, hence if either one of the lifting ropes breaks, the end of the lever it is attached to will drop down. The shaft *H*, which runs under the car from one side to the other, carries at its end a lever *L'* that when raised lifts the wedge *N* and forces it into the space between the guide *M* and the side of the jaw of the safety plank. Whichever way the lever *C* may be tilted by the breaking of one of the lifting ropes, it will rotate the shaft *H* and lever *L'* in the proper direction to throw up the wedges *N* and thereby lock the car against the stationary guides *M*. The levers on the end of the shaft *H* are long enough to strike the guides *M*, when raised high enough, and are sharp at the ends so that they will cut into the guides.

It might be thought that if the wedge *N* is only raised far enough to catch in the space between the guide *M* and the safety-plank jaw it would be forced upward so tightly as to stop the car without further assistance. This would be the case if the wedge had a sufficiently long taper, but if it were so proportioned, it would require an enormously strong jaw to resist the bursting strain; moreover, the car would be so tightly wedged that it would require a greater force to release it than could be easily obtained. With the wedges of the proportions used, it is necessary to make the lever that lifts the wedge so that it will dig into the guide, and as the car moves down through, say, a foot or two in coming to a stop, the lever shaves the side of the guide, thereby not only forcing the wedge tighter against the guide, but producing an additional retarding force. When a car is caught by the safety, all that is necessary to release it is to start in the upward direction, and the force exerted by the lifting cylinder is enough to overcome the friction of the wedges against the guides.

In the foregoing we have shown how this safety acts providing one of the ropes breaks. Elevator cars, however, seldom drop when one of the ropes breaks, but frequently attain very high velocity when the ropes do not break, and on that account it is necessary that the safety be arranged so as to act when the speed reaches a certain point, no matter what causes the increased velocity. This result is accomplished in the safety shown in Fig. 25 by means of the Otis safety governor seen mounted on one of the overhead beams. This governor is driven by the rope *L* which is fastened to one end of the lever *G'*, as clearly shown at *G*, Fig. 26. The spring that holds *G'* is strong enough to keep the lever in the normal position and rotate the safety governor; hence, the latter will rotate at a velocity proportional to the speed of the car. A drawing of the governor is shown in Fig. 28, and it will be seen from this that the governor can be adjusted by means of the spring on the spindle to act at any desired velocity. The governor driving rope passes through the clamping jaws *H H'*, and when the governor speed becomes great enough to lift the rod *Z* and throw the jaws together, the rope will be clamped. Then, as the rope cannot move, the outer end of the lever *G'* on the safety plank will be held stationary as the car descends; hence, the shaft *H* will be rotated, throwing the safety wedges *N* into action to stop the car.

It is evident that the car can descend only as far as the upward movement of the end of the lever *G'* and the compression of the spring on *L* will permit before the rope will have to slide through the clamps

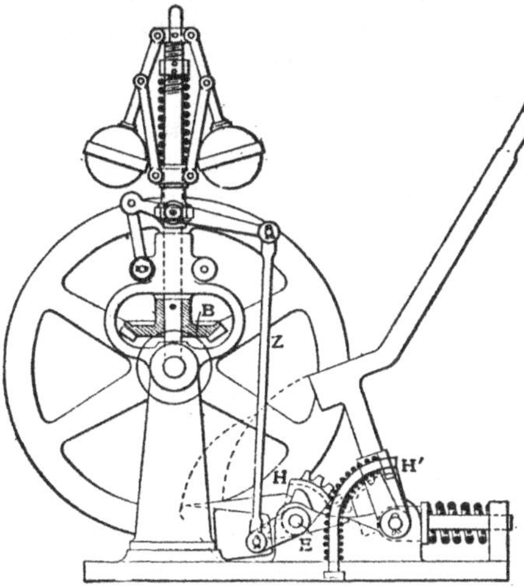

FIG. 28

H H' of the governor. Now as the distance through which the spring can be compressed, plus the movement of the end of *G'*, is only a few inches, it follows that unless the car is stopped very short, the rope *L* must break if it cannot slide through the clamps *H H'*. The distance in which the car will stop is always considerably more than the compression of the spring plus the movement of the end of *G'*; hence, while it is necessary for *H H'* to clamp the rope tight enough to move *G'*, the pressure must not be so great as to prevent the rope from slipping. For the same reason, in order to make the safety governor reliable it is necessary that the operating rope shall be in just as good condition as the elevator lifting ropes. The failure to inspect this rope properly and make sure that it is at all times in perfect condition has been a prolific cause of accidents.

The jaws of the safety plank and the wedge *N* should be kept clean and in proper adjustment at all times. As the guides *M* have to be kept well lubricated, it can be easily seen that if the safety jaws are neglected they will soon become clogged with a mixture of grease and dust, and this may give considerable trouble by causing the wedge to stick to the side of the guide and thus go into action when everything else is running in proper condition. The wedge *N* and the adjusting wedge on the opposite side of the guide will wear away gradually; therefore, the latter must be set up as often as required to keep the clearance between the guide and the safety jaw of the proper amount. If the clearance is too great the wedge *N* is liable to not catch firmly when called into action, and if the clearance is too small, the safety is liable to act without cause.

The operating valve shown in Fig. 25 is the same in general principle as the one shown diagrammatically in Fig. 7, but has several details of construction that are not illustrated in the latter. Its actual design can be readily understood from Fig. 29, which is a sectional elevation of the valve and the casing. The casing is made in three parts, marked *7*, *8* and *9*. The first forms the top, and provides a dome into which the rack *6* on the end of the valve rod can rise as the valve is lifted by the rotation of the pinion on the end of the shaft *A*. This shaft carries at its outer end the hand-rope sheave shown at the side of the valve in Fig. 25. The parts *7* and *8* are divided at the center of the shaft *A* and form a bearing for the latter. The lower part *9*, which is the valve casing proper, has ports *10* and *11* for connection with the lower end of the circulating pipe and the lower end of the cylinder in the manner indicated by Fig. 7.. The part into which the circulating pipe is connected forms a separate casting in Fig. 25, and the casing *9* is bolted to it. The port *12* in part *9* of the valve casing is for the purpose of connecting with the pressure-water supply if for any reason it is not desired to have this connection made in the circulating pipe. The valve casing is lined with brass tubing *4* and *3*. The former is simply for the purpose of providing a smooth surface for the cup packing of *V'* to slide against, but the latter is provided for the additional purpose of making ports of such a character that the cup packings of *V* may be able to slide over them readily. If the ports were large openings, the packings could not pass over them because on the up movement they would be caught by the edges of the ports. With the brass linings this trouble is overcome by perforating the brass with a large number of small holes, about one-quarter of an inch in diameter. The combined area of the holes is much larger than would be required in a single port, this increase in opening being provided so as to reduce the friction of the water running through the holes by reducing the velocity of flow.

The pressure of the water tends to force the valve piston *V'* upward, and the other piston *V* downward; both pistons being of the same diameter, the valve is balanced. The force necessary to move the valve

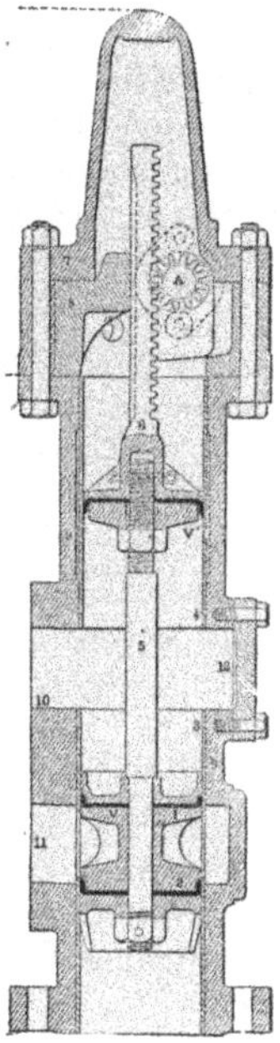

FIG. 29

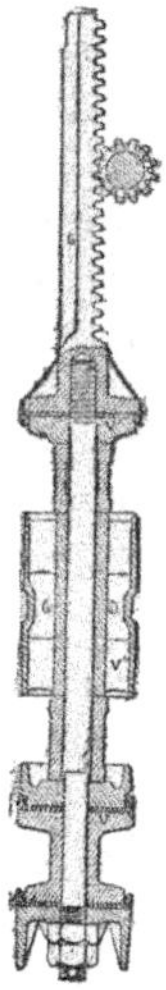

FIG. 30

is considerable, however, as the friction of the cup packings is great, being produced by the pressure of the water acting upon the entire surface of the leather in contact with the brass linings of the valve casing. On this account the pinion on the shaft *A*, through which the valve is moved, is made very small, while the hand-rope sheave is large —about 20 inches in diameter—so that while the valve travels a few inches in either direction the hand rope has to be pulled through a distance of from two to four feet, according to the size of the valve and the speed of car. For high car speeds the hand rope movement is increased, so that the automatic top and bottom stops may be able to arrest the movement of the car without making the stop abruptly.

In looking at the lower end of valve piston *V* it will be noticed that the lower head that clamps the packing *2* is made tapering; this is done so that in moving the valve down, to stop the car on the up trip, the outlet from the lower end of the cylinder may not be closed so quickly as to produce a violent stop. Even with this precaution it is possible for the operator to close the valve too rapidly; hence, in addition to this tapering of the main valve, a check valve is inserted in the passage that connects the valve casing with the cylinder. This check is directly under the lower end of the circulating pipe, so that if the operator closes the valve too suddenly the descent of the piston in the cylinder will not be arrested instantly, but will continue its movement, and force the water under it to pass through the relief check valve into the circulating pipe and thus to the upper end of the cylinder.

If the operator moves the hand rope so quickly on the down trip as to produce a violent stop, the piston will continue to rise in the cylinder and the water above it, which cannot pass to the lower end of the cylinder on account of the valve being closed, will be forced back through the inlet pipe *I* to the pressure tank. In this case, as no water can pass into the lower end of the cylinder, the continued upward movement of the piston causes it to leave the water and thus form a vacuum. This vacuum, combined with the pressure in the tank, soon arrests the movement of the car—in fact, in a very few inches—but the stop is not so sudden as to jolt the passengers, as would be the case if there were no relief for the water imprisoned in the cylinder.

One objection to having the connection between the cylinder and the pressure tank through the inlet pipe *I* is that if for any reason the pressure in the tank should drop, as by the springing of a bad leak, the water in the upper end of the cylinder can immediately run out, and with such freedom that if the car were at the top of the building, as in Fig. 25, it would attain a dangerous speed by the time it reached the bottom. This danger can be entirely obviated, however, by placing the pressure tank on the roof, so that the water in the cylinder has to run out against a head, due to the elevation of the tank. To this head is added the pressure of the atmosphere, because as the valve is closed no water can pass into the lower end of the cylinder, and as the piston runs up a vacuum is formed under it, and these combined pressures are sufficient to prevent the car from attaining a dangerously high speed in its descent.

When the pressure tank is placed in the basement the danger above referred to is avoided by using a valve of the type shown in Fig. 30. The difference between this valve and that of Fig. 29 is that it is provided with an additional piston *V″*, which is called the throttle valve. When this valve is used, the inlet pipe from the pressure tank is attached to the port *12*. When the elevator is stopped, the throttle valve *V''* is directly opposite the port *12*, and thus obstructs the flow of water from the port *10*. It will be seen that a groove is turned in *V''* at the center line; in addition the valve is not made a perfect fit in the valve casing, and the clearance afforded by these two features is sufficient to permit water to pass by in as large an amount as may be necessary to prevent too sudden a stoppage of the car, if the operator should close the valve

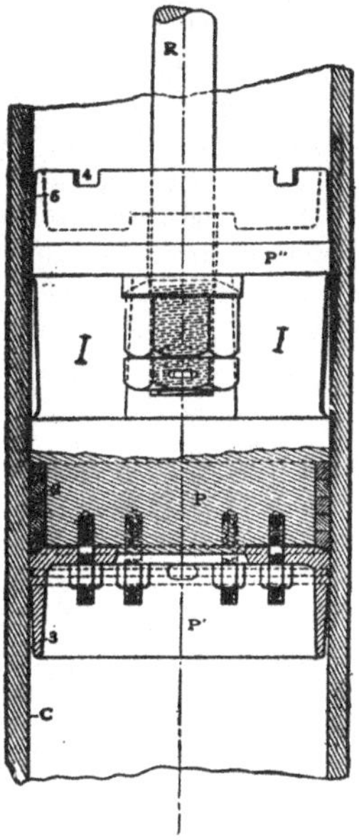

FIG. 31

too quickly; but it will not allow the water to flow through fast enough to enable the car to descend at a high velocity if the pressure in the tank should fail.

When the valve is moved in either direction to set the car in motion the water passes from the port *12* to the port *10*, there being sufficient opening around the throttle valve, even when the operating valve piston *V* is only slightly open.

The pistons used in vertical hydraulic elevators are made in several designs, some being arranged so as to be packed from the upper end, and others so as to be packed from the lower end. Fig. 31 shows one of the latest designs of pistons arranged to be packed from the lower end of the cylinder, which appears to be the favorite type now. The drawing shows a section through the complete piston, with packing in place, also a section of the cylinder *C*. Ordinary square packing is used, and this is held in position by a follower secured by six bolts. The parts *P* and *P''* are made to fit the cylinder, but the intervening section is cut away on opposite sides, so as to afford space for the ends of the piston-rods and their fastening nuts. The top and bottom parts of the piston are connected by the pillars *1* and *1*.

In packing these pistons it is necessary to be careful not to press the packing in too tight, as there is danger of bursting the cylinder by so doing, and even if this much damage is not done the friction caused by the excessive pressure may be so great as to prevent the car from attaining its full velocity. If a hard packing is used, and this is forced into place dry and very tight, the chances are that when it becomes well soaked it will expand enough to burst the cylinder. Bursting hydraulic-elevator cylinders is not a very rare occurrence, and when it does occur it is due to too great pressure of the piston packing against the sides of the cylinder.

CHAPTER VIII

LOW-PRESSURE VERTICAL TYPE ELEVATOR LEVER CONTROL

In Chapter VII we discussed the vertical-cylinder elevator of the type used in buildings from five to seven or eight stories high. The same style of elevator could be used for higher runs, by increasing the gear to three or four to one, but it is not desirable to do so, owing to the fact that as the hight of the building increases the car speed is increased, and hand-rope control is not satisfactory when the car runs faster than about 250 feet per minute. When a car runs at a low velocity the operator can control it perfectly by means of the hand rope; but as the velocity increases, the difficulty of making accurate stops at the floors of the building becomes so great that only the most experienced elevator operators can control the movement of the car in a manner to give satisfaction to the passengers. Furthermore, the difficulty of controlling a high-speed car by means of a hand rope is due not only to the increased velocity, but also to the fact that more effort is required to move the valve, as the latter must be increased in size, owing to the fact that more power is required to develop the higher speed. The effort to move the hand rope can be kept down somewhat by increasing the distance through which it must be moved to open or close the valve, but this does not help matters very much.

Even if the difficulty of car control could be entirely overcome, the hand rope would still be objectionable for high speeds, because it would be necessary to adjust the automatic top and bottom stops so as to retard the car too rapidly to make a satisfactory stop or else give the hand rope so much movement that the operator could not make at intermediate floors as quick stops as are often required. For these reasons the hand-rope control is not used in high-speed elevators intended for first-class service.

In place of the hand rope an operating lever is used, and as the movement of the end of this is only through a distance of about one foot, it would be next to impossible for the operator to move the operating valve by it if a direct connection were made, for it would require an effort that even an athlete could not exert for more than a few minutes. To get around this difficulty, a small valve operated by the car lever is provided, and this valve controls the flow of water in and out of a motor cylinder, the piston of which moves the main operating valve. This small valve is called a "pilot," and elevators in which it is used are said to be provided with "a pilot-valve and car-lever control." An elevator of this type, as made by the Otis Elevator Company, geared four to one, is shown in Fig. 32.

It will be noticed in this illustration that there is an extension at the top of the valve chamber, on the right side. This extension carries the pilot valve and also a sheave *1*. This sheave moves the pilot valve, and

FIG. 32

OTIS VERTICAL HYDRAULIC ELEVATOR FOR FIRST-CLASS PASSENGER SERVICE

the movement of the latter controls the flow of water into and out of the upper end of the main valve chamber, which, as will be seen, is of larger diameter than the lower portion. In this enlarged end of the valve chamber is located a piston that moves the main valve. The sheave *1* is rotated through a small angle, in either direction, by sheave *2*; the latter carries on one side two smaller sheaves over which pass ropes *n* that run up the elevator well and at the top of the building pass over two similar sheaves which are stationary. These ropes are connected with a lever mounted on the rear end of a shaft running under the elevator car, and which carries at its front end a lever that projects up through the floor of the car. This is the operating lever, and by turning it in one direction the sheave *1* is caused to rotate in a corresponding direction, say in the direction that will move the pilot valve so as to cause the car to ascend. Moving the car lever in the opposite direction will cause the car to descend.

The pressure water enters through pipe *A* and is discharged through pipe *D,* in the lower end of the cylinder. At *B'* is placed a speed-regulating valve, the office of which is to prevent the car from running at too high a velocity if the operator throws the main valve wide open when the load is light. At *B''*, between the main valve and the lower end of the cylinder, is located a valve which stops the car automatically at the top and bottom floors of the building. This valve is actuated by the rope *5*, which passes through the end of arm *7* projecting from one side of the traveling-sheave frame. The rope passes over a sheave *6* and thence runs down and around a sheave mounted on the back end of a shaft carrying a pinion which meshes into the gear on the automatic stop valve *B''*.

The illustration shows a car running on iron guides, which are the only kind permitted in fireproof buildings. For metal guides the wedge safety shown in Fig. 25 cannot be used, as the friction of wood against iron is too uncertain, being considerable if there is any grit between the surfaces, and slight if there is no grit and the surfaces are well lubricated. The safety shown in Fig. 32 is of the type known as a clamp, or brake safety. It is also called a drum safety, or a toggle-joint safety. It is given all these names because it is a brake made in the form of heavy shoes that are clamped tightly against the elevator guides by means of the toggle joint, or equivalent devices, operated by the rotation of a drum placed in the safety at a point between the two ends.

The construction and operation of the several parts of Fig. 32 can be better explained in connection with the line drawing Fig. 33, and in other line drawings which follow. In Fig. 33 the speed-regulating valve *B'* is not shown, and the rope connections for transmitting the motion of the car lever to the pilot-valve sheave *1* are not the same. There are many modifications of these rope connections, but all of them are classified under two generic systems known as the "running rope" and "standing rope." The arrangement of Fig. 32 is of the running-rope type, while that of Fig. 33 is of the standing-rope.

THE STANDING-ROPE SYSTEM.

The operation of the standing-rope system can be more easily understood from Fig. 34, which shows the rope connections complete and

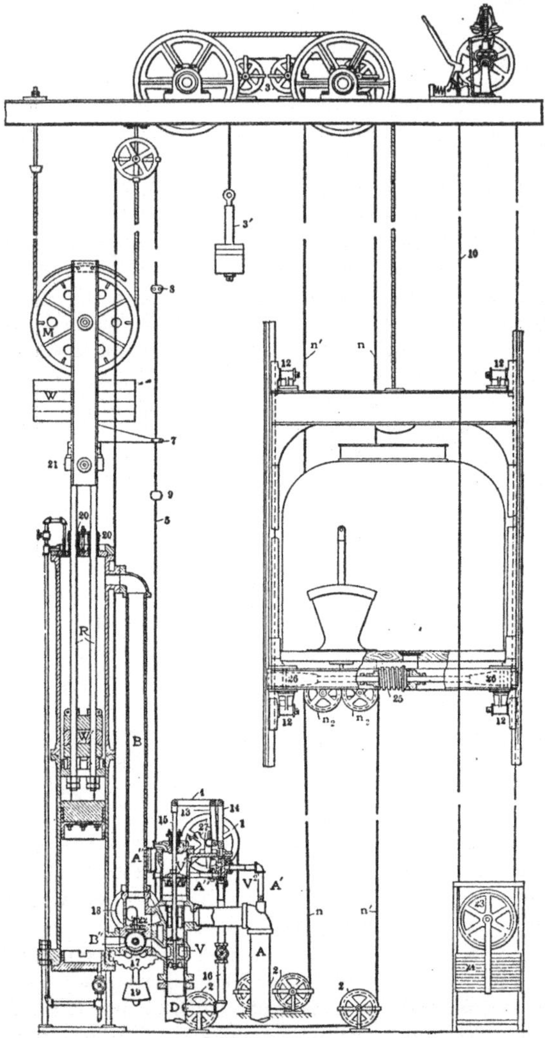

FIG. 33
OTIS VERTICAL HYDRAULIC ELEVATOR

free from other complications. The rope *n'* comes down from top sheave *3* on the right, and passes under the right-side sheave carried by the cross lever connected with the car-operating lever. From the under side of the right-side sheave the rope passes to and around the top of

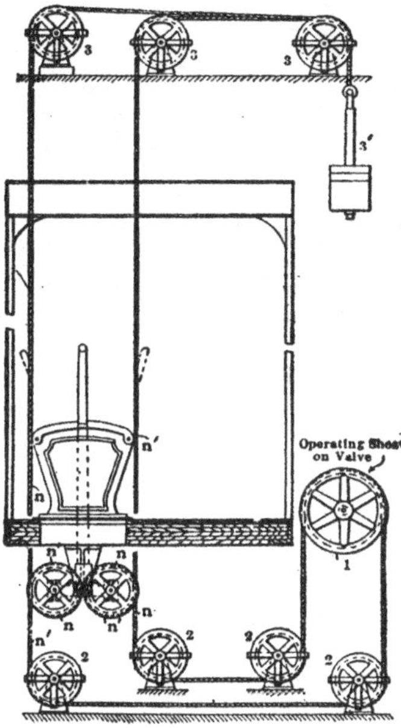

FIG. 34
ILLUSTRATING OPERATION OF STANDING-ROPE SYSTEM

the left sheave, and thence down and under the two lower stationary sheaves *2*, *2*, then up and around the pilot-valve operating sheave, and under the two central stationary sheaves *2*, *2*, thence over the right and under the left sheaves carried by the car lever, and so on up over top sheave *3*, where the two ends of the rope are connected and from their end is suspended the tension weight *3'*. If the car lever is moved to the right, the ropes *n* will slacken up while the ropes *n'* will draw up, from the fact that the right-side sheave will be depressed and the left-side one elevated. This change in the ropes will rotate the pilot-valve sheave clockwise. If the car lever is moved to the left, the action will be just the opposite and the valve-operating sheave will be rotated counter-clockwise. Thus, through the taking up of one rope and the letting out of the other, the pilot-valve operating sheave is made to move in accord with the movement of the car lever. This system, which as already explained is made in several modified forms, is called the standing rope, because the ropes remain stationary while the car is running. In running-rope systems, the ends of the ropes are fastened to the car lever, or two ends to the lever and two to the car, and the ropes run when the car runs.

Returning to Fig. 33, it will be seen that when the sheave *1* is rotated clockwise, the lever *4* is pulled down by the connecting-rod *13*, the lower end of which is connected with the crank on the sheave shaft. This downward movement of *4* will depress rod *14*, and thus move downward the pilot valve V^2, simply because it is easier to move this end down than the opposite end which is connected with the top of the main valve-rod. When the pilot valve is moved down, pressure water will flow from pipe *A* through the small pipe *A'* to and through the valve to pipe *A''*, and thus to the upper end of the main valve chamber, above the motor piston *V'*.

The pressure water is at all times in the space between valve *V* and the under side of *V'* and also in the space above *V'*, but so long as water cannot pass in or out of this latter space the main valve cannot move, for it could not move upward without compressing this water, and it could not move downward without forming a vacuum above *V'*. As there is pressure water above and below *V'*, this valve taken by itself is in a state of perfect balance, but the main valve *V* is unbalanced, because the pressure water acts against its upper side, while only the pressure of the discharge tank acts against its lower side. As a result of this state of things, as soon as water enters the space above *V'*, this piston will move down, carrying the main valve with it, thus connecting the top and bottom of the cylinder, so that the water may circulate and the piston run upward. It will be noticed that as soon as *V'* begins to move down by the admission of water above it, the pilot valve is moved upward, because rod *13* will be held stationary by sheave *1*. When *V'* moves down a certain distance, it will have moved the pilot valve upward to the central or stop position; that is, the position in which the flow of water into the space above *V'* is stopped; hence, when this point is reached *V'* will descend no farther.

If the sheave *1* is turned counter-clockwise, rod *14* will be lifted and thus pull up the pilot valve V^2, and this movement will connect pipe *A''* with pipe *16* and permit the water in the space above *V'* to run out into the discharge pipe D, and the main valve to ascend and connect the lower end of the cylinder with the discharge pipe; so that the piston may run down and pull up the elevator car. In this movement it will again be seen that as soon as *V'* begins to move upward, it will move rod *14* downward, and thus depress the pilot valve until it reaches the stop position and closes off the connection between A'' and the pipe *16*. To fully understand this movement of V' it must be remembered that the pressure in pipe *16* is much below that in the pipe *A*, so that when A'' is connected with *16* the pressure above V' is greatly reduced, and the latter being unbalanced will be forced upward by the pressure under it, which will be greater than the downward pressure on *V*, owing to the difference in the diameters of the two pistons.

The distance through which the sheave *1* is rotated depends upon the distance through which the car-operating lever is moved, so that if the latter is moved a short distance the sheave *1* will rotate through a small angle. In this case the pilot valve will be shifted but a short distance from the central position, and, therefore, the main valve will only have to move a small portion of its full stroke to return the pilot to the central position. If the car lever is moved as far as it will go, the rotation of the sheave *1* will be through a wide angle, and the pilot will be shifted a considerable distance from the central position. In this case the main valve will have to move its full stroke to return the pilot to the central position. Thus it will be seen that in every case the opening of the main valve is directly proportional to the distance through which the car lever is moved, and the control the operator in the car has over the movement of the main valve is as complete as it would be if he stood by the side of the valve and effected its movement by means of a lever attached directly to the piston-rod *15*.

CHAPTER IX

PILOT-VALVE CONTROL—REGULATING VALVES

All modern hydraulic elevators intended for first-class passenger service are provided with a pilot-valve control, and although the design in the various types may differ considerably, the principle of operation is the same as that of the valve here shown. This design, however, is the best one from which to gain a thorough knowledge of the principles of operation, because it shows them up very clearly. The stop screw *27* is a safety feature provided to prevent the operator from opening the valve wide enough to permit the car to run away when running up with a light load. When the speed regulator *B'*, shown in Fig. 32, is used, this stop simply becomes an additional precaution, as *B'* alone is capable of taking care of the car speed, if properly adjusted.

The construction of the main valve of Fig. 33 can be easily understood from the drawing Figs. 35 and 36. Fig. 35 is a vertical elevation in section, showing the valve in the stop position. Fig. 36 is a horizontal section taken on line *A—B,* just above the port with which the pressure pipe *A,* Fig. 33, is connected. The lower portion of the valve is the same as that shown in Fig. 30, and so is the corresponding part of the valve chamber.

The upper part differs from the last-named valve simply in that it is of a larger diameter. In the simple hand-rope valve the pistons *V* and *V'* are made of the same diameter, so that the pressure of the water pushing up against *V'* may be just equal to the pressure pushing down upon *V,* to produce a perfect balance and make the valve as easily movable upward as downward, disregarding the weight of the valve itself. In the valve here shown it is necessary to make *V'* of a larger diameter than *V,* so that when the pressure above *V'* is reduced the valve may become unbalanced and thus be forced upward by the pressure acting on the under side. This valve is provided with the throttle valve *V''*, which will prevent a runaway if for any reason the pressure in the tank drops to zero, or nearly so, and it also prevents making a violent stop on the up trip if the operator moves the lever to the central or stop position too suddenly.

The construction of the pilot valve is well shown in Figs. 37 and 38, the first being a sectional elevation of the valve and casing, showing the valve in the stop position, and the second a sectional elevation of the casing alone, taken at right angles to Fig. 37. The valve consists of a sleeve mounted upon the spindle V_3, which serves to hold in place cup packings *40*, *41* and *42*, as clearly shown. The pressure water enters through the port *45*, and passes to the pipe A'' through the port *46*. The discharge water from the upper end of the main valve chamber enters through the port *46* and discharges through the lower end *47* into the pipe *16* (see Fig. 33). The valve chamber is provided with brass linings,

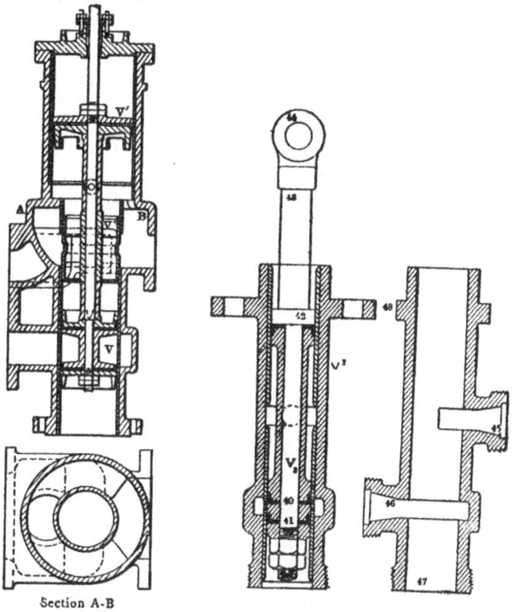

Section A-B
FIGS. 35-36 FIG. 37 FIG. 38

and the lower lining has numerous holes drilled through it opposite the port *46*, so that the cup packings may pass over them on the up stroke without being caught.

In Fig. 37 it can be seen that the packings only lap a short distance over the port holes drilled in the brass lining; therefore, the valve does not have to be moved very far in either direction to uncover the ports and start the elevator. The amount of lap of the valve packings is made sufficient to cause the valve to close when the car lever is some distance from the central position. If the valve were made with less lap, the operator would have to move the lever with great accuracy to stop the car, for if it were not moved far enough the car would slow down to a very low speed, but would not stop; and if the lever were carried a trifle further the car would stop and then begin to run in the opposite direction. To obviate this trouble the valve packings are made to lap enough to require a considerable movement of the car lever to either side of the center to open the ports, but not any more than is actually necessary for the proper operation of the car.

In the simple hand-rope control elevator the car is stopped automatically at the top and bottom landings by the closing of the main

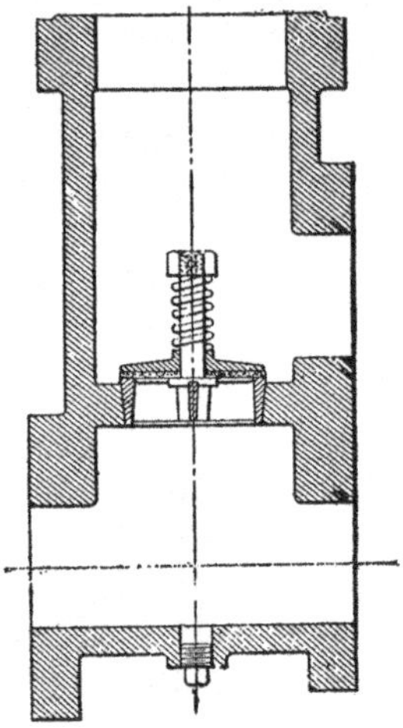

FIG. 39

valve through the movement of the hand rope by the motion of the car itself, the latter striking stop balls fastened to the rope at the proper points. In the pilot-valve control, as can be seen in Fig. 33, a separate stop valve *B″* is placed between the valve chamber and the lower end of the cylinder. This valve is actuated by the rope *5* which has stop balls fastened to it at proper points, as shown at *8* and *9*. These balls are struck by the arm *7* when the car comes within proper distance from either end of its travel for the automatic valve to start to close. The closing of this valve is gradual, so that it does not entirely check the flow of water until the car has traveled a distance that may vary from about three feet up to six, eight or more feet, depending upon the speed at which the elevator runs. The distance within which the valve is closed is determined by the size of the sheave *18*, the gear *17*, and a pinion that drives *17*.

Looking at Fig. 33 it can be seen that if the car is going up the arm *7* will descend and at the proper time will strike the stop ball *9*, and thus carry the rope *5* with it so as to rotate the sheave *18* in a clockwise direction, and the pinion of the sheave-shaft will rotate *17* and therefore

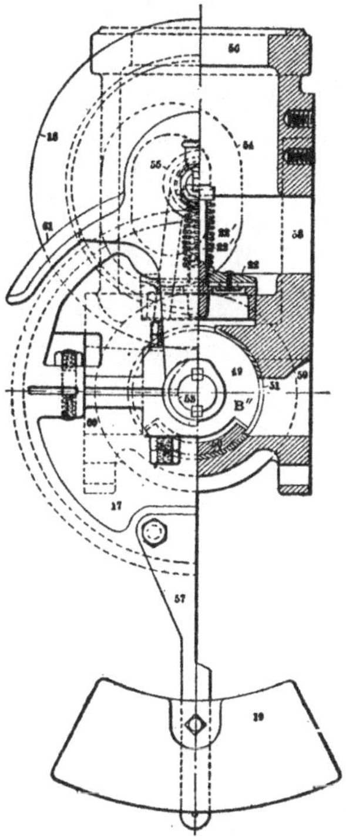

FIG. 40

the valve *B″* in a counter-clockwise direction, so that the valve will stop off the flow of water from under the piston into the pipe *D*, by swinging up and over the port on the right side of the circular valve chamber. If the car is running up at full speed, and is stopped entirely by the action of *B″*, it may shut off the flow of water faster than the momentum of the moving parts is overcome by the action of gravity, so that the car may continue moving after the valve is closed. In such a case the car ropes would tend to slack up, as the piston could not move any further, owing to the incompressibility of the water. In a case of this kind there would be no such occurrence with the valve *B″* as constructed, because the relief valve *22* would be raised, and the water imprisoned under the piston would find an outlet into the lower end of the circulating pipe, and thus to the upper end of the cylinder.

It is very seldom that the valve *22* is called into action through the too sudden stopping of the car by the action of the valve *B″*, as this

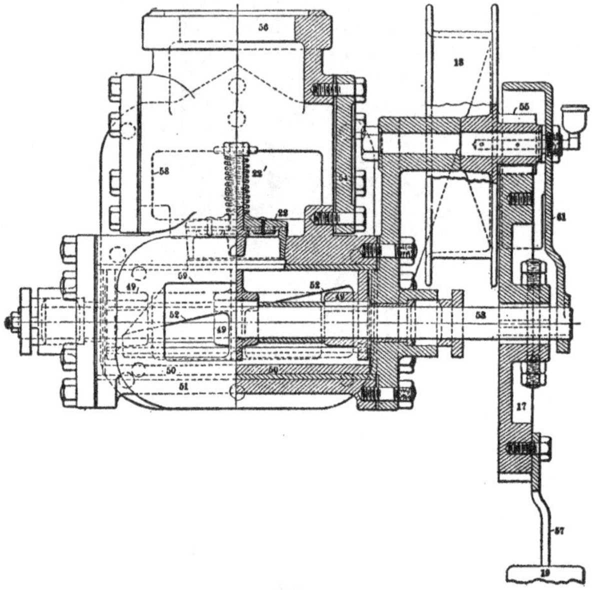

FIG. 41

valve is always adjusted so as to not act so quickly as to produce such an effect, but sometimes valves can get out of adjustment. The principal value of the relief valve *22* is to prevent too sudden stops when the operator pulls the lever to the stop position too rapidly. This valve is used even with the simple hand-rope control, as was mentioned in Chapter VII, the casting which forms the foot of the circulating pipe being of the form shown in Fig. 39, which differs from the casting shown in Fig. 33 only in not having a circular chamber to accommodate the valve *B″*.

AUTOMATIC STOP VALVE.

The construction of the automatic stop valve is fully shown in Figs. 40 and 41, the first of which is a view taken at right angles to the axis of the shaft, with the right side in section. The other drawing is a view parallel with the valve-shaft, and also shows the right-side half in section. The face of the valve is formed by the segment *50*, which is held in position and moved by three-flanged hubs, *49*, these latter being firmly secured to the shaft *53*. The gear *17* is mounted on this shaft, and is driven by the pinion *55*, which is cast on one side of the sheave *18*. The segment *50* rotates within a brass lining *51*, which has ports cut through it on both sides, the upper edge of the ports being on an inclined line, as shown at *52* in Fig. 41; this construction being for the purpose of closing the port gradually so as to not stop the elevator too

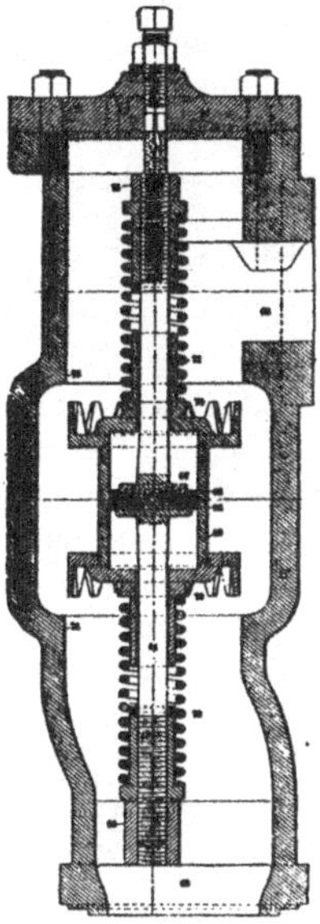

FIG. 42

suddenly. The segment *50* is not rigidly secured to the hubs, *49*, for the reason that, if so made, it would not be possible to start the elevator from either end of the well.

This valve moves into the position to close the outlet from the lower end of the cylinder, at the end of every trip, and is held in that position by the arm *7* bearing against one or the other of the stop balls, *8*, *9*; therefore, unless the piston moves away from the end far enough to permit the stop ball to move the valve, *B″* will remain closed, so that the opening of the main valve would not have any effect in starting the elevator. With the segment *50* made so as to have considerable play in the supporting flanges of the hubs *49*, as is clearly shown in Fig. 40, there is no difficulty in making a start, because when the elevator is stopped, the current of water flowing through *B''* is in such a direction as to press the segment *50* against the casing *51*, and when the main valve is moved so as to run the elevator in the opposite direction, the pressure of the water acting on *B''* will be toward the center of the circle and thus will push *50* away from *51*, and permit a sufficient amount of water to leak through to start the car gradually. As soon as the car begins to move, the arm 7 permits the stop ball to return to the normal position, and then the weight *19* draws the gear *17*, and thereby the valve *B''* into the open position. From this it will be seen that this valve not only acts as an automatic limit stop, but also it provides means whereby the car is prevented from starting off with a violent jerk if the operator moves the lever too rapidly to the full-speed position.

SPEED-REGULATING VALVE.

The construction and principle of action of the speed-regulating valve *B'*, shown in Fig. 32, will be more fully understood by reference to the line drawing Fig. 42, which is a section parallel with the axis of the valve-rod. The port *63* is bolted against the upper end of the cylinder, while the circulating pipe is screwed into the lower end *62*, thus whether water is passing from the pressure tank into the upper end of the cylinder, or the water is simply circulating from the upper end to the lower end of the cylinder, it must pass through this speed-regulating valve. Suppose the elevator is running upward, then water will pass from the pressure tank, up through the regulating valve to the upper end of the cylinder. This current of water will strike against the lower end of the valve *70*, and push it upward against the tension of the spring *71*. If the elevator is running down, the stream of water will flow from the port *63* down through the regulator and into the circulating pipe. In this case the force of the water will act against the top of the valve *70* against the tension of the spring *72*. By properly adjusting these springs the device can be arranged so as to act at any desired velocity. The action depends entirely upon the velocity of the water flowing through the valve; therefore, it is not affected by any variations that may occur in the pressure of water. The operation of the valve is so self-evident as to require no very extended explanation. It can be seen that on the down stroke of the piston, if the valve is forced upward far enough by the impact of the stream of water, the upper end will enter the neck *73*, and thus gradually reduce the opening through which the water must pass, and thereby reduce the volume that passes, hence the velocity. On the up-stroke of the piston, the valve is the neck *74*, of the valve chamber. The interior valve ose of preventing too sudden a movement of the valve as a dash-pot. The packings *68* make a tight joint, and t of leakage to give the desired retardation is obtained necessary difference between the diameter of the rod through the ends of the valve. The springs *71* and *72* d more or less, as may be required, by means of the at their ends.

CHAPTER X

SAFETY DEVICES

The safety device shown on the car in Figs. 32 and 33 is of the clamp, or brake, type. It is known as the "Otis wedge-clamp safety," and its construction and operation can be fully understood from Fig. 43 and the line drawings that follow. The lower part of Fig. 43 shows the under side of the elevator car, with the safety attached thereto. The center portion shows the roof of the car and the device placed thereon to hold the governor rope. The upper portion shows the safety governor, whose office is to actuate the safety. For the purpose of showing clearly the construction of the safety, the side of the framing that is in front of these parts is drawn as if it were transparent, and while this type of illustration serves admirably to accomplish the object intended, it is liable to be misleading if it is not kept in mind that this side is made of a channel beam, so that in the actual apparatus the internal parts cannot be seen at all. The operation of the safety is as follows:

The end pieces *26* form powerful clamps that are pressed against the same elevator guides that the guide-shoes *12* run on. At the inner ends of these clamps rollers *80* are mounted, and between these wedge-shaped pieces *79* are forced by the rotation of the drum *25*. The forcing of these wedges between the rollers *80* spreads the inner ends of the clamps *26* and forces the outer ends against the elevator guides, thus developing a frictional resistance that increases gradually as the pressure of the clamps is increased, until the car is brought to a stop. The rotation of the drum *25* causes the shafts *78* to move outward, and thus force the wedges *79* into the space between the rollers *80*. The drum *25* is rotated by the rope *11*, and this rope is actuated by the safety governor in the following manner: The governor rope *10* is in effect endless, as shown in Figs. 32 and 33, its ends being fastened to the piece *10'*. This piece has conical projections on its sides that fit into depressions in the ends of the levers *10"*. These levers are pivoted at 10_1, and their back ends are forced apart by a spring 10_3. The tension of this spring is sufficient to hold the levers *10'* firmly enough to drive the safety governor; so that in the normal running of the elevator the parts remain in the position in which they are shown, and the governor is driven at a speed that corresponds with the velocity of the car.

FIG. 43

The stand 10_2 holds two small sheaves, in addition to the levers *10″*, and the rope *11* passes over the lower sheave and around the top one, its end coming forward and being firmly secured to *10′*, at the point *11″*, as shown in the illustration. If the elevator car attains a velocity higher than that for which the governor has been adjusted, the balls will swing outward so far that the lifting-rod will throw the rope clamps into action,

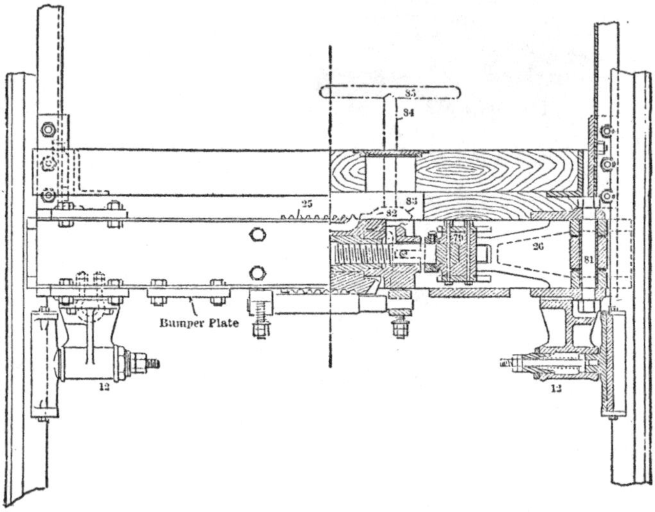

FIG. 44

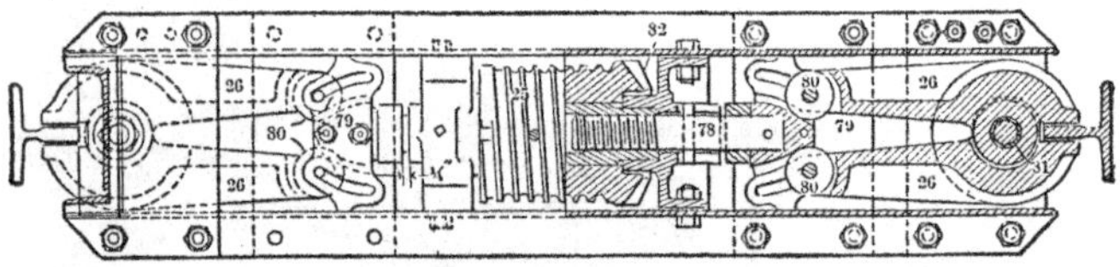

FIG. 45

and thus immediately stop the movement of the latter. As the car still continues to move, the piece *10′* is pulled away from the ends of levers *10″* and then the rope *11* is drawn upward and the drum *25* is rotated. As soon as the drum begins to rotate, the pressure of the clamps *26* is applied to the elevator guides, and the farther the car moves after this action, the more the drum rotates and the tighter the clamps are applied, until the braking force becomes sufficient to stop the car.

A very good feature of all these clamp safety devices is that they cannot stop the car suddenly, but in every case the retardation is gradual, so that the car may not be brought to a state of rest in less than three or more feet. If the governor rope *10* and the drum driving rope *11* are sufficiently strong, and in good condition, the safety is sure to stop the car, because the farther it moves the greater will be the pressure of the clamps *26* against the elevator guides, so that after a while the braking force must become great enough to overbalance the momentum of the moving mass.

The construction of the several parts of this safety can be better understood from the line drawings Figs. 44 and 45, the first being a side

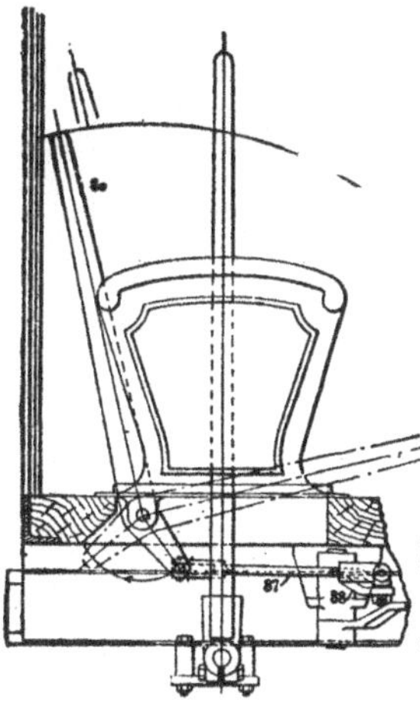

FIG. 46

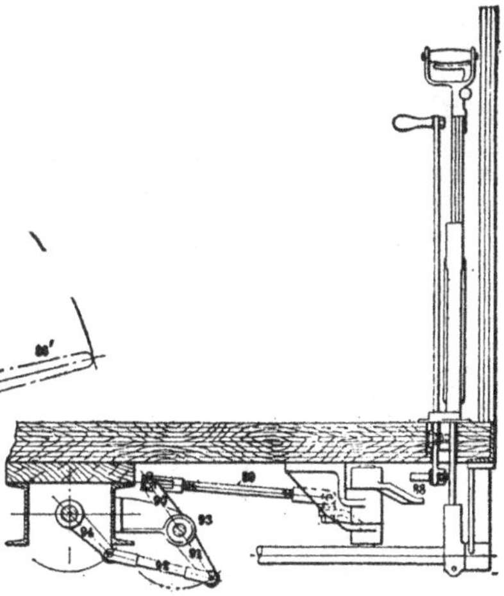

FIG. 47

view and the second a plan. The brake clamps *26* are hinged on a strong steel stud *81*, and carry at their inner ends rollers *80*. The wedge-shaped pieces *79* are mounted on the ends of the shafts *78* and are prevented from getting out of line by side guides. The shafts *78* are prevented from turning by pins that slide in slots in the bearings in which they are held, as is shown in Fig. 45. The inner sleeves of the drum are threaded to fit the screw ends of the shafts *78*, with right- and left-hand threads, so that when the drum rotates the shafts are forced outward, and the wedges *79* are driven in between the rollers *80*, thus forcing the outer ends of the clamps *26* together, with an ever-increasing force. The bevel gear *82* at the end of the drum is for the purpose of releasing the clamps after the safety has gone into action. This gear is used in connection with a wrench that consists of a pinion *83*, shown in broken lines in Fig. 44, mounted on the lower end of a stem *84* that is constructed at the top with a handle *85*. A trap-door is provided in the car, and by removing this the wrench can be put in position and by means of it the drum can be turned backward to the normal position so as to release the clamps *26*.

Fig. 44 also shows the construction of the elevator guide-shoes *12*. As will be noticed, these are provided with adjusting screws, and as they work on a swivel, they can be set so as to have but little play.

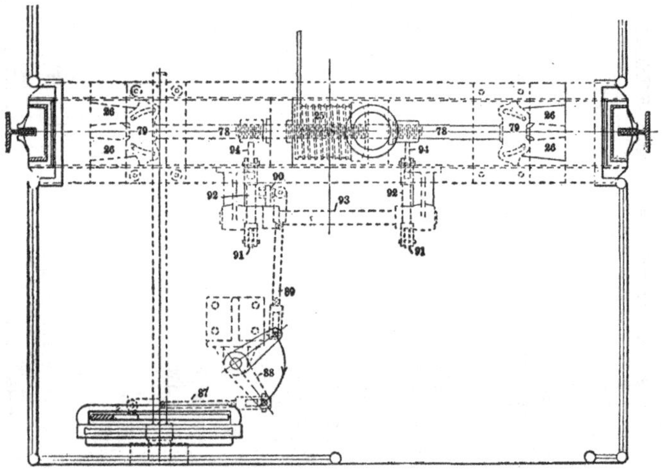

FIG. 48

On this account the brake clamps *26* can also be set close without fear of their rubbing against the elevator guides and thus getting so loose that the clamp will not act soon enough, or with sufficient force when called into action.

In Fig. 32 it will be noticed that there is a second lever in the car on the right side of the operating lever. This is an emergency lever used in most first-class elevators to enable the operator to retard or even stop the car, if for any reason he cannot control it by means of the regular lever. The construction of this device (which is simply an addition to the clamp safety) and its operation can be understood from the drawings Figs. 46, 47 and 48, the first of which is an elevation giving a side view of the main and the emergency levers. Fig. 47 is an elevation at right angles to Fig. 46, looking at the latter from the left side, and gives an edge view of the two levers. Fig. 48 is a plan of one-half of a car, showing the safety device and the connecting-rods and links connecting in with the emergency lever, in broken lines. In this arrangement, the shafts *78* are made in two parts and are joined by the hubs of the cranks *94*, the latter having right- and left-hand threads in their bore, into which the ends of *78* are screwed. When the emergency lever *86* is moved toward the position *86'*, it acts through the connecting levers and rods to turn the cranks *94* in the direction that will force the ends of *78* out, spread the wedges *79* and apply the brake clamps *26*. The several cranks and connecting-rods are numbered consecutively, so that following the numbers the way in which the motion of *86* is transmitted to *94* can be easily traced. This device can be used to stop the car if the lever *86* is moved far enough, but it would be better to apply it with enough force to simply retard the velocity of the car to, say, half speed or less, and thus descend safely to the bottom floor of the building

CHAPTER XI

GOOD FEATURES OF MAGNETIC VALVE CONTROL; OBSTACLES THAT PREVENT ITS GENERAL USE—DESCRIPTION OF CONTROL SYSTEMS

The running-rope and standing-rope arrangements for moving the pilot valves of hydraulic elevators enable the operator to control the movement of the car perfectly, i. e., to cause it to run at any desired speed, from maximum to a barely perceptible motion, and to start and stop as slowly or rapidly as he may desire; but, structurally considered, these arrangements are not entirely free from objection. Comparing them with the simple hand rope, it can be seen that they have several additional parts, and this alone increases the first cost and makes them more expensive to maintain. In practice, however, their operation has proved entirely satisfactory, with the single exception of their liability to get out of adjustment through the stretching of the ropes. When in proper adjustment, the car lever will stand vertical when the valve is closed. If when the ropes stretch they both lengthen out the same amount, the adjustment will not be disturbed, and the fact that the ropes have stretched will not be noticed, as the tension weight will take up the slack; but if one rope lengthens out more than the other, then the car lever will not stand vertical when the valve is closed, but will incline to one side or the other. If the difference in the stretch of the ropes is not great, the "stop" position of the lever will vary so little as to make no difference in the operation of the car; but in some cases the lever will stand several inches from the central position when the car is stopped, and then it becomes necessary to readjust the ropes. This is accomplished by taking up the rope that has stretched the more, until the car lever is brought to the central position, when the valve is closed. When the ropes are new they may require frequent adjustment, but after a few months' operation practically all the stretch will have been taken out of them, after which slight trouble will be experienced.

As the standing-rope and running-rope systems comprise considerable mechanism, it has been thought that some sort of magnetically operated device could be used that would be far more simple and compact, and certainly easier to handle, as a small electric switch could be substituted for the operating lever. This would be a decided gain in buildings where many passengers are carried, since considerable room is required for the proper manipulation of the operating lever, while for the movement of a small electric switch very little space is needed. Another advantage would be that the ropes and sheaves, which take up quite a lot of space, would be replaced by small wires which could be located in any convenient position, the connection with the car being made through a flexible cable similar to that used for conveying current for incandescent lamps.

Many inventors have endeavored to produce an electric controller, but up to the present no one has brought out anything that is of sufficient merit to enable it to replace the mechanical devices generally applied to first-class elevators in large office buildings. The principal objection to magnetic controllers is that they do not enable the operator to control

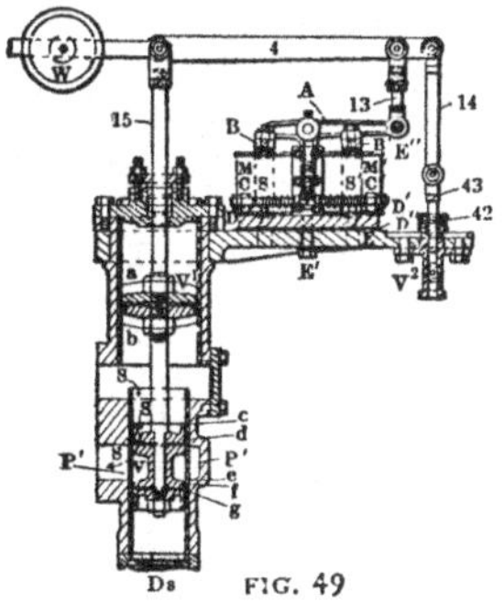

FIG. 49

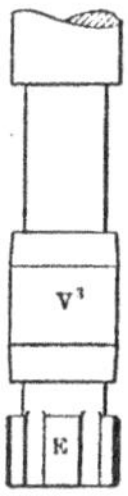

FIG. 50

the movement of the car as closely as can be done with either the standing-rope or the running-rope method. Another objection is that they are more liable to get out of order. For these reasons, then, these controllers are not generally used for passenger elevators in large buildings. Nevertheless, they have a field of their own in which they are decidedly superior to mechanical devices, and in which they have practically superseded all other methods of control.

In many buildings owned by insurance companies and other large corporations, elevators and dumb-waiters have been installed for the exclusive use of officers or employees. Such elevators are not in constant use, and therefore are not provided with operators, but each passenger operates the elevator himself. As every one who desires to use the car cannot be expected to be an experienced elevator operator it is necessary to provide a controller so simple that any one can use it, and this simplicity can be easily obtained in the magnetic controllers, which require only a switch or push-button at each floor, and similar devices in the car.

MAGNETICALLY CONTROLLED PILOT VALVE.

A magnetically controlled pilot-valve gear, such as is used under such circumstances, is shown in Fig. 49. The operation of this device is as follows: The lever *A* is actuated by the pull of the magnets *M* and *M'*, which attract the armature *B* and *B'* when an electric current is passed through the magnet coils. The arm *A'*, extending downward from *A*, is held between the two horizontal rods *C* and *C'*, which are pressed against it by the springs *S* and *S'*. The nuts *D* and *D'*, together with the threaded joint in rod *13*, comprise means for adjusting the

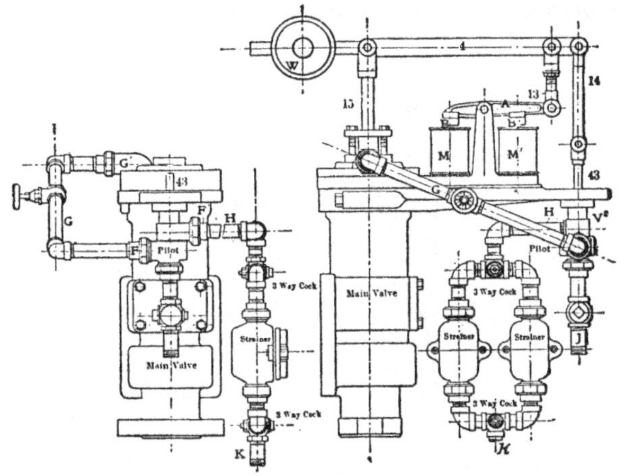

FIG. 51 FIG. 52

position of the pilot valve, the rod being set when the mechanism is put together, and the nuts being used for the final adjustment.

The magnets are of the horse-shoe type, there being two coils each in *M* and *M'*. The armatures *B* and *B'* are long enough to cover both poles in the magnets. By comparing Fig. 49 with the mechanically operated pilot valve, it will be seen that the magnets *M* and *M'* displace the rope sheave, and the lever *A* displaces the crank on the end of the sheave shaft. In all other respects the two arrangements are the same; the pilot valve, however, is solid in the magnetically operated system, instead of being provided with cup packings. This difference is made in order that the valve may move more easily, and therefore not require the use of excessively large magnets. This solid pilot valve is hardened and ground to insure a perfect fit. Its actual construction is not clearly shown in Fig. 49, but will be fully understood from the enlarged illustration, Fig. 50. The lower end E is of the same diameter as the body of the valve V^3, in order to prevent springing and the consequent wearing of the edges of the port recessed in the valve chamber. To provide an opening for the water to escape through the lower end of the chamber into the discharge tank, the enlarged end E is cut away at four places, as clearly shown.

As the pull of the magnets is limited, it is necessary to guard against the entrance of chips or other solids into the valve, because these might be caught between the latter and the edges of the ports in the valve chamber and lock the valve. For this purpose strainers are provided through which all the water passes to the pilot valve. The way in which these strainers are connected, and the general arrangement of the entire valve-actuating piping, are fully shown in Figs. 51 and 52. Two strainers are provided, so that it may not be necessary to stop the elevator to clean them. Only one of the strainers is in use at a time, the two three-way cocks being provided for the purpose of connecting either strainer with the piping. The one out of service can be opened and cleaned at any time. The lower end of the bottom three-way cock is connected with the pressure tank through the pipe K, and the lower end of the pilot-valve chamber is connected with the discharge tank through pipe J. The rest of the piping is so clearly shown as to require no explanation.

CHAPTER XII

MAGNETIC CONTROLLER FOR BATTERY CURRENT

The controller shown in Fig. 49 is intended for use in buildings where there is a direct-current incandescent-light circuit; but in cases where such a circuit is not available the magnets must be operated by current derived from primary or storage batteries, and as such currents are expensive it is necessary to modify the construction of the controller so that it may be operated with less current. The way in which this is accomplished is shown in Fig. 53, which shows the Otis magnet controller for battery current. This controller is also used for alternating currents, the magnets being of the solenoid type. The principal difference between Figs. 49 and 53 is that in the latter are two secondary pilot valves *D'* which are much smaller than the regular pilot valve and therefore can be moved with much less effort. These valves control the flow of water into and out of the cylinder *Z*, so as to push the piston up or down, as may be required. The movement of the piston, through the rod *13* and the lever *4*, actuates the main pilot valve. The weight of *W* is for the purpose of balancing the weight of the lever *4* to the right of rod *15*, and also the pilot valve and connection rods. The weight *W'* is for the purpose of counterbalancing the weight of *W*, the lever *4* and the pilot valve and connections.

The construction of the secondary pilot valves *D'* can be understood from Fig. 54, which is a section through the valve chambers taken at right angles to the view in Fig. 53. The two pairs of solenoid magnet coils act independently of each other on the two valves D' and D_2, and when they are not energized the valves rest in the position in which they are drawn in Fig. 53, so as to connect the ports *a* and *b*. The ports *a a'* are connected with the opposite ends of the auxiliary cylinder *Z*, while the ports *b* are connected through the central passage in the valve casting with *a''*, and thus with the discharge pipe. If the magnets *M M*, Fig. 53, are energized, the valve *D'* will be drawn down and then port *a* will be connected with *c*, which, as shown in Fig. 55, is connected with the supply pipe; therefore, pressure water will pass to *a* and thus to the upper end of *Z*, and force the piston down against the tension of the spring *S'*. Thus the pilot valve *43* will be depressed and the pressure

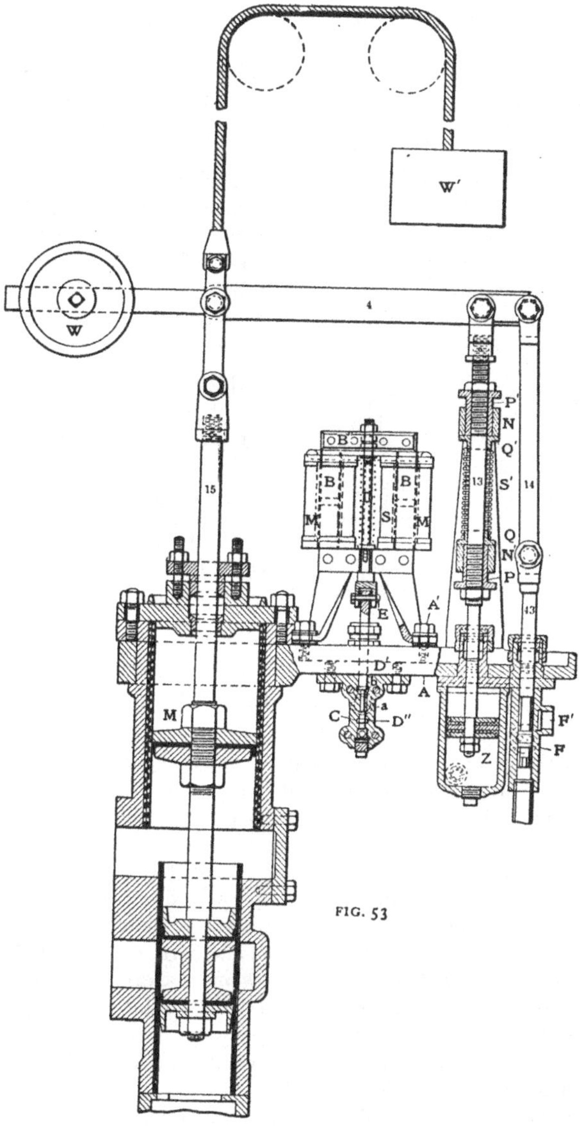

FIG. 53

water will pass through the pipe *E*, to the upper end of the main valve chamber.

If the magnets *M′ M′* are energized, the valve D_2 will be pushed down, and then the pressure-water port *c* will be connected with *a′*, and thus with the lower end of *Z*, thereby forcing the rod *13* upward against the tension of the spring *S′*, Fig. 53. This spring is made stiff enough to force the compression rings *Q Q′* against the stops and thus bring the pilot valve to the central position whenever the current is cut off from either pair of magnets. The sleeves *P P′* limit the movement of the rod *13* and thus the stroke of the pilot valve *43*. The pilot valve and valves $D' D_2$ are made to fit tightly by grinding, but not too tightly, so they may move as easily as is consistent with a good fit. $D' D_2$

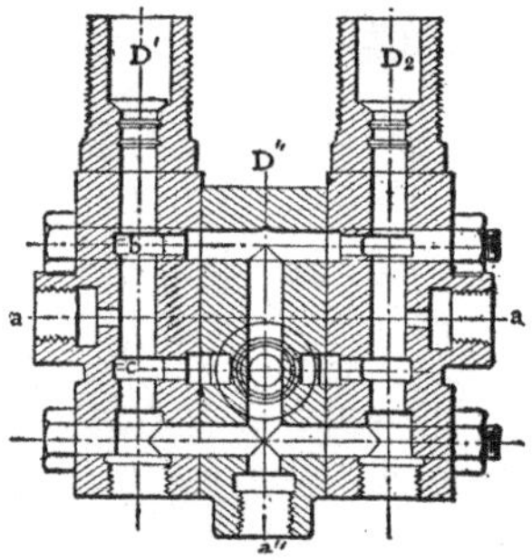

FIG. 54

do not have to fit as closely as the pilot valve; for even if they leak considerably the only effect is to waste a small amount of water, because when in the stop position they both connect with the discharge pipe. Hence, after filling both ends of the cylinder *Z*, the additional leakage would pass to the discharge. Owing to this fact these valves can be made so as to move very easily, the principal resistance being in the stuffing-boxes; these should not be screwed down any more than is necessary to make them tight, and a soft packing should be used in them.

Figs. 56 and 57 are external views of the parts shown in Figs. 53 and 49, respectively, and will serve to show more clearly the arrangement, particularly of the pipe connections and the strainers. In Fig. 57 the two magnet spools in front are removed so as to show clearly the construction of the lever *A* with the extending arm *A′*; also the rods *C C′* and the tension springs *S S′*.

Magnetically operated valves have to be arranged so that they can be easily handled by persons who are not skilled elevator operators. A great many controlling devices of this kind have been invented, and in the following paragraphs will be described two which are quite extensively used.

Fig. 58 is a wiring diagram of a simple controller, by means of which the car can be operated from any floor of the building as well as from within the car. The controller circuit is connected with the supply wires *P* and *N* by means of a two-pole switch shown at the left side of the diagram. If this switch is closed and the controller switches are all

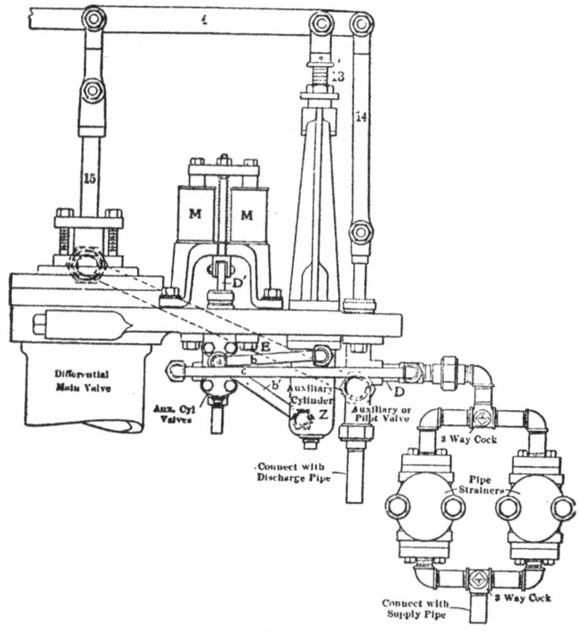

FIG. 55

open, as is the case when the car is not running, the current can pass from *P* to the wire *A*, then up the elevator well to the several floors of the building, and return by the wire *B*, passing through the door-contact switches *B'* at each floor, and thence to the wire *B''* and to the lever of the floor switch. One of these floor switches is located at each floor from which it is desired to operate the elevator. As the floor switch is open, the current can go no farther. From the wire *A* it will be seen that a wire *A'* runs upward; this runs to a point about half way up the elevator well and there enters a flexible cable, which connects with the elevator car and reaches the up and down push-buttons; but it can pass no farther, as these push-buttons do not connect with either *C″* or *K′*.

If a person at any floor desires to use the car, he operates the floor switch, turning it so as to set the car in motion toward that floor. Suppose the car is above him, he then turns the floor switch to the left so as to cause the car to run down. The current from *B″* will pass to *C′*, to *K*, to *J*, to and through the pair of down magnets on the valve to

FIG. 56

the wire *E′*, and thence to the supply wire *N*. When the car reaches the floor, the passenger opens the floor switch, then the elevator inclosure door, and steps into the car. The opening of the inclosure door releases one of the floor-contact switches *B′* and thus breaks the circuit, so that the car cannot be moved until the door is closed again. After the passenger has entered the car and has closed the landing door, he can cause the car to run either up or down by simply pressing the proper push-button and keeping it pressed in until the car reaches the floor at which he desires to stop. If he desires to go down he presses the lower button, when the current passes from *A′* to *K′*, thence to *K* and, through the route as before explained, to the wire *N*.

This arrangement is very simple, but has two objections, one of which is that the passenger has to hold the floor switch (or the car button) closed until the car reaches the floor, and has to use his judgment in determining just how far the car may be from the floor when he opens the switch, so that it will stop even with the landing. Another objection is that when a passenger is already in the car some one else,

FIG. 57

at one of the floors, can interfere with its operation by turning the floor switch so as to produce the opposite movement of the car. In the same way a passenger in the car can interfere with any movement directed from a landing.

The wiring diagram Fig. 59 shows a more complicated arrangement than that just explained, but is far superior to it, as it is entirely non-interfering; and at the same time it simplifies the work of the passenger, for if he is at any floor and desires to use the elevator, he has only to press a button for an instant, when the car will be set in motion, whether above or below him, and will come to a stop when it is even with his floor. After entering the car and closing the landing door, all he has to do is to press, for an instant, the button corresponding with the floor to which he wishes to go, and the car will at once start off and stop automatically at that floor. The arrow heads on the lines of the diagram show how far the circuits are complete when the car is not

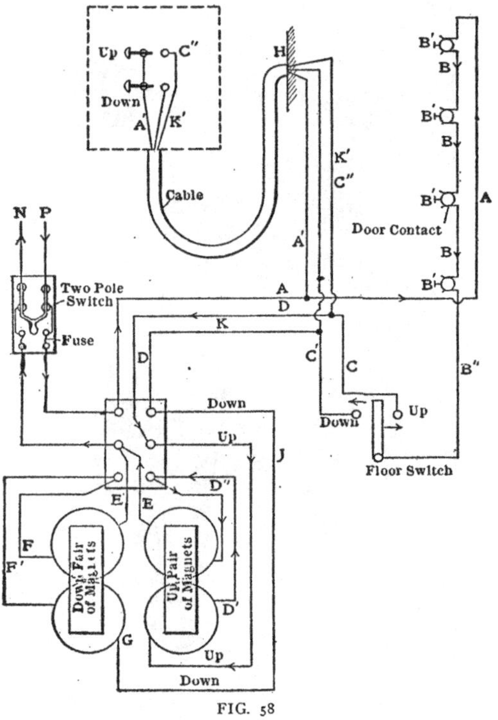

FIG. 58

running. The floor controller is a change-over switch arrangement, the object of which is to reverse the connections between the wires connecting with the push-buttons at the several floors and in the car, and with the brushes *E E'*, as the car passes above or below the floor; so that by means of a single button the car may be made to run down, if above the floor, or to run upward, if below the floor. The relay magnets are for the purpose of rendering it possible to keep the elevator in motion until the proper floor is reached, without having to keep the push-button depressed. It will be noticed that a wire runs from *f* up to the contacts opposite the four relays. If any one of the push-buttons is depressed, the current will flow through the corresponding relay magnet, the contacts at the right will be drawn up and a direct connection will be made with point *f*. Thus, suppose button *2 P* is pressed; the current will at once pass through the relay *22* and the contacts on the right side will be closed, so that the current may flow through from the junction point at *f*. The floor controller revolves, being driven by

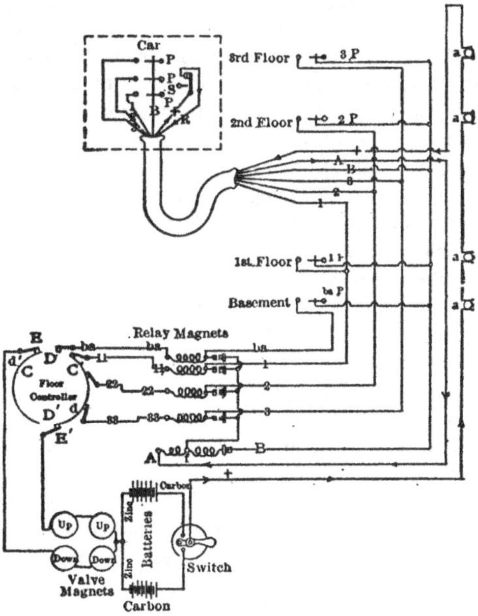

FIG. 59

the elevator, and is so geared that it makes the upper end of the contact arc *C′* travel from the position in which it is drawn (when the car is at the bottom of the well) to the point *d*, when the car is at the top floor. From this it will be seen that if the button *2 P* is pressed, the car will run until the upper end of the contact curve *C′* passes from under the brush *22*; and then the car will stop, and it will be even with the second floor.

The two coils between the wires *A* and *B* form a differentially wound magnet which exerts no effort when the current passes through both coils, and hence does not draw apart the contacts at the right side; but if one of the buttons is pressed, the instant the circuit from the junction *f* through the relay is closed, the right-side coil of the differential magnet becomes inactive, and then the switch contacts are drawn apart and no current can pass to the wire *B*, so that after this time it makes no difference how much the push-buttons may be manipulated, they can have no effect upon the running of the car. The circles *a a a a* represent inclosure-door switches at the landings, which are provided as safety devices to prevent absent-minded persons from moving the car away from the landing without first closing the door.

While Fig. 59 is shown arranged to be operated by a battery, Fig. 58 is connected with supply wires, which may create the impression that it is for use in cases where the current is derived from a lighting circuit. As a matter of fact, however, both arrangements can be used with current derived from any source, providing it is not alternating current; for the latter, slight modifications are made.

An examination of these wiring diagrams, as well as the drawings 49 and 53, will show at once that these magnetic valve-operating devices open the valve wide to set the car in motion, and close it entirely in one movement when the car is to be stopped. The person operating the elevator cannot in any way vary the running speed, nor can he cause the car to get under headway either quickly or gradually at will, nor make a quick stop or a slow one; all he can do is to open the valve to start, and close it to stop. The speed the car will take will depend entirely upon the way in which the valves are adjusted, and this is true as to the rate of acceleration in starting and of retardation in stopping. This lack of flexibility is the reason why magnetic valve-control has not come into general use for passenger elevators. Some magnet controllers have been devised that partially accomplish the results obtained with the mechanical devices, but they are complicated and cannot be regarded as entirely practical.

CHAPTER XIII

DOUBLE-POWER HYDRAULIC ELEVATOR SYSTEM

In Fig. 24 we presented a diagram illustrating the double-power system of hydraulic elevators, an arrangement devised to reduce the power consumption, and make it more nearly proportional to the actual amount of work done. With a steam or electrically driven elevator, the power required is directly proportional to the load lifted, and it cannot very well be otherwise; in other words, it is "the nature of the beast" to put forth an effort equal to the work it has to do and no more. With hydraulic elevators this is not the case, for the simple reason that to run the elevator car from the bottom to the top of the building requires one cylinder full of water, regardless of whether the car is empty or fully loaded. The lifting capacity, however, must be sufficient to run the car up with the maximum load, and this means that the pressure maintained in the pressure tank must be high enough to do this maximum work. This being the case, all the water pumped into the tank must be forced against a pressure great enough to elevate the car when fully loaded. It may be asked, where does the power go to when the car is run up light, since the lifting capacity of the piston is sufficient to raise the maximum load? The answer is simply that generally the surplus power is absorbed in forcing the water at a high velocity through the contracted opening in the operating valve, the latter not being opened wide under such conditions. If the valve should be opened wide the car will run at a higher speed than when loaded, and then the surplus power will be absorbed in forcing the water through the piping, as well as the valve, at an increased velocity. As it is contrary to the nature of the hydraulic elevator to use energy in proportion to the work done, special arrangements have to be devised to accomplish the result, and although these arrangements do not accomplish it perfectly, they do it well enough to be of decidedly practical value.

The Otis double-power vertical elevator is illustrated in Fig. 60. The valves are shown in the stop position. An inspection of the drawing will show that the main valve and valve chamber are longer than the low-pressure valves described in previous chapters, and in addition there is an extra valve piston V^2. The upper outlet port of the chamber connects with the lower end of the circulating pipe, and the lower outlet port connects with the bottom of the cylinder. The valve chamber is made longer so as to provide space for the port of the high-pressure pipe, and the valve is made with the additional piston V^2, so that when the valve is in the position shown, or lower down, the high-pressure water may not be able to pass into the cylinder.

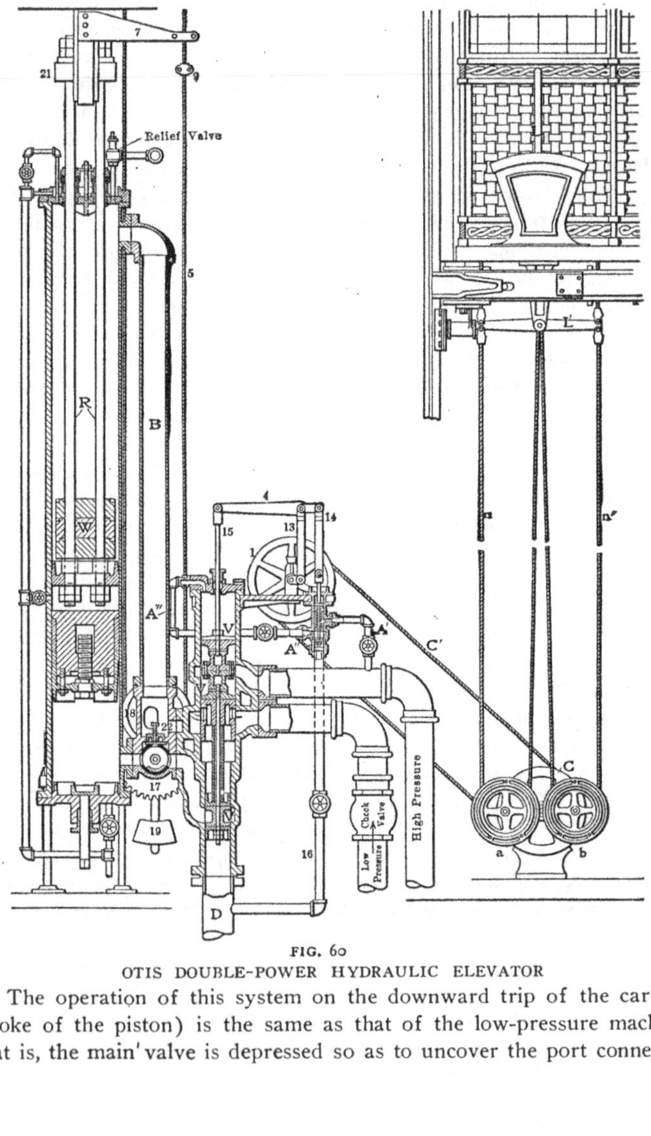

FIG. 60
OTIS DOUBLE-POWER HYDRAULIC ELEVATOR

The operation of this system on the downward trip of the car (up stroke of the piston) is the same as that of the low-pressure machine; that is, the main valve is depressed so as to uncover the port connecting with the lower end of the cylinder, and then the water above the piston passes down through the circulating pipe, through the valve chamber to the under side of the piston, and the latter ascends in the cylinder. On the up trip of the car, the main valve is raised until the port connecting with the lower end of the cylinder is uncovered partly or in whole, according to the speed required, and then the water under the piston escapes into the discharge pipe *D*, and pressure water from the low-pressure pipe passes through the valve into the circulating pipe and to the upper end of the cylinder, thus forcing the piston down. If the load is less than one-half the maximum, the valve will not have to be raised any farther than is necessary to uncover the ports connecting with the low-pressure pipe, as this pressure is sufficient to raise this load. If the load is more than one-half the maximum, the low-pressure water alone will not lift it, unless it is only slightly in excess, and even then the velocity of the car will be too slow to be satisfactory. In such a case the operator moves the operating lever farther, so as to lift the main valve higher, and by so doing a portion of the ports of the high-pressure valve is uncovered; that is, the piston V^2 passes above the lower margin of the ports, and then water passes into the valve chamber from the high-pressure, as well as from the low-pressure pipe; as a result the pressure acting on the piston is higher than the pressure in the low-pressure pipe. If the load is only a trifle more than the low-pressure water alone can lift, only a small amount of water will be drawn from the high-pressure pipe, but if the load is nearly equal to the maximum, most of the water will be supplied by the high-pressure pipe, while when the load is fully up to the maximum, no water will be drawn from the low-pressure pipe.

From the foregoing explanation it can be seen that no matter how small the load may be, it will require just as much power to lift it as it would for the heaviest load that the low-pressure water can raise, but for any load between the latter and the maximum amount the elevator is designed to raise, the power will not be very far from proportional to the load. The high-pressure water is generally double the low-pressure, the usual pressures being 200 and 100 pounds, respectively, so that the expenditure of energy is nearly in proportion to the work done for all loads greater than one-half the maximum.

When this system was devised, it was expected that in its operation the high-pressure water upon entering the valve chamber would not only flow into the cylinder, but would also back up into the low-pressure pipe, and to prevent this action a check-valve was placed in the low-pressure pipe, as shown in the drawing. The actual operation of these machines, however, has shown that the high-pressure water does not run back into the low-pressure pipe, unless the load is so great as to require a pressure nearly as great as that of the high-pressure water, so that for any load under, say, 75 per cent. of the maximum, the apparatus would operate perfectly if the check-valve in the low-pressure pipe were removed. It is not very easy to explain how water can run into the valve chamber from a pipe in which the pressure is 100 pounds, when water is running in from another pipe in which the pressure is 200 pounds, but it is nevertheless a fact that it does run in. If it did not run in the speed of the elevator car could not be kept up when the valve is raised just far enough to slightly open the high-pressure ports. To make this point clear, suppose the load is slightly in excess of what the low-pressure water alone can lift at full speed; then, if no high-pressure water is used, the car will rise at a slow velocity; if now the valve is raised so as to open the high-pressure ports, say, 5 per cent. of their full area, the speed will be at once reduced very decidedly, if the high-pressure water backs up into the low-pressure pipe and prevents water coming in from that source, because the opening of the high-pressure ports is so small that sufficient water to produce a high speed cannot pass through them. In actual practice, however, it matters little how slight the opening of the high-pressure ports may be, the effect is to give an increased speed; and if the car lever is moved very slowly over the portion of its movement that begins to open the high-pressure ports, the speed will be noticed to increase slowly and evenly as the lever advances. This could not be the result if the opening of the high-pressure ports had the effect of stopping off the flow from the low-pressure pipe, for if such were the action until the high-pressure ports were opened enough to permit water to pass through them in greater quantity than it previously entered through the low-pressure ports, the speed would be lower than with the low-pressure alone. Elevator engineers who are familiar with this problem have theorized upon it in an endeavor to find a plausible explanation of the action, and while no one appears willing to advance a decided opinion upon the subject, the general conclusion is that there is some kind of inspirator or injector effect produced, or probably both effects exist at once, the high-pressure water serving to suck the low-pressure water into the valve chamber as it rushes down past the low-pressure ports, and then, turning around just above piston *V*, it may rush into the ports leading to the circulating pipe, and draw in the low-pressure water with it.

In Fig. 60 it will be noticed that there is a relief valve at the top of the cylinder, while no such valve is provided with the low-pressure machines. This difference is made necessary by the presence of the check-valve in the low-pressure pipe. This valve is not required in low-pressure machines; therefore, if the car is stopped too rapidly on the down trip the water in the upper end of the cylinder can flow back through the throttle valve into the pipe. With the double-power system **there** is **no** escape for the water either through the low-pressure or the

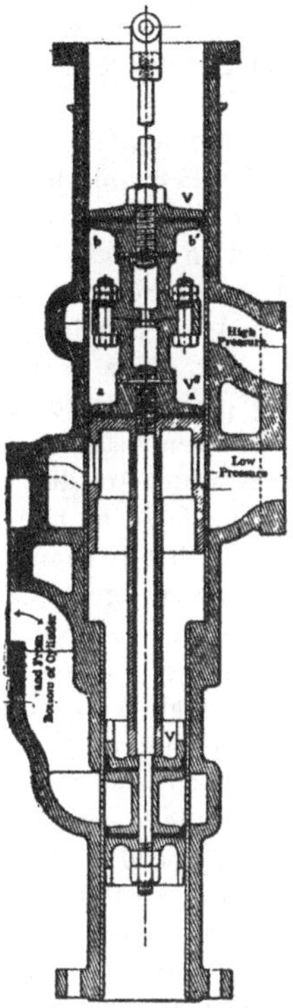

FIG. 61

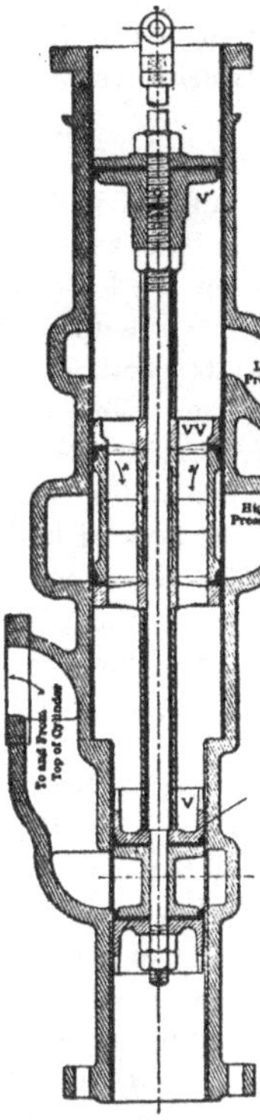

FIG. 62

high-pressure pipes, hence a relief valve has to be provided at the top of the cylinder.

NON-CIRCULATING VALVES.

Double-power hydraulic elevators have been made by the Otis company with non-circulating as well as circulating valves. The former are used in cases where the varying weight of the lifting ropes is compensated for by the hight of a column of water rising in a stand-pipe connected with the bottom of the cylinder, the principle of which is fully explained in Chapter V.

Fig. 60 shows the circulating type. The valve and valve chamber of this type are shown in Fig. 61. It will be seen that on the right side the top port connects with the high-pressure, and the bottom one with the low-pressure pipe. On the left side there are two ports, one to connect with the lower end of the cylinder, and the other, which is just above it, to connect with the circulating pipe. The valve has the cup packings *a* and *b* turned toward each other, so as to hold the high-pressure water between them when the valve is in the central position, as drawn in Fig. 60.

The valve chamber and valve for the non-circulating machine are shown in Fig. 62, and as will be noticed differ considerably from the circulating type. In the valve chamber it will be seen that the high-pressure port is below the low-pressure, and on the left side there is only one port which connects with the top of the cylinder. This is the same port, in so far as its position is concerned, as that in Fig. 61, which connects with the lower end of the cylinder. The port which in the latter figure connects with the circulating pipe is closed up in Fig. 62. Looking at the valve in Fig. 62 it will be seen that the central valves *V V* that shut off the flow of high-pressure water are different from the similar valves in Fig. 61. The latter are made so as to hold the high-pressure water between the two pistons, so that it cannot pass to the cylinder until the valves have been raised high enough to carry V^2 above the lower edge of the high-pressure port. In the non-circulating valve the action is the reverse of this. At all times the low-pressure water has free access to the entire valve chamber, through the central opening in the high-pressure valve *V V*, the arrows *a a* indicating the path of the stream. In starting the elevator on the up trip the valve is depressed until the top of the lower valve *V* uncovers the lower port in the valve chamber, then water can flow into the upper end of the cylinder and the piston will be depressed. If it becomes necessary to use some of the high-pressure water, the valve is further depressed, until the upper end of the valve *V V* uncovers the upper part of the high-pressure port in the valve chamber, then high-pressure water will flow in and passing down through the valve *V V* will go into the cylinder together with the low-pressure water.

The non-circulating valve is not used very extensively, as it is required only in cases where the weight of the lifting ropes is compensated for by means of a stand-pipe. This arrangement is far more elegant than compensating chains dangling in the elevator well from the bottom of the car, but is considerably more expensive, and that in all probability fully accounts for its not being more generally adopted. In New York this arrangement is in use in the St. Paul building and one or two others.

CHAPTER XIV

OTIS RUNNING-ROPE SYSTEM

The rope connections, shown in Fig. 60, for operating the pilot valve by means of the car lever *L*, represent one form of the running-rope system. The standard running-rope arrangement used by the Otis Elevator Company is shown in Fig. 63. The horizontal rock-lever *L'* is fastened to the same shaft as the lever *L*, and to its ends are secured the running ropes *n n'*. The upper ends of these ropes are made fast at the upper end of the elevator car, as shown at *M M'*. The *n* rope passes under a sheave *a*, attached to the side of sheave *C*, and then runs up the elevator well to the top of the building and passes over the sheave *a'*, thence running down to the hitching point *m* on the car top. The *n'* rope passes under the sheave *b* attached to the side of sheave *C*, opposite to that on which the sheave *a* is mounted, and thence running up to the top of the building passes over the sheave *b'* and down to the top of the car to *m'*. If the lever *L* is moved in either direction it will lift one of the ropes and depress the other, and thereby cause the sheaves *a* and *b* to rotate and carry with them the sheave *C*. The movement of the latter sheave, through the rope *C'*, turns the actuating sheave 1 on the pilot valve stand.

This rigging of the ropes is not quite the same as that shown in Fig. 60, as the latter is arranged so that both ends of each running rope are attached to the ends of the rocking lever. Owing to this difference the ropes in passing up the elevator well have to cross over, so that the one that passes under the sheave on the right side at the bottom must pass over the left side sheave at the top. This way of arranging the ropes is clearly illustrated in the diagram Fig. 64, in which the solid lines show the position of the ropes when the car lever is in the central position, and the broken lines represent the position of the ropes, and lower sheaves, when *L* is shifted to the position *A'*. An examination of this diagram will show clearly why it is necessary to cross the ropes in the elevator well when all the ends are connected with the rocking lever *L'*, and why they should not be crossed when the top ends are connected to the upper part of the car frame, as in Fig. 63. It will also be seen that in the arrangement of Fig. 64, if the operating lever is shifted to the position *A*, points *a* and *b* will be shifted to *a'* and *b'*, and the lower sheaves *c* and *d* will be shifted to *c'* and *d'*; that is, these

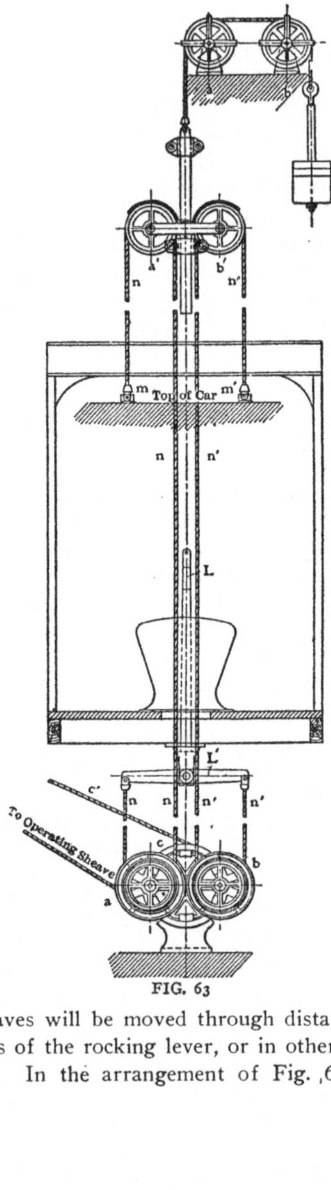

FIG. 63

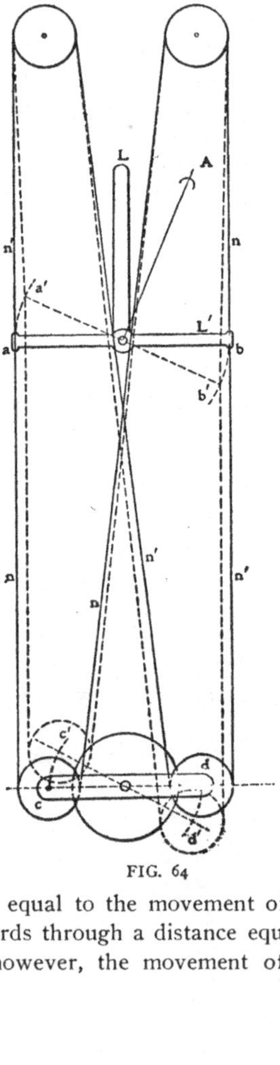

FIG. 64

sheaves will be moved through distances equal to the movement of the ends of the rocking lever, or in other words through a distance equal to *aa'*. In the arrangement of Fig. ,63, however, the movement of the lower sheaves is equal to one-half the distance through which the ends of the rocking lever are moved. In both cases it must be understood that the upper sheaves *a'* and *b'* are stationary.

A PUZZLING PILOT VALVE.

The pilot valve shown in Fig. 60 has puzzled quite a number of engineers, for the reason that the valve chamber has an outlet which is generally plugged up, and the object of this outlet is not easy to trace. The construction of the valve, the position of the outlet, and its object can be made clear by the aid of Fig. 65, which shows on the left side a vertical sectional elevation of the valve chamber with the valve in the

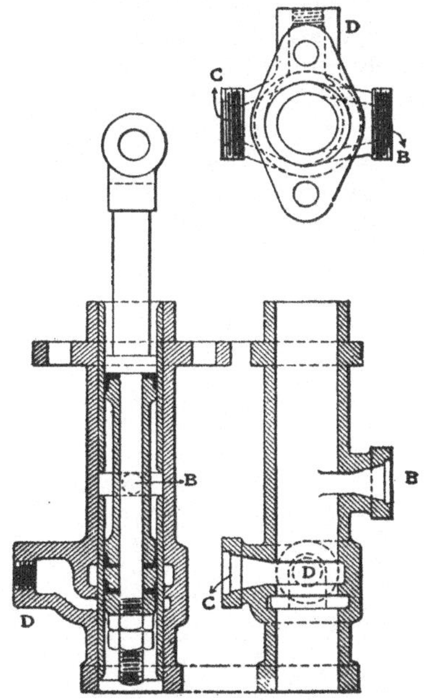

FIG. 65

stop position, and on the right side another vertical section of the chamber taken at right angles to the first-named, while at the top a plan view is given showing the relative position of the three outlets of the chamber. The outlet that will almost invariably be found plugged up is at *D*. This opening was originally made for the purpose of connecting a pressure-regulating device intended to make the double-power system automatic in action, in so far as the use of high-pressure water is concerned. When this device is used, *D* is connected with the upper end of the valve chamber, and in the pipe connection there is a valve that is normally closed, but is controlled by a pressure regulator that in turn is connected with the upper end of the cylinder, and is, therefore, actuated by the pressure therein. The regulator is adjusted so as to act with a pressure considerably above the low-pressure water. The operation is as follows: If the pressure at the top of the cylinder, which is due to the load in the car, is more than can be maintained by the low-pressure water, flowing in at the required velocity, and is also great enough to actuate the pressure regulator, so as to open the valve in the pipe connection from *D* to the upper end of the valve chamber, then water will escape from the upper side of the motor piston *V'*, through this *D* connection, and thus permit the valve to rise higher and open the high-pressure valve wider, so as to let in more high-pressure water, and thus increase the car speed. In practice it has been found that this regulator is not as much of an improvement as might be expected, and if it gets out of adjustment slightly is liable to make the operation of the car far from satisfactory. If it is set to act at a pressure only slightly above the low-pressure water, it is liable to set the car in operation if an unusually heavy load is taken on. In the regular running of the car, also, it is liable to produce an irregular motion, like an oversensitive engine governor. For these reasons it has been practically discarded, and the opening *D* plugged up.

CHAPTER XV

PRINCIPAL CAUSES OF DISORDERED MECHANISM AND HOW PREVENTED OR REMOVED—DIRECTIONS FOR PACKING THE PISTONS AND VALVES

The hydraulic elevator, like every other type of machine, must be kept in perfect order to give satisfactory results. Some of the effects produced by a disordered condition of the mechanism can be easily detected; in fact, they make themselves conspicuous by causing the elevator car to run erratically. Other defects, however, can be discovered only by careful inspection of the entire apparatus. Disorders of the first-named kind do not as a rule produce any permanent injury or deterioration of the running parts, and are of necessity soon rectified; hence, they are not as serious as those of the second type, whose only effect is to deteriorate the apparatus. The principal disorders which produce noticeable effects are leaks in the lifting-cylinder piston, the stuffing-boxes and the valves. These cause the car to settle slowly after being stopped.

Sometimes the car bounces up and down a number of times when stopping, especially if a sudden stop is made. This action, however, is not produced by an actual defect in any part of the apparatus as a general rule, but is more likely due to an accumulation of air in the lifting cylinder. This air, being elastic, will compress in the act of stopping, on a downward trip of the car, and will expand on an upward trip, owing to the fact that the momentum of the car and other moving parts resists the stopping, thus increasing the pressure within the cylinder on the downward trip and reducing it on the upward trip. After the valve has been closed the momentum of the moving parts is fully absorbed either by compressing the air to a pressure greater than is necessary to balance the load, or by expanding it to a lower pressure. As a result, in the first case the extra pressure of the air will force the piston down and lift the car, while in the second case the piston will move upward to compress the air to the balancing point, and the car will run down.

Owing to the elasticity of the air, the expansion and compression will be repeated several times, imparting to the car a teetering motion which is supposed by many people to be produced by the stretching and contracting of the lifting ropes. As this teeter effect is far more noticeable in high buildings, the belief that it is due to the elasticity of the ropes is strengthened; but as a matter of fact the explanation is that in high buildings the car speed is much greater; therefore, the stops are necessarily more sudden, and the momentum of the moving parts produces a greater expansion or compression of the air confined in the cylinder.

Theoretically, stretching of the lifting ropes actually takes place every time the car is stopped, but, although ropes when very lightly loaded will elongate considerably with a slight increase in weight, when strained to even 5 or 6 per cent. of their ultimate strength they will not stretch much more than solid rods, so that the rise and fall of a car due to the stretching of the ropes is probably not more than a small fraction of an inch, even for a length of two hundred feet or more.

The accumulation of air in the lifting cylinder is not due to a disordered condition of the apparatus, but to some extent it may be excessive on account of improper arrangement or proportions of the discharge and pressure tanks. All water contains air in very small bubbles, the amount being greater when the water is agitated than when it is at rest. If the pressure is increased the bubbles come together and form a smaller number of larger size, with increased buoyancy, and so the air is "squeezed" out of the water. If the water is compressed into a tank where it can remain quiet for a time, considerable air will be forced out of it. It therefore follows that with a large pressure tank less air will pass into the cylinder than with a small one, but the difference is hardly enough to warrant the use of an extra large tank for this reason alone, and it is seldom taken into account by designers of elevator installations.

In some cases the discharge tank may be so arranged as to cause an undue amount of air to get into the water, and if so it can be easily improved. If the discharge pipe is above the surface of the water in the tank, the stream falling into the tank will carry with it considerable air, and if the suction pipe is nearby, some of this air will be drawn into the pump. By placing a few division boards within the tank so as to space off the suction outlet, the water will have a chance to settle and free itself of most of the air before it is drawn into the pump.

TO REMOVE AIR FROM LIFTING CYLINDER.

To remove air from the lifting cylinder, when the air has accumulated in sufficient quantity to cause the car to teeter noticeably when stopping, an air-cock is provided at the upper end. If this cock is left open while the elevator is running, all the air will be expelled after a few trips have been made. Most of the air, however, is forced out while the car is running down. On the up trip of the car, water rushes into the upper end of the cylinder from the circulating pipe, and the surface of the water is kept well churned, so that water, as well as air, passes out through the air-cock; but on the down trip the piston moves upward, forces the water up against the air, and out into the circulating pipe. As the water surface remains comparatively quiet, air alone passes through the cock, and a far greater amount is expelled than in the other case. When all the air is out of the cylinder, the bouncing of the car will stop, and then the air-cock should be closed. The car would continue to run satisfactorily with the cock open, but there would be an

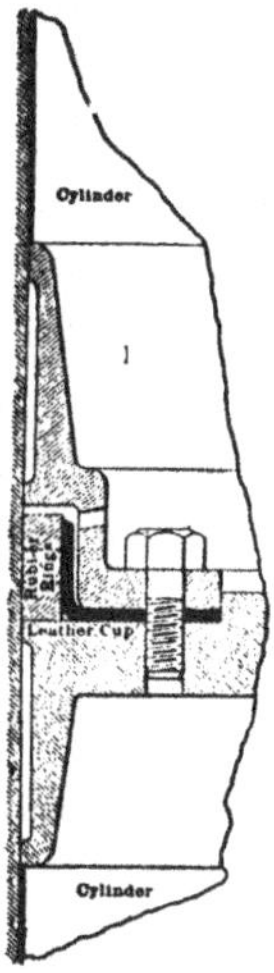

FIG. 66

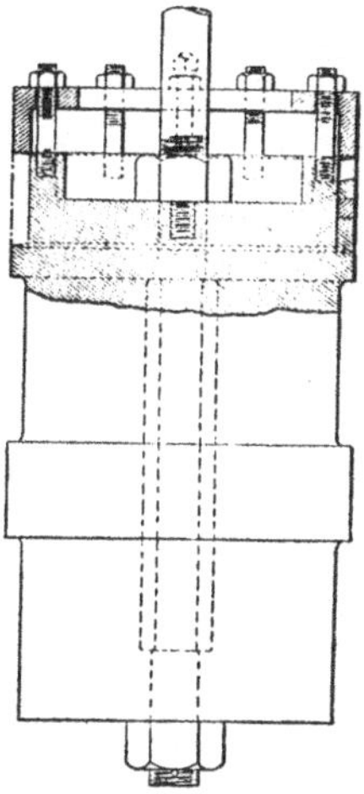
FIG. 67

unnecessary waste of water, and the car would settle after stopping at a floor; hence, it is important to make sure that the cock is closed.

SETTLING OF THE CAR.

Whenever the elevator car settles after stopping at a landing it is because there is a leak in either the piston or the valves. To determine where the leak is located, stop with the car above the lower floor, close the gate valve in the supply pipe, and move the main valve into position for the upward trip. If the car still settles, the chances are that the leak is in the piston, because the packings of the main valve will be above the port connecting with the lower end of the cylinder, and to circulate through the valve the water would have to pass by both packings. It is improbable that both packings would leak at the same time. While this is not a positive test, it is the best available, and four times in five will prove reliable.

If it becomes necessary to tighten the piston packing, the way to do it will depend upon the type of packing and piston. Some pistons are designed to use a combination packing, consisting of a deep leather cup with an outside wearing surface of square-section rubber rings. The water presses the leather cup against the rubber, making a water-tight joint, at the same time forcing the rubber rings against the cylinder surface. The construction of this packing will be understood from Fig. 66. This style of packing was formerly used almost exclusively and is to be found in nearly all the elevators which were installed fifteen or more years ago.

In modern elevators the packing used for pistons is the same as that employed for steam-engine piston-rod stuffing-boxes, and the piston is made with a packing space, and a compressing ring to press the packing into it. The construction can be readily understood from Fig. 67, which shows one of the latest designs of pistons for Otis low-pressure, vertical elevators, although in this design the piston is packed from the upper side, while in the design most generally used it is packed from the under side. These pistons being provided with regulation stuffing-boxes, ordinary hemp packing can be used, but unless the packing is done with good judgment the results may be serious. This is true with regard to almost any kind of packing, but particularly with those that are apt to swell appreciably when they become wet. If the stuffing-box is packed very tightly with a dry packing, when it begins to absorb water it will begin to swell, and may continue to swell until it bursts the cylinder.

The bursting of hydraulic-elevator cylinders is not an unusual thing, and in the great majority of cases it is due wholly to excessively tight packing; there are elevator engineers who insist that this is always the cause. Even if the packing is not tight enough to burst the cylinder, it may be tight enough to absorb so much of the power of the water as to cause the car to run much too slowly, not to speak of the waste of power.

PACKING THE PISTON.

If the piston is made so as to be packed from the upper end of the cylinder, the course of procedure in packing is as follows: The car is run to the bottom of the elevator well; then the valve in the supply pipe is closed, to prevent water from flowing in from the pressure tank while the packing is being done. The next step is to remove the water in the cylinder above the piston, and this is accomplished by opening the air-cock in the upper cylinder-head, and also a valve that connects the drain pipe with a point on the side of the cylinder below the bottom of the piston, when the water will be drawn off into the drain pipe. After this much is accomplished, the cylinder-head is to be removed, but before doing this the position of the part which contains the piston-rod stuffing-boxes and the location of the outer clamping ring must be carefully marked, so that when the head is replaced the parts may be returned to their original positions.

It must be borne in mind that in removing the old and putting in the new packing, it is possible to impart to it a rotary movement that would throw the piston-rods out of line. When the cylinder-head has been removed, the stuffing-box gland-screws are removed, and the gland is raised out of the way; then the old stuffing is taken out and new put in, as in any other stuffing-box. It is to be observed, however, that as it is a far more laborious job to pack an hydraulic piston than a steam-engine or pump stuffing-box, greater care should be taken to make sure of not having to open up the cylinder a second time.

As before stated, not only is there danger of bursting the cylinder if the packing is pressed in too tightly, but even if this damage does not result, there will be excessive friction; therefore, every care must be taken not to get the packing too tight. At the same time it is necessary that it be tight enough to prevent the water from passing, and, while it is not possible to give specific directions or signs by which to determine when the packing is tight enough, men of experience will have no difficulty in determining this point.

To avoid trouble in replacing the packing-gland, it is advisable to mark the position in which it goes before removing it, so as to replace it in the same position. As it is necessary that the gland be screwed down evenly all around, that is, so that its outer surface may be parallel with the cylinder, the distance from its upper surface to the top of the cylinder should be accurately measured at four points equidistant from each other.

If, when the cylinder-head is removed, it be found that the piston is too far down to be reached conveniently, it can be raised by placing a clamp on the lifting ropes some distance above the car and drawing it down by means of a tackle attached to the upper frame of the car. In doing this care must be taken that the clamp is attached only to the lifting ropes, and not to those that run to the independent counter-balance. In cases where the machine is geared high—four or six to one—and the independent counterbalance is light, this method may not prove satisfactory, unless the car is loaded so that it cannot be lifted by this effort to raise the piston. The piston can also be raised directly by pulling up the traveling-sheave frame; but if this is done it will be necessary to be careful to return the ropes to their proper position when the frame is lowered, and in some cases this is not easily done on account of the sheaves not being in a position where the ropes can be readily reached. If the piston is raised so high as to bring the packing opposite the porthole connecting with the circulating pipe, care must be taken that in lowering it the packing is not caught, also that the packing is not forced out into the port while placing it in the stuffing-box.

After the piston is packed and the cylinder-head is replaced, one should see that the air-cock is open, and that the main valve is in the stop position. Then slowly open the valve in the supply pipe, and keep the air-cock open until water runs out of it, indicating that all the air is expelled; then close the air-cock and the elevator will be ready to run.

If the piston packing is of the leather-cup and rubber-ring type, the packing process is the same as described, with the exception that after removing the follower *F*, Fig. 66, the next step is to remove the leather cup and clean out thoroughly the packing space, and also the holes in the follower through which the water passes to press out the leather cup. If the cup is found to be in good condition, which is probable, it may be used again, but if much worn it must be replaced, and to get the new one in position it will be necessary to cut it carefully on a diagonal line, so as to pass it over the piston-rods; then it must be sewn up neatly (which is not an easy job), so that it may be of the same shape and size as before cutting. The rubber rings are to be replaced by others of the same size; the cross-section is generally 5/8 inch square. They should be cut somewhat longer than the circumference of the circle, say from 1/4 to 1/2 inch, according to the diameter of the cylinder, so that the ends may be pressed tightly against each other. Three rings are usually required, and they should be of such thickness as not to fill up the space tightly, for if they do they may not be forced out against the cylinder wall by the leather cup. If they fit loosely they will be pressed out so as to make a tight fit.

If the piston must be packed from the bottom, the packing operation is the same as that just described, but the piston must be run down to the lower end of the cylinder, which means that the car must be run to the top of the building. In order that the piston may be brought low enough to be easily reached, it must be run down until it strikes the lower cylinder-head, and to do this in the case of a hand-rope-operated elevator it is necessary to slide up the stop-ball on the hand rope. In the case of a pilot-valve machine, the stop-ball on the rope that actuates the automatic stop-valve must be moved. In either case, the car is first run up as far as it will go, then the stop-ball is shifted to a point where it will prevent the car from moving far enough to get up a speed that would cause the piston to strike the cylinder-head hard, even if it were not run up carefully. In running up the car, however, the valve should be opened just enough to cause the car to move.

When the car is run up until the piston rests on the lower cylinder-head, it must be secured to the overhead beams by means of ropes several times as strong as may be considered necessary to hold it safely, say ten times as strong. This being done, the valve in the supply pipe is closed,

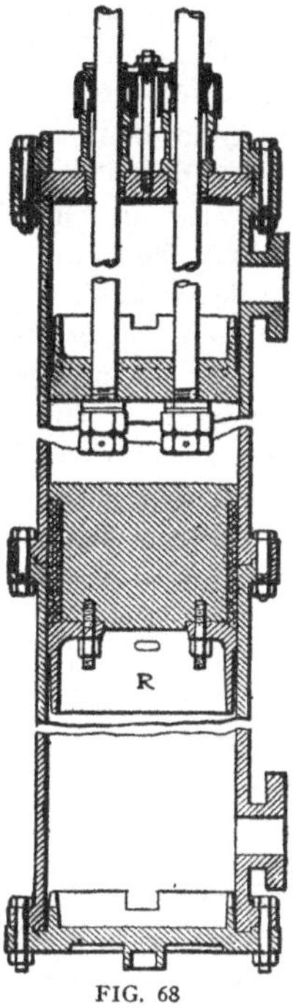

FIG. 68

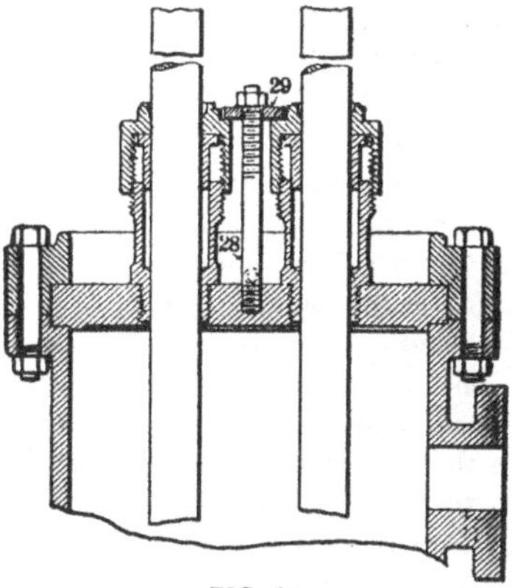

FIG. 69

and the car lever is turned for the down trip, so as to connect the circulating pipe with the lower end of the cylinder. The next step is to open the valves connecting the drain pipe with the side and bottom of the cylinder, and also the air-cock at the top, so as to drain out the water. If the discharge tank is higher than the bottom of the cylinder, the valve in it must be closed, so that the tank water may not run into the cylinder, as it would otherwise surely do, owing to the fact that the cup packings in the valve are not set so as to be tight against water running in from the discharge end of the valve cylinder. Having drained the cylinder, the lower head is removed and the packing is done in the same manner as for top-packed pistons.

In the latest designs of bottom-packed pistons, one of which is shown in Fig. 68, the packing chamber is made deep and the clamping ring *R* is screwed tightly against the bottom of the piston. This renders it unnecessary to measure the distance from the ring to the lower end of the cylinder to make sure that it is in its true position, because when screwed up tightly it will be true; but as there is no provision for adjusting the depth of the packing, the latter must be gaged in quantity so that it will be just tight enough when the clamping ring is screwed up. Owing to the greater amount of packing there is no difficulty in doing this, generally, without splitting any of the packing. When the piston is repacked, and the cylinder-head in place, move the main valve to the up-trip position, so as to just open, then open slightly the valve in the supply pipe, having first closed the valves connecting with the drain pipe. Leave the air-cock open and the cylinder will fill slowly with water, and the air will be expelled. When all the air is out, close the air-cock and return the main valve to the stop position; then open the supply-pipe valve wide, release the car and run it down a short distance; then reset the stop-ball, and the elevator will be ready to run.

The piston-rods pass through stuffing-boxes of the same type as those used on steam engines, as can be seen from Fig. 69, and they are packed in the same way. To keep the glands from working loose the locking piece *29* is provided, and this is secured in place by the stud *28*. In repacking the piston-rod boxes, one should always be sure to replace the lock-clamp *29*, especially if the cylinder is located in the elevator well at one side of the car, because if the cap of one of the boxes should work loose, it might permit the water to spray out in sufficient quantity to give the passengers in the car an undesirable shower bath. Before starting to pack the piston-rods, the car should be run to the bottom of the building and the valve in the supply pipe closed, so as to remove the pressure in the cylinder, and thus prevent water from spurting out of the stuffing-box. It is also advisable to open the air-cock so as to be sure of relieving the pressure entirely. Flax packing is recommended for this service, and it should be about ¼ inch in diameter. About eight turns will be sufficient.

TO PACK THE VALVES.

To pack the valves they must be removed from the valve chamber, so they may be taken apart to remove the old cups and put new ones in their places. There is no difficulty in removing the valves from machines provided with a pilot-valve gear without a throttling valve, as there is nothing to interfere with the free withdrawal of the leather cups. This may be seen at once from Fig. 53, which is a vertical section of such a valve. It will be evident that both the small lower chamber and the large upper one are smooth, so that as soon as the valve-chamber head is removed and the connecting-rod is free, the valve can be drawn upward. In Fig. 29, which is a vertical section of a simple valve of the type used with hand-rope control, the cups of the lower valve have to pass over the space *12*, and thus are liable to catch on the upper lining *4*. This difficulty seldom occurs, however, because the inner edge of the lining is rounded off, in the manner illustrated in Fig. 70, so that the cup

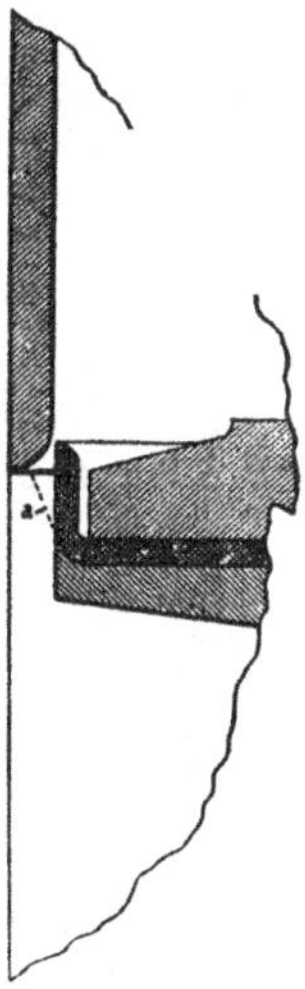

FIG. 70

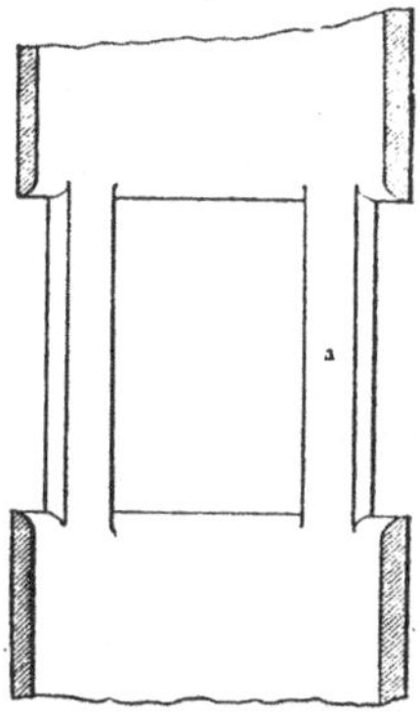

FIG. 71

can spread out considerably before it will be caught, as indicated by the dotted line *a*.

These valves are not all made in this way. In many of them the linings *3* and *4* are combined in one, with ports cut through where the space *12* comes, the construction being as illustrated in Fig. 71. If the valve chamber is constructed as in Fig. 29, and the cup appears to catch, it can usually be freed by gently rotating the valve, at the same time pulling it up just hard enough to make it lift if the cup slides back into position. If this method fails to free the cup, then the proper thing to do is to drop the valve back far enough for the cup to enter the lining *3*, and allow the valve to remain in this position for a few minutes, so as to allow the leather to set, and then try to lift the valve again, but with greater care.

Differential valves provided with a throttle generally work in chambers constructed as shown in Fig. 71, and with these there is no difficulty in removing the valve, because if the packings spread out and catch they can be pressed back by simply rotating the valve, as the longitudinal connecting bars *a* will flatten down the raised edge of the cup.

To pack the main valve it is necessary to run the car to the top of the building and secure it there, the same as for packing the piston from the lower end of the cylinder, the only difference being that it is not necessary to lift the car any farther than it ordinarily runs. After firmly securing the car, the valve in the supply pipe is closed, and then the water is drawn off the circulating pipe by opening the valve that connects its lower end with the drain pipe. If the discharge tank is higher than the bottom of the valve this must also be shut off, so that the water may not run out of the tank through the valve chamber when the valve is removed.

If the valve is of the rack and pinion type used with hand-rope control, all that is necessary after securing the car and closing the supply and discharge valves is to remove the bolts and take off the cap and the intermediate block which contains the bearing in which the hand-rope sheave rotates. This having been done, the valve can be removed. If the valve is of the differential type, then the upper end of the valve-rod must be disconnected from the cross lever that connects it with the pilot valve, and the pipe connection with the upper head must be broken; then the head can be lifted off and the valve drawn out.

The valves should be packed at a time when it is not necessary to hurry in getting the elevator back into operation, in order that the valves may be allowed to soak for an hour or two after the new cups have been put in place. To make sure that the cups are tightly clamped between the valve body and the head, it is advisable to drop a little oil on the flat part of the leather before putting it in place, so as to soften it. After the repacked valve has been replaced and the valve chamber properly closed, all connections having been correctly made, the valves in the supply and discharge pipes must be opened, also the air-cock in the top of the cylinder, so as to fill up the circulating pipe and dispel the air from the upper end of the cylinder; then close the air-cock and the elevator is ready to run.

To replace the cup packings in the pilot valve it is necessary to close the valve in the supply pipe, after running the car to the bottom of the well. If the discharge tank is higher than the valve, the discharge connection must also be closed. The valve, as may be understood from Fig. 65, can be drawn out of the chamber as soon as its upper end is disconnected from the connecting-rod. The cups at the lower end of the valve will have to pass up over the break in the lining at *B*, but as the inner corners are rounded off, there should be no difficulty in passing this point. Before removing the pilot valve it is advisable to draw just enough water from the circulating pipe to remove the pressure, in order that none may spill out through the top of the pilot-valve chamber. If the large valve at the upper end of the main valve, which is called the motor piston, is not tight, the pilot valve cannot be removed until all the water that is above the level of the pilot-valve chamber is drawn from the circulating pipe; otherwise, this water would pass by the motor piston into the upper end of the main valve chamber, and thence through the pipe connection to the inlet *D* of the pilot valve.

CHAPTER XVI

AUTOMATIC DEVICES USED FOR STOPPING CARS AT TOP AND BOTTOM LANDINGS; THEIR CARE AND THEIR VALUE AS SAFETY APPLIANCES

The automatic devices employed to stop elevator cars at top and bottom landings are the most valuable safety appliances used, and they must be kept in perfect working order. For systems in which the cars are operated by the ordinary hand-rope method, the automatic stopping device is very simple. It consists of two stop-balls fastened to the hand rope at the top and bottom in such manner that they will be caught by the sleeve through which the rope slides, a little before the car reaches the stopping point. These stop-balls are so adjusted that the sleeve strikes them at the proper distance from the floor level to cause the car to come to a state of rest three or four inches beyond the floor, if running at normal speed. As the careful operator usually pulls the hand rope to stop, the stop-balls are not often struck by the sleeve. When the car is stopped properly by the operator the reduction in speed is more gradual than would be the case if the stop-balls were relied upon.

Although it adds to the comfort of the passengers not to have the car stopped by means of the stop-balls, it makes it more imperative to examine the stop-balls frequently and carefully in order that there may be as little danger as possible of their getting loose and shifting out of place. If the car were frequently stopped by the stop-balls, any slight displacement of them would become apparent by the car running a little beyond the point where it should stop, indicating that the stop-ball had shifted. By taking advantage of this warning, the stop-ball can be returned to the proper position before serious damage results. If, however, the automatic stop is rarely used, the ball may become loose and shifted out of position a distance that would be dangerous. This is especially likely to occur with the bottom stop-ball, as gravity will cause it to slide down on the hand-rope after it becomes loose and it will not be in place to stop the car when needed. With the upper stop-ball there is not so much danger, because if it slides down it will only come into action too soon, and stop the car before it reaches the top floor. Yet it is possible for the upper stop-ball to become just loose enough to be pushed up on the rope when struck by the sleeve, and, although it may not slide up far enough to permit the car to rise too high at the time, it will in all probability be high enough to fail to act when next called upon; therefore, the upper stop-ball requires nearly as much watching as the lower one.

STOP-VALVE FOR PILOT-VALVE CONTROL.

In the case of pilot-valve elevators, the automatic stopping device is entirely separate from the operating lever in the car, and is actuated by the movement of the cross-head of the lifting piston. It is much more complicated than the stop-ball device, as it comprises a number of parts, and there is more chance for it to get out of order if neglected. A description of the automatic stop-valve and connections used with pilot-valve control is given in Chapter IX, but the action of this valve and the way in which it should be adjusted will be more fully explained herewith, in connection with the elevator elevation drawing, Fig. 33. It might be assumed that as this drawing represents only one design of automatic stop-valve for use with pilot-valve control, the explanation will be of service only in connection with this particular machine, but such is not the case; all hydraulic elevators provided with pilot-valve control are constructed upon the same principle, being different only in detail modifications, so that what is true of the general principles of operation for one is true of all, and all that is necessary is to take into account the changes in form of the various parts, which in most cases is trifling.

In Fig. 33 the automatic stop-valve is located at *B″*, and is actuated by the rope *5*, which runs upward by the side of the hydraulic cylinder to a point several feet above the highest position reached by the arm *7* attached to the crosshead of the machine. The rope *5* passes over a sheave at the upper end, as clearly shown in the drawing, and down around the sheave *18*, which is mounted in a bearing attached to the chamber of the automatic valve *B″*.

The arrangement and location of the sheave *18* can be better understood from Fig. 72. On the shaft of *18* there is a pinion that meshes into the gear *17*, which is mounted on the end of the shaft that carries the center *49* of the valve *B″*. The ratio between the diameters of the gear *17* and the pinion on the shaft of the sheave *18* is such that *17* makes a quarter of a turn between the instant when the arm *7* strikes either one of the stop-balls *8* and *9* and the time when the car comes to a state of rest. During this interval the flow of water through the valve *B″* is reduced so gradually as to bring the elevator to a stop without a noticeable jar if the velocity is not above the normal. The weight *19* provides the necessary momentum to return the stop-valve to the open position when the main valve is turned so as to move the car in the opposite direction.

In Fig. 33 it will be noticed that the arm *7* acts on stop-balls placed on the right-hand rope, therefore if the elevator is going upward, as *7* will be moving downward, the ball *9* will be struck, and thus the sheave *18* will be rotated clockwise, and the movement of *17* and the valve *B''* will be counter-clockwise; the curved plate *50*, Fig. *72*, will swing around to the right side and cover the outlet from the lower end of the cylinder

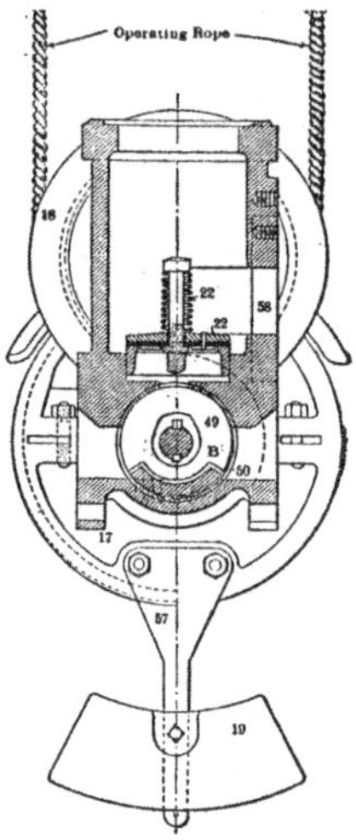

FIG. *72*

to the main-valve chamber, stopping the motion of the piston, and, consequently, the car. The curved piece *50* is held in the center carrier *49* loosely, so that, while the pressure of the water trying to escape from the lower end of the cylinder will force it against the valve seat, to make a tight joint, if the main valve is depressed and pressure is brought on the outside of *50*, it will be forced away from the seat, and water from the circulating pipe *B* will pass through the small opening slowly and produce a gradual start of the elevator.

As soon as the piston begins to move upward the arm *7* will also move, when gravity will draw the weight *19* downward to its lowest position. When the car is sufficiently near the lower floor for the automatic stop-valve to act, the arm *7* will strike the ball *8*, and by pulling up the right-hand side of the operating rope will rotate the sheave *18* counter-clockwise, swing the plate *50* around to the left and prevent the flow of water into the lower end of the cylinder.

Again, it will be seen that as long as the main valve is depressed the pressure will hold *50* against the seat, and prevent the escape of water from the lower end of the cylinder; but if the main valve is now raised, so as to connect the port from the valve *B″* with the discharge pipe, the pressure against the inner side of *50* will be removed, and the water in the lower end of the cylinder will force *50* away from its seat and escape, permitting the piston to move downward slowly. As the piston moves down, the arm *7* will also move down and free the stop-ball *8*, so that gravity again acting on *19* will swing it down to the central position and move *50* entirely out of the way of the flowing water.

If the plate *50* were not held loosely in the carrier *49*, the elevator could not be started after it had reached either end of the well, because the flow of water would be effectually stopped, and the opening of the main valve for either direction of movement would not have any effect. As the water used for operating elevators is generally clean, or at least free from bodies large and strong enough to wedge the plate *50* in the closed position, there is very little danger of an elevator becoming stalled through this cause. If, however, the car should refuse to start from either end, and no other reason could be found for its not moving, it could be inferred that the plate *50* was held firmly against the seat in some unexpected manner. Whether such were the case could be easily ascertained by placing the main valve so as to just open for the desired direction of movement, and then move the stop-ball away from the arm *7* to allow the weight *19* to drop to the central position. Upon doing this, if the elevator were set in motion, it would show that *50* had been caught, and then the valve chamber should be opened to ascertain and remedy the cause.

TO REMEDY DEFECTS CAUSED BY FRICTION.

While it is improbable that anything will cause the plate *50* to remain against its seat when the main valve is reversed, it is possible for the weight *19* to be incapable of overcoming the friction of all the moving parts of the automatic valve-gear, and when the main valve is moved into position to run the car up from the lower landing, or down from the top, the weight *19* may not return to the central position, but remain at the side. If this should occur it would prevent the car from attaining a speed anywhere near normal; in fact, it would not run with much more than 10 per cent. of the normal velocity. To remedy a defect of this kind it is necessary to find out where the valve mechanism sticks, and put it in proper condition, which in most cases can be done by thoroughly cleaning, readjusting and oiling all the bearings and relieving any undue friction in the stuffing-boxes.

The refusal of the automatic stop-valve to return to the open position when the elevator is started from either end, due to the sticking of the moving parts, is not the most serious thing that can occur with it. Suppose the valve mechanism works so hard that when the arm *7* strikes one of the stop-balls it simply slides it along on the rope, what would be the result? Simply that the car would not stop, but would pass the floor, and if at the bottom of the building it might strike the bumpers hard enough to injure passengers, unless the ropes were rather short, in which case the piston would be drawn up against the top cylinder-head before the car struck and damage the cylinder-head, piston or other parts. If the stop-ball were to slip when the car was running upward, the piston would strike the bottom cylinder-head, with more or less serious results. From the foregoing it will be apparent that it is of the utmost importance to keep the automatic stop-valve mechanism in perfect condition all the time, the bearings and gears running freely and well oiled, and the stuffing-boxes tight enough to prevent leakage, but not too tight to prevent the valve-rod from revolving freely. The stop-balls must be firmly secured in position, and the fastening of the weight *19* must be perfect.

To secure correct running of the elevator the valve *22* must be in good order. If it leaks, the car will settle when standing at a landing, and if the spring *22'* becomes clogged and fails to compress freely, smooth stops will not be made when running upward or when the car is stopped at the top floor by the automatic valve, if the car is running at high speed. In such cases it generally happens that the operator closes the main valve too abruptly, and the automatic valve may close too soon; but if the valve *22* is working properly the excess of pressure developed in the lower end of the cylinder by the compressing action of the piston, due to its momentum, will force the valve up, allow the water to escape from the lower end of the cylinder into the circulating pipe, and permit the piston to move down some distance farther before coming to a stop.

The reason why the automatic stop is the most valuable of all elevator safety devices is that if the car for any reason runs down with abnormally high velocity, and all the usual safety appliances fail to act and stop it before it reaches a point near the lower landing, the automatic stop will begin to act, and by gradually stopping the flow of water will arrest the motion of the car, so that if it does not stop before reaching the bumpers its velocity will have been reduced and a disagreeable jolt will be the worst result. If elevator accidents were due to the actual breaking of the lifting ropes, the automatic stop would not be of such importance, ere are only a few cases on record where cars suspended from a er of ropes have dropped through the breaking of the latter. In every case the elevator attains a dangerously high velocity through disarrangement of the machinery, or, as it is generally expressed, evator "runs away." Since such is the fact, it is evident that if the atic stops are kept in perfect working order every runaway can evented from doing serious damage, although if the car comes at a sufficiently high velocity the escape of water through the relief *22* will continue long enough to permit it to strike the bumpers hard.

CHAPTER XVII

EFFECT OF STRETCHING OF THE ROPES

Elevator lifting ropes stretch continually from the first day they are used until they are replaced, and the position of each automatic stop-ball has to be changed from time to time to compensate for the elongation. When the ropes are new they stretch rapidly, but after a few months' service they apparently stop stretching and remain at practically the same length until they begin to reach the point where they should be removed, then the stretching slowly increases. To guard against keeping in service a rope that is sufficiently worn to be in danger of giving out, it is necessary to make frequent inspections, but these inspections cannot be relied upon as infallible because even the most experienced men are not sure to detect every flaw. Examining grease-covered ropes in service is very different from inspecting new ones. The thoroughness of the operation can be greatly increased if the ropes are well cleaned with kerosene, but even when this is done it only enables the inspector to determine external conditions; the condition under the surface can only be conjectured. If the outside strands are badly worn or broken, it is time to put on a new rope. The fact that the ropes begin to stretch rapidly when reaching the end of their usefulness affords a means of determining their condition that should not be overlooked. Even if from external appearance they appear to be sound, they should be mistrusted if there is a noticeable increase in the rate of stretching.

If an elevator is provided with pilot-valve control the stretching of the ropes will cause the car to stop below the floors when arrested by the automatic stop-valves. Slight elongations of the ropes can be compensated for by shifting the stop-balls *8* and *9* downward. A considerable elongation cannot be compensated for in this way, because the effect of changing the position of the stop-balls is to cause the car to stop even with the floors, and to shift the piston out of position in the cylinder so that it will run closer to the lower end. As the clearance at each end of the cylinder is considerable, the position of the piston can be changed several inches without doing harm, but beyond this point the lower cylinder-head will be struck by the piston and possibly cause damage.

If an elevator is provided with hand-rope control the stretching of the ropes will not affect the stopping of the car, because this is controlled by the relative position of the stop-balls on the hand-rope, and this will evidently remain unchanged regardless of the length of the lifting ropes. In such elevators the effect of stretch in the lifting ropes is to cause the piston to run nearer to the lower end of the cylinder and finally to strike it if the ropes are not shortened. The way in which the ropes are shortened can be easily understood from Fig. 73, which shows the overhead beams on which the main sheaves are supported, the traveling

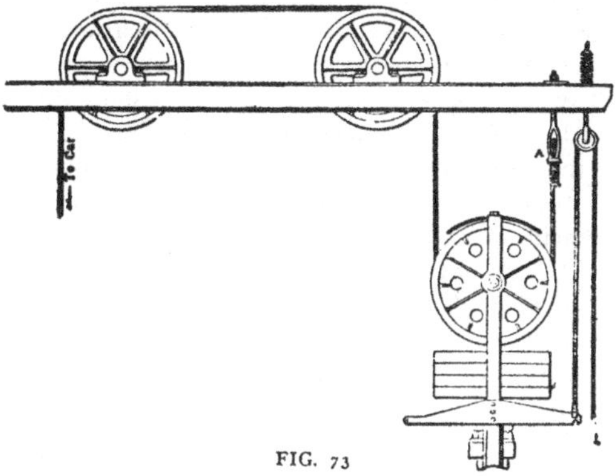

FIG. 73

sheave and the method of securing the end of the rope. In elevators geared two to one, as in this illustration, the ropes are fastened to the overhead beams, but with higher gears they are secured farther down, generally to the beams that carry the intermediate sheaves. The end of the rope is firmly fastened in a long shackle-bolt, and when the rope is first installed its length is made such that the nut on the bolt is at the end. By drawing up the bolt, the rope can be shortened several inches. When the rope cannot be taken up any more by the bolt, it becomes necessary to cut a piece off the end of the rope.

To be able properly to keep track of the elongation of the ropes it is advisable to place a mark on some stationary surface in line with lowest position of some part of the traveling-sheave frame; then if the car is operated by a hand rope the stretching of the lifting ropes, as well as the amount of stretch, can be detected by noticing how far the traveling sheave runs below the datum mark. In the case of elevators controlled by pilot valves the distance the sheave runs below the mark will serve to show how near the piston comes to striking the lower cylinder-head, and the ropes may be shortened before any damage is done.

HOW TO SHORTEN THE ROPES.

The first thing to do is to ascertain how much to reduce the length. This can be determined by measuring how far the traveling sheave runs below the proper position and multiplying this by the gear of the machine. For example, if the sheave runs three inches below the datum mark and the gear is four to one, the ropes will have to be shortened one foot. The gear can always be ascertained, as it is equal to the number of times the ropes run up from the traveling sheaves. Having decided how much to take out of the ropes, the next step is to run the car down until it rests upon the bumpers, moving the automatic stop-ball out of the way for the purpose. Then a strong clamp is firmly fastened to the ropes several feet below the ends, and this is drawn up by means of a tackle-block firmly fastened to the beams that sustain the shackle-bolts, until the latter have been raised enough to be free. The shackles are then removed, the ropes are shortened the proper amount and the shackles replaced, after which the ropes are drawn up by means of the tackle to raise the shackle-bolts high enough to run the nuts on their ends. There should be check-nuts on all the bolts, and in addition a safety-pin should be passed through a hole in the end of the bolt that there may be no danger of the bolt working off. This danger is greater than might be supposed, as the impression naturally would be that there is no movement of a rotative kind which would tend to loosen the nuts. On the contrary, there is a strong force all the time tending to twist each rope. As can be seen, when tension is placed upon the ropes it acts to untwist them, and if the shackle-bolts were held in frictionless bearings they would rotate, the number of turns increasing as the strain is increased. On removing the strain, the ropes would twist up again and rotate the shackle-bolts in the opposite direction. When the traveling sheave is moving down to lift the car, the strain on the ropes is greater than when it is moving upward to lower the car; therefore, the shackle-bolts are actually subjected to strong forces that tend to turn them in one direction when the elevator ascends, and in the opposite direction when it descends. On this account, if single nuts were used they would work loose in a short time, and even check-nuts are liable to work free, hence the necessity of using safety-pins. The best plan of all, however, is to fasten the shackle ends by means of a piece of wire passing from one to the other, so as to prevent rotation as far as possible.

The course of procedure when shortening the ropes is as follows: If the shackle-bolt is of the design shown in Fig. 73, an enlarged view of which is given in Fig. 74, the ends are straightened so the clamp *A* may be removed; then the rope can be withdrawn and the wire binding taken off. The old bend in the rope is then straightened and another bend is made far enough down to shorten the rope the required amount. As the rope is stiff, strong clamps will be needed to bend it snugly around the center piece. While still clamped, the binding wire is wound on as tightly as possible, and if the distance between the end and the upper position of traveling sheave permits there is no objection to making the wire binding six inches, or even a foot, long. The ends of the wire should be turned outward and upward, as shown in Fig. 74, otherwise the rope may be drawn through the binding wire. If the end

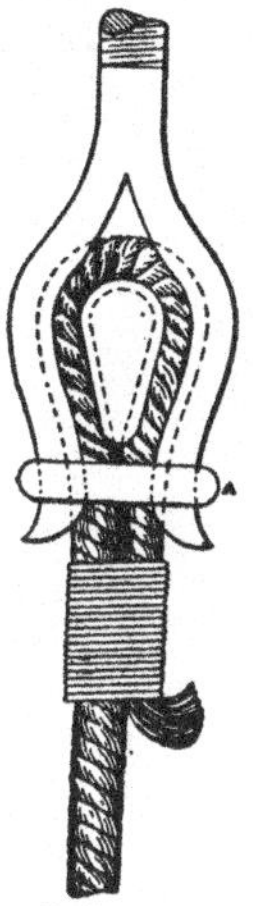

FIG. 74

FIG. 75

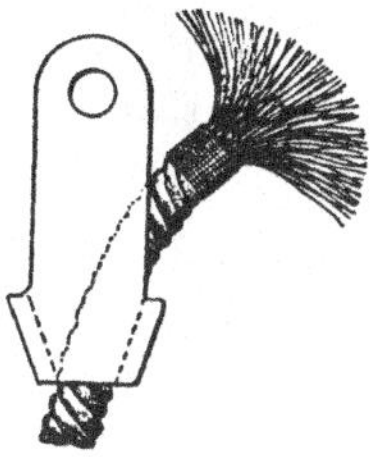

FIG. 76

is unraveled and wound around the body of the rope, and then covered with the binding wire, the job will be much stronger, although it may not look so well.

TYPE OF SHACKLE USED.

The shackle shown in Fig. 74 is not as generally used as the type in which the end of the rope is fastened in a conical ring, with the ends of the wires bent over to form an enlargement to prevent pulling the rope through the ring. This kind of fastening, when properly made, is best of all, and will hold more securely than the rope itself. The shape of this shackle is shown in Figs. 75 and 76. In some cases the bolt is made part of the shackle, and in others it is connected by means of a pin, as in the illustrations. To secure the end of the rope a binding band of wire is first put on at a distance from the point where the rope is to be cut off that will leave ends long enough to bend over and fill the cone cup to the top, as illustrated in Fig. 77, *a a* representing the point at which the rope is to be cut off, *b* being the binding band. This band is made of soft, iron wire (about No. 20), and to secure the ends of the wire firmly and neatly the winding should be done as shown in Fig. 78, the starting end *a* being run along in the space between the strands of the rope, the first turn of the band passing over the end, as shown at *c*, all the other turns being wound to the left. The band should be about three-quarters of an inch long, and the ends *a* and *b* are to be twisted together, and tucked under the band in one of the spaces between the strands.

It is necessary to put this band on before cutting the rope off, to prevent untwisting. It is also advisable to place a temporary band just

FIG. 77

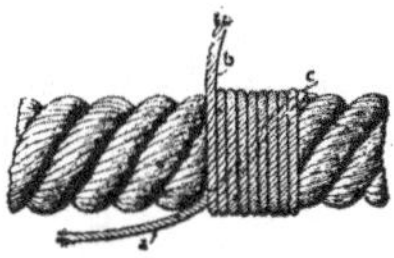

FIG. 78

below the cutting-off point, to facilitate the cutting. For the latter purpose a good hack-saw is best, although a half-round file can be used, especially where there is no way of holding the rope firmly. After the rope has been cut off, it is passed through the shackle in the manner shown in Fig. 76 and the ends of the wire spread out. Next, the ends are bent over in the manner illustrated in Fig. 79, but not as in Fig. 80. The latter way is easier and may on that account be resorted to, but it will not hold the rope safely under a heavy strain, as can be easily realized by looking at Figs. 81 and 82. In Fig. 81 it can be

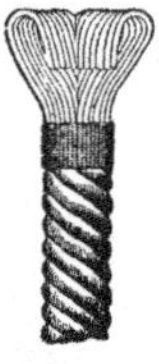

FIG. 79

FIG. 80

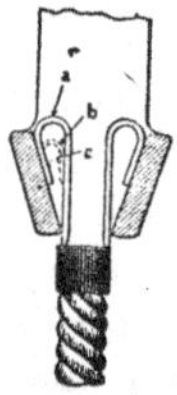

FIG. 81

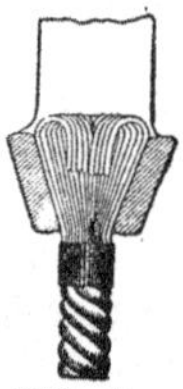

FIG. 82

seen that the ends of the wire will be held against the sides of the cone by friction and the tension of the rope will draw the wires through, changing the position of the bend, as indicated by the dotted curves *b*, *c*, until the rope is pulled entirely out of the shackle. With the turned-in ends of Fig. 82 there is no possibility of pulling out. After the ends have been properly bent, the rope is pulled into the shackle by means of tackle, or screws, and the spaces between the bent ends of the wire are filled with molten babbitt metal, or zinc. Sometimes lead is used, but it is too soft to be reliable.

When new ropes are put on they are fastened at the ends in the manner stated. The easiest way is to remove and replace one rope at a time, first running the car down to the bottom of the building and shutting the supply-pipe valve. In putting on the ropes that connect with the independent counterbalance, they must be made of such length that the car will rest on the bumpers before the counterbalance strikes the overhead beams, and when the car runs up to the top of the building the counterbalance must strike its bumper before the car reaches the beams. The proportionate lengths of the ropes are shown in Figs. 83

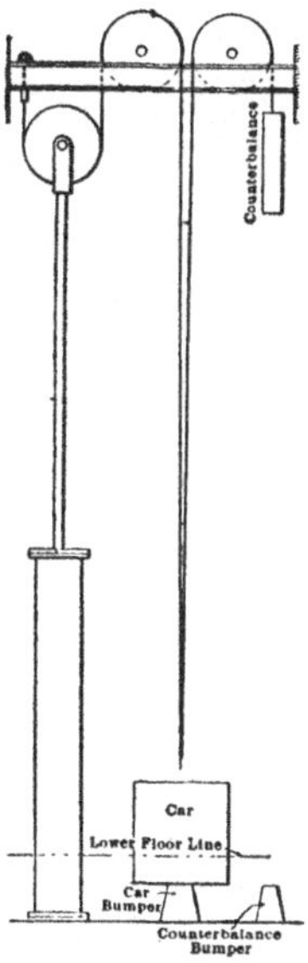

FIG. 83

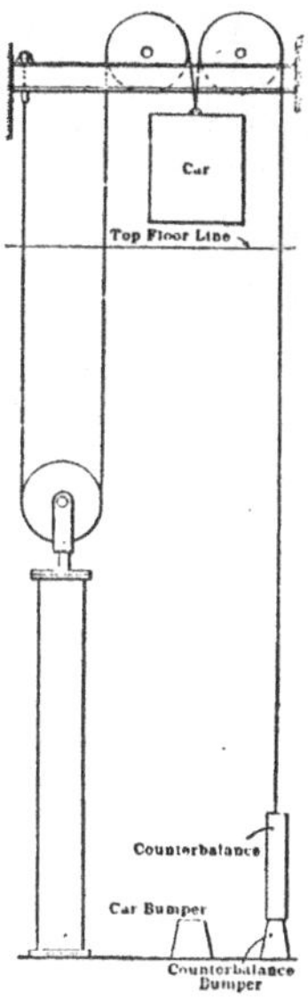

FIG. 84

and 84. In Fig. 83, if the ropes were so short that the counterbalance would strike the beams first, the ropes would undoubtedly be pulled apart. In Fig. 84, if the car ran into the overhead beams before the counterbalance struck the bumper, the ropes would be likewise pulled apart by the enormous force due to the momentum of the counterbalance. In the case of Fig. 83 the broken ropes would drop down the elevator well onto the top of the car, possibly with serious results to the passengers. In the case of Fig. 84 this also would generally be the case, as in most buildings the counterbalance runs in the elevator well at the side of the car.

CHAPTER XVIII

CONSTRUCTION OF TRAVELING-SHEAVE FRAMES, FOUNDATIONS AND SUPPORTS, TO AVOID PISTON AND CYLINDER FRICTION

In previous illustrations the traveling sheave has been shown as if moving unguided in its path. This construction was universally employed some years ago, and most vertical-cylinder elevators in actual use are so arranged; but nowadays very few machines are installed with unguided

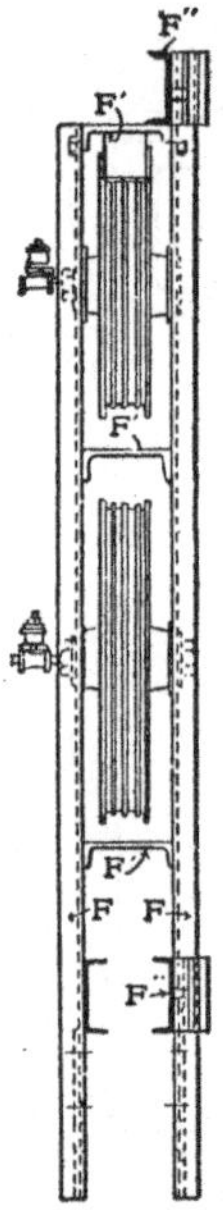

FIG. 85

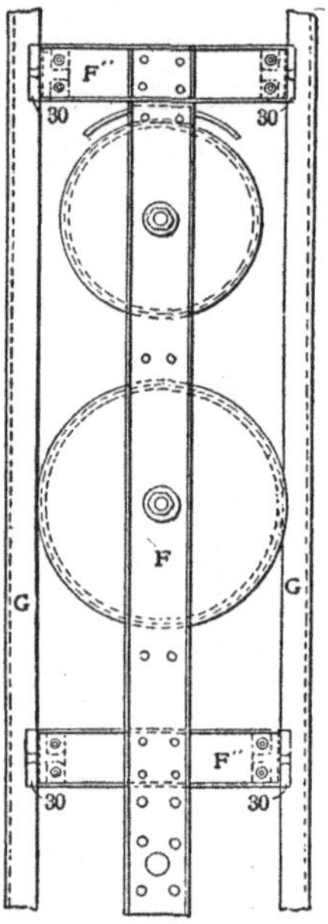

FIG. 86

sheaves. The modern practice is to construct a substantial frame, provided with shoes that usually run on iron guides, to hold the traveling sheaves. A sheave frame of this kind, as made by the Otis Elevator Company, is shown in Figs. 85, 86 and 87, the first two being end and side elevations and the third a plan view. This frame consists of sides

F F made of channel beams about nine inches wide, with connecting pieces *F′* above, below and between the sheaves, guide-holding channel beams *F″* being placed at the top and bottom. On the ends of the channels *F″* are secured guide shoes *30*, which run on the guides *G G*. The lower beam *F″* and a similar one opposite, as clearly shown in Fig. 85, form a support for the counterbalance weights, and the distance between the beams and the lower sheave is sufficient to permit the use of as many weights as may be required. The lower ends of the side frames *F F* are reinforced by flat plates riveted to them, so as to afford proper bearing for the trunnions of a crosspiece through which the upper ends of the piston-rods pass.

The construction of this piston-rod connection is shown in Fig. 88, which also shows a section through the lower ends of the side frames *F F*, from which the position of the reinforcing plates can be seen. The

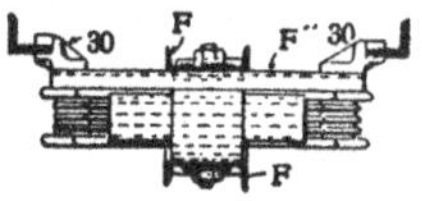

FIG. 87

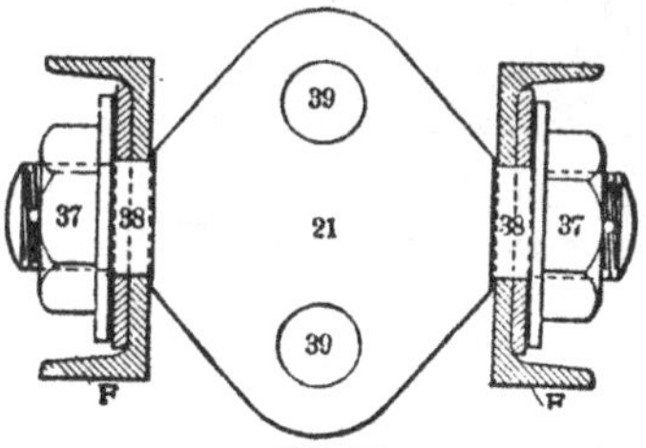

FIG. 88

trunnion ends of the piece *21* are so made that the nuts *37* can press the washers against a shoulder before binding on the side frame, the bearing *38* being considerably larger than the screw. Thus the piece *21* is free to swing and thereby equalize the strain on the two piston-rods, these passing through the holes *39*. The equalizing of the strain is not perfect, however, for if the piece *21* is tilted much, the nuts on the high side will have a bearing on the outside edge, while the nuts on the low side will bear on the inner edge; therefore there will be considerable difference in the leverage. Owing to this fact, whenever it is found that the crosshead *21* is out of square with the piston-rods, it should be trued up by screwing down the nuts on the high side.

The studs on which the sheaves revolve are constructed as shown in Fig. 89. They consist of two parts, a steel center stud and a brass outside sleeve. The stud reaches from outside to outside of the frames *F F*, while the brass sleeve is a trifle longer than the width of the sheave-hub; hence, when the nuts are tightened, the sleeve is prevented from rotating, as a consequence of being tightly clamped between the side frames. A thin strip is set in one side of the stud *32*, and keyways are cut on opposite sides of the diameter of the brass sleeve to slide over this strip. Through the center of the stud an oil duct extends from end to end, so that an oil-cup may be mounted (see Fig. 85) on either side of the sheave. From this central oiling channel radial ducts are run, as shown in the enlarged view of the stud, Fig. 90, and in the bore of the brass sleeve are grooves to match the holes in the stud; from these grooves ducts extend to the outside. It will be seen that, whatever

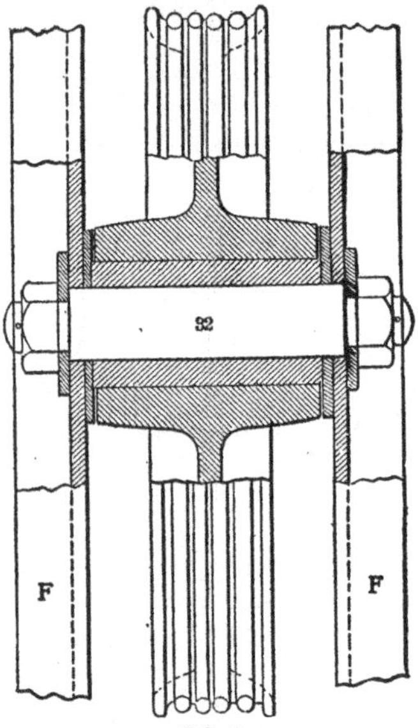

FIG. 89

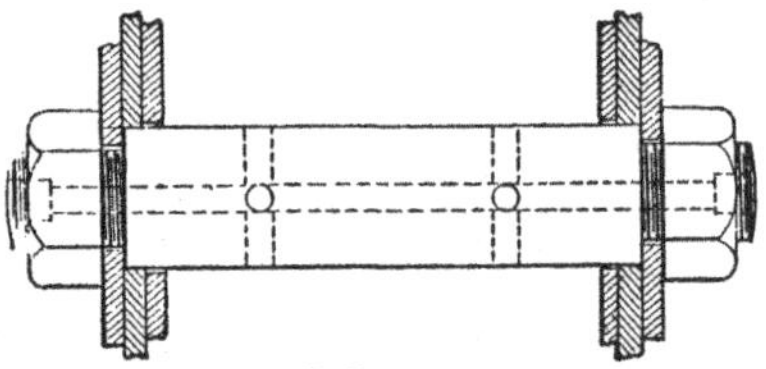

FIG. 90

the position of the stud, there will be a clear passage from the oil-cup to the outer surface of the brass sleeve which forms the sheave bearing. The sleeve has two keyways, so it may be reversed when one side is worn out, the pressure on the sheave-bearing being always in the same direction; and it makes no difference how much space there may be on the upper side; in fact, space at this point is a benefit, if anything; as it serves to collect the lubricant so it will be spread evenly over the bearing surface.

COMMON METHOD OF ANCHORING CYLINDERS.

When water under pressure is admitted to the upper end of the cylinder to lift the elevator car, it not only acts to force the piston down, but by pressing against the under side of the upper head acts to lift the cylinder, and unless the latter is held down firmly it will go up in the air and the car will remain stationary. Thus it will be seen that the cylinder must be well anchored to keep it from lifting, particularly if the elevator is geared high, say with a ratio of four or six to one. The most common way of anchoring vertical cylinders is by bolting them to a foundation of such size that foundation and cylinder will weigh more than the pressure of the water on the under side of the top cylinder-

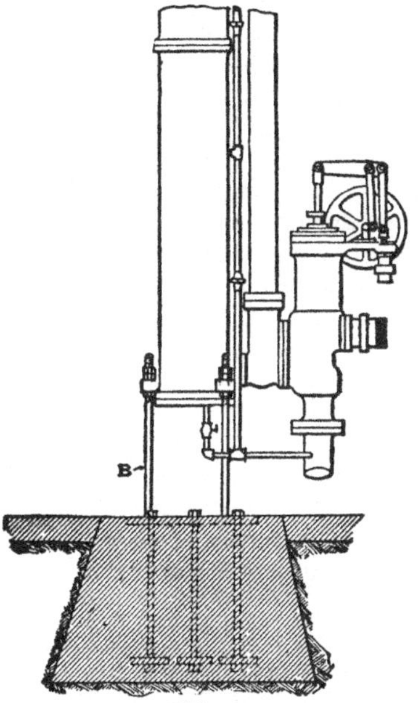

FIG. 91

head. The cylinder has three strong lugs at the lower end, as can be seen in Fig. 33. On the foundation is bolted a cast-iron plate, 1½ inches thick. The bolts extend nearly to the under side of the foundation and are secured in anchor plates of sufficient size to enable the bolts to lift the entire mass of masonry without breaking it. Three large connecting bolts are tapped into holes in the foundation plate, and the other ends pass through the cylinder lugs. The general arrangement of all these parts is shown in Fig. 91. The bolts *B* are about 1¼ inches in diameter and long enough to hold the lower head of the cylinder about two feet above the floor. By means of nuts, the bolts can be adjusted alike. To prevent the cylinder from swaying, it may be steadied from the walls or floor, according to which may be the more convenient. Sometimes the cylinder is kept from lifting by being secured to the framing of the building, where the latter is of steel construction. In such cases the three bolts *B* simply serve to hold the cylinder in position, but are not counted upon to act as part of the anchorage, because it is not good engineering to depend upon two separate supports when the construction is such that it is not possible to determine just what portion of the strain is carried by each.

For quite a while after an elevator is installed the foundation bolts should be examined frequently to see whether they have become loose, in which case they should be tightened. The clamps which hold the cylinder against the floors of walls should also be tightened if there is lost motion at any point. The cylinder and the guides on which the traveling-sheave frame runs, should be in line with each other. As the piston rods are of small diameter and project a considerable distance above the stuffing-boxes, when the sheave frame is in the lowest position there will not be much side strain, even if the cylinder and the guides of the sheave frame are out of line as much as half an inch; but the stuffing-boxes will eventually wear away on the side where the rods rub hard, so that it will be difficult to keep the packing tight. On this account, if it is found that the alinement is not perfect, the upper end of the cylinder should be shifted. Usually this can be done easily by slacking up the clamps and putting in or taking out a liner. It may be necessary, however, to cut away part of the flange of the girder that the cylinder rests against, in order to get the parts into line. If this should be the case, it is always preferable to shift the sheave guides, unless it is certain that the girder flange can be cut away without weakening it, however slightly. As a rule, if the part to be cut away is very close to the wall or other support of the girder, there will be no harm done by cutting into the flange, say, a quarter or half of an inch, but if the cutting would have to be done at a point near the center of the span it would be better not to attempt it unless assured by the designer of the building that it can be done safely. These girders are very accurately calculated by designing engineers with regard to withstanding strains, and, although a large margin of safety is allowed, the removal of a portion of a flange might reduce the stiffness to an injurious extent.

HOW TO STRENGTHEN SHEAVE SUPPORTS.

Not only must the lifting cylinder be properly anchored, but the stationary intermediate sheaves must be secured so that the rope strain will not pull down their supports. This will be understood from Fig. 92,

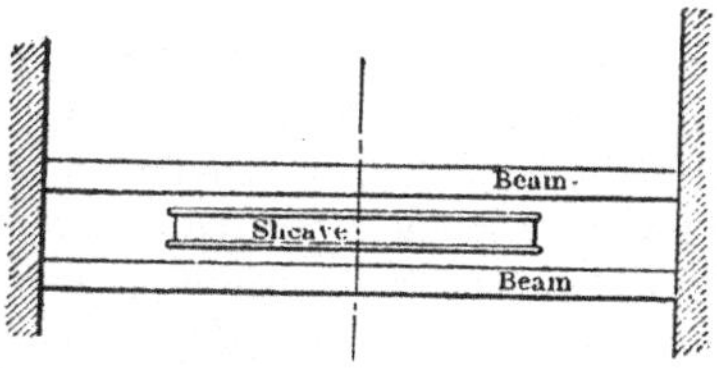

FIG. 93

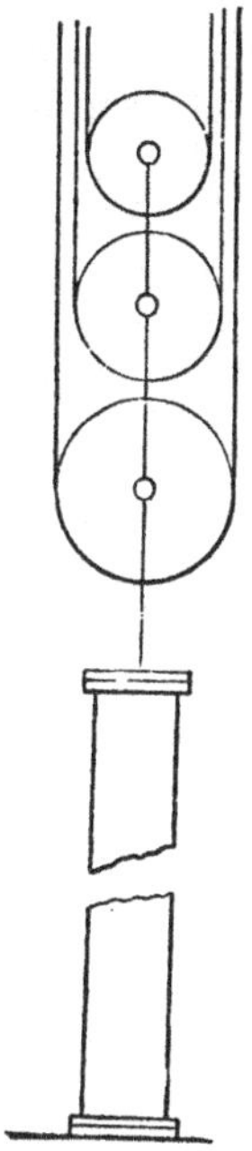
FIG. 92

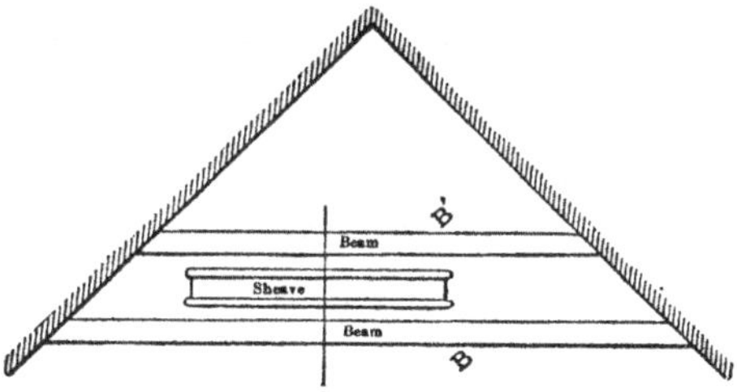

FIG. 94

in which it will be seen that only a portion of the total downward pull of the piston acts on the car-lifting rope *R*, the balance being impressed directly upon the supports of the stationary sheaves *S* and *S'*. In this diagram the machine is geared six to one, so that one-sixth of the pull of the piston acts on the lifting rope *R*, and the other five-sixths on the supporting beams *B B'*, two-fifths coming on the first-named and three-fifths on the second. In all first-class buildings, the supports for these sheaves are generally strong enough, but occasionally they are not, either through faulty calculation on the part of the elevator engineers or cheap

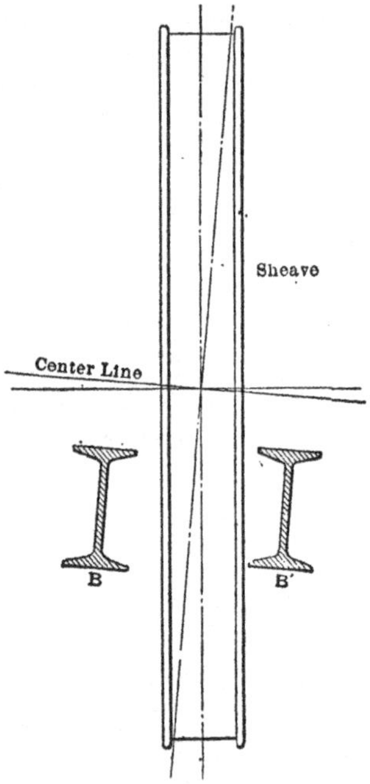

FIG. 95

construction. The principal elevator manufacturers employ engineers who can calculate accurately every part of an elevator structure. Others, however, employ men who have no technical knowledge, but work out everything by "practical experience" and the "rule of thumb," and, although such men generally err by making things stronger than necessary, in their desire to be on the safe side, they sometimes make serious mistakes and get things entirely too weak. As to the strength of the building, there is not so much certainty. First-class office buildings are sure to be of ample strength at every point, but cheaply constructed apartment houses and family hotels, with cast-iron columns and bolted joints, cannot be counted upon as being reliable in every particular. Where the floor beams of such buildings are not stiff enough to support the beams *B B'*, the construction can be improved by means of braces extending from the beams that sustain the supports *B B'* to the floor below, or, if necessary, to the second or third floor below. In extreme cases, where it does not appear safe to trust to the building itself, a perfectly rigid construction can be obtained by running braces from the supports *B B'* to the cylinder, so that the strains will be limited to the elevator apparatus. It is understood that this construction has been adopted by elevator builders in certain Western cities, for cases where buildings are cheaply built.

The building may be strong enough to afford proper support to the beams *B B'*, yet these may not be of proper size to give the required stiffness if improperly located. If they are sustained by parallel beams or walls, as indicated in Fig. 93, little harm will be done through lack of stiffness; but if they are supported by walls or beams at right angles, as in Fig. 94, there may be trouble, for as can be clearly seen the beam *B* will spring more than *B'*, owing to its greater length, and this will have the effect of throwing the sheave bearings out of line, as indicated in Fig. 95. When the support *B* sags more than *B'*, if both beams become twisted so as to remain parallel with the sides of the sheave, there will be little harm done; if, however, the beams do not twist, but remain in the vertical position, as shown, the sheave-shaft will be out of alinement with the bearings, and will run hot or score, until the corners wear away enough to give a good bearing. The easiest way to surmount this difficulty, if the support *B* does not sag much, is to line up the bearings parallel with the sheave-shaft. If the sagging is considerable, the beam should be propped up, if this can be conveniently done, or it may be stiffened by riveting a plate on the side of the web, taking the strain off the sheave while the plate is being put on.

CHAPTER XIX

WHY THE PISTON IS WEIGHTED

In each of the sectional drawings of vertical cylinders already shown the piston is represented with a number of counterbalance weights resting upon it, and no doubt many readers have wondered why this construction is used, inasmuch as it increases the length of the cylinder and therefore its cost. The reason can be made perfectly clear by the aid of Fig. 33. Suppose that the piston has just reached the point where the automatic stop valve begins to swing around to the right side to cut off the outflow of water, and that after the piston passes over one-half the remainder of the stroke the automatic valve closes the outlet completely, then, if the piston still persists in going down, the water under it must lift the relief valve *22* in order to escape. If the car is running upward very fast at this instant, as is likely to be the case if the load is light, the momentum will be great and the car will tend to keep up its motion after the stop valve has closed the outlet. The momentum of the traveling sheave and piston will also be considerable, but not as great as that of the car, owing to the lower velocity. The sheave, however, will move down as rapidly as the car goes up, but the piston might not do so, because to descend with a velocity equal to that of the traveling sheave it would not only have to overcome the friction of all its parts, but it must also have sufficient energy left to force the water through the valve *22*. The momentum of a light piston would be insufficient to do all this work; therefore it would lag behind and the piston-rods would slacken up; in extreme cases the rods might be forced up through the crosshead in the traveling-sheave frame.

Thus far I have considered only what would take place if the car were arrested by the action of the automatic stop while ascending, but it can be readily seen that if at any time during the up trip the operator closes the main valve too quickly when the speed is high, the same effect will be produced. To prevent such an occurrence the piston is weighted. It may be suggested that it would be preferable to make the piston heavier, instead of adding counterbalance weights, and this is very true; but heavier castings would have to be handled and there is room for an honest difference of opinion whether it would be advisable to save pieces by increasing the weight of a part which has to be finished in a lathe, and which in the process of erection has to be handled with great care.

TO PREVENT FRICTION.

As the weight of the piston alone causes it to descend, it is evident that it is necessary to restrict the resistance to motion to a minimum at all times, otherwise the stopping and starting on the up trips may be very unsatisfactory. If the piston packing is too tight sufficient friction may be developed to retard it unduly when the car stops on down trips, even if the stops are not abnormally sudden. If the cylinder is not well lubricated, also, the friction may increase enough to interfere with the movement of the piston; therefore, this point should not be overlooked. An oil-cup is provided at the side of the cylinder, and if this cup is filled every ten days or two weeks and the cocks opened for about half an hour, the oil will run in and, being lighter than the water, will rise to the surface and collect under the bottom of the piston, whence it will gradually spread over the cylinder surface. After the oil has drained out of the cup, the circulating cocks must be closed; if they are not, water will escape and the car will settle while stopped at the landing. Cylinder-oil is the proper lubricant for this purpose. It is seldom necessary to oil the cylinder oftener than once in ten days, because the water itself acts as a lubricant. If the cylinder requires lubrication it can be detected by a rumbling and also by unevenness of motion. These symptoms, however, are not always an indication that the cylinder requires oiling, as they may be produced by inequality of tension in the piston-rods, which will cause one side of the piston to sag and bind in the cylinder. This condition can be detected by examining the cross head plate in the traveling-sheave frame, to see if it is square with the piston-rods, and also by noticing if the rumbling and chattering are unequal on the up and down trips. If there is more noise when the piston is going up, it indicates that the rods are not strained equally.

In addition to keeping the cylinder properly lubricated, and the piston packing properly adjusted, it is necessary to keep the relief valve *22* in proper working order. A light spring is used to hold this valve to its seat because very little tension is required, but if it becomes clogged it may not move freely enough to allow water to pass through the valve as fast as necessary to avoid holding the piston back, causing the piston-rods to slacken up. When the car is going down, as the water in the cylinder simply circulates from the top to the bottom of the piston, the pressure above and below the valve *22* is nearly equal, hence slight pressure is needed to hold the valve to the seat. When the car is going up the lower end of the cylinder is connected with the discharge pipe, so that practically all the pressure is removed from under the valve *22*, hence no spring pressure is required to keep it closed. On an up trip, when the elevator is stopped either by moving the main valve or the automatic valve, the momentum of the descending piston will act to compress the water in the lower end of the cylinder and thus force it up through the valve *22* into the lower end of the circulating pipe. If the valve *22* lifts easily and opens wide enough, the water will be forced through it as fast as may be necessary to prevent holding the piston back; but if the valve moves hard, or does not lift enough, the piston will be held back and the piston-rods will slide down through the holes in the upper end, until the momentum of the car and traveling sheave is absorbed, then the weight of these parts will cause them to descend and pull the piston-rods back until they reach their bearing, when the car will stop with a jolt.

The automatic-valve chamber is provided with openings opposite the valve *22* to permit of easy access to the latter. To open up this valve chamber it is necessary to draw the water from the circulating pipe and, unless the piston is perfectly tight, also from the cylinder. Of course it will be understood that the valve in the supply pipe must be closed first. It is considerable of a job to get at the valve *22*, but fortunately it is not often necessary; for unless the elevator behaves unsatisfactorily in making stops on the up trips it may be taken for granted that there is nothing the matter with this valve. Then, too, if such stops are not as perfect as they should be, the fault may not be in the valve *22*, but in the cylinder, hence the latter should be put in proper condition before attempting to clean out the chamber of the valve *22*.

CHAPTER XX

THE ELECTRICAL FEATURES OF VERTICAL-CYLINDER ELEVATORS, WITH MAGNET CONTROL, OPERATION AND CARE OF PILOT VALVES AND CONNECTING MECHANISMS

This chapter will be confined to a consideration of the purely electrical features of vertical-cylinder elevators and the mechanical devices that operate in connection with them. The general arrangement of the various parts of a magnetically controlled vertical-cylinder hydraulic elevator may be seen by reference to Fig. 96, which shows the arrangement of the pilot valve and the secondary pilots used for operation by battery, or alternating current, the electrical features being the same as in those designed to be operated by current taken from an incandescent-lighting circuit. The magnets that move the pilot valve are located at *A*. They are connected with the batteries and the floor controller *D* by means of wires carried in the cable *B*. The arms *C* of the floor controller are connected with relay magnets, contained in the magnet box shown, by wires in the cable *E*, while other wires in the cable *F* connect the relay magnets with the push-buttons located at the sides of the elevator doors on the several floors. From a point *G* located half way up the elevator well is suspended a cable *H* which carries the wires connecting with the push-buttons in the car. The manner in which the relay magnets are connected with the floor and car push-buttons, and also with the contact rollers on the floor controller, is clearly shown in the wiring diagram, Fig. 59.

The purpose of the floor controller is to control the direction of movement of the car whenever a push-button is depressed. If the car is above the floor with which the button corresponds it will run down, and if below the floor it will run up to that floor. From this it will be seen that the floor controller is virtually a reversing switch, so made that the point at which it effects the reversal is controlled by the position of the car in the elevator well.

Two views of the floor controller are shown in Fig. 97, and from these it will be seen that the contact arcs *d d'* of Fig. 59 are replaced in the actual machine by the spiral strips *a a'*, Fig. 97. This construction is used to permit the contact rollers *b b* to pass over the break between the ends of *a* and *a'*, moving with sufficient velocity to prevent injurious sparking. The break *D*, Fig. 59, is shown at *d* in Fig. 97. In order

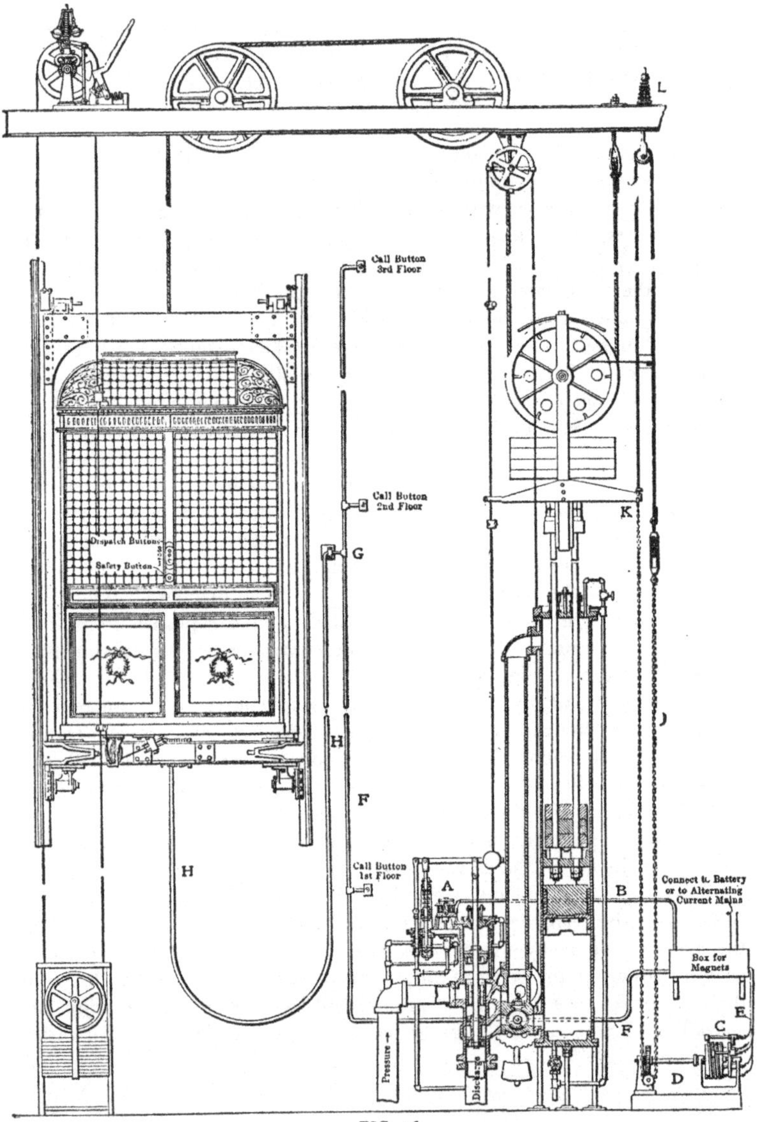

FIG. 96
GENERAL ARRANGEMENT OF VARIOUS PARTS OF A MAGNETICALLY CONTROLLED ELEVATOR

that the contact rollers *b b* may pass from the contact strip *a* to the strip *a'*, or *vice versa,* at the right time, the drum of the floor controller is rotated by means of a sprocket-wheel and the chain *J,* Fig. 96, and the gearing is proportioned to allow the roller corresponding to each floor to pass the break *d* at the time the car is passing the floor. The break *d* consists of a "dead" piece of such length that the circuit between the roller and the contact strip is broken shortly before the car is even with the floor, whether going up or down. The distance from the floor

FIG. 97

at which the break occurs is just enough to cause the car to come to a stop on a level with the floor.

NEED OF KEEPING CONTACTS BRIGHT.

The current passing through the floor controller is very weak (about one ampere), so that it cannot cause much sparking when the rollers *b b* pass off the ends of the contact strips; nevertheless, if the ends of the contacts are permitted to become badly burned the duration of the sparking will become increased, and this will not only cause increased burning at the ends of the strips with each succeeding break, but it will also cause the car to travel slightly farther before stopping. Thus it will be seen that to prevent rapid deterioration of the ends of the contact strips *a a'*, and of the rollers *b b,* as well as to secure accurate stopping of the car, it is necessary to keep these parts bright all the time. It is also essential to keep the surfaces of the contact strips bright all over, for, as will be seen in Fig. 59, the current that actuates the valve-lifting magnets passes through these strips, and also through the relay magnets, and if sufficient corrosion or dirt should accumulate to stop the flow of current, or even to reduce its strength so that it would not hold the relay magnets closed, the circuit would be broken and the car would stop. The sprocket-chain *J* must be kept properly adjusted so as not to slip off the teeth of the sprocket-wheel, but at the same time it should not be so tight as to lift on the controller shaft sufficiently to cause the bearing to run hot. The drum of the controller is arranged to run on a thread so that it may travel endwise as it rotates, and thus keep the contact rollers *b b* on the contact strips *a a'*. The pitch of the thread on which the drum runs is the same as that of the strips *a a'*. The floor controller shown in Fig. 97 is slightly different from that shown in Fig. 96, in that it has a large driving sprocket on the end of the shaft, in place of a small sprocket and a worm-gear reduction; in every other respect the construction is identical.

CARE OF THE RELAY MAGNETS.

In the magnet box shown in Fig. 96 are placed the relay magnets, and as the successful operation of the elevator depends on the proper working of these, they should be examined frequently and kept in good condition. All they require ordinarily is to be kept free from dust, and the contacts should be kept bright to prevent injurious sparking. The best way to keep contact surfaces bright, on both the floor controller and the relay magnets, is by the use of fine sandpaper, about No. 0. To prevent injury to the insulation, the parts through which electric currents pass must be kept well protected from water, and even moisture. As a rule, when the apparatus is installed all the electrical parts are placed where they are not exposed, but sometimes in making changes in a building the necessity of keeping these parts perfectly dry is lost sight of, and then trouble begins. The only parts of the electrical apparatus that cannot very well be removed from where they are liable to be damaged by water are the two sets of magnets that actuate the pilot valve; these must necessarily be placed at *A*, Fig. 96. To protect them as much as possible, they are insulated as a whole from the valve casing in the manner shown in Figs. 53 and 49, the former being the type of controller shown in Fig. 96 and the latter the type used when the operating current is taken from a lighting circuit. In Fig. 53 it will be seen that the bolts *A'* that hold the magnet stand to the valve plate *A* are well insulated, and the valve-rod *D* is insulated at *E;* so that if a leak should develop in the insulation between the wire coils and this part of the magnet frame it would do no damage.

In Fig. 49 an insulating plate *D''* separates the magnets from the plate E, and the bolts *E'* are also insulated from it. The magnet rock-lever *A* is insulated from the connecting-rod *13* at the joint *E''*. If the plate *E* is kept clean there is little danger of impairing the insulation *D''*, because the surface surrounding the stuffing-box of the pilot valve is lower than that on which the magnets are mounted; there is also a drain-pipe to carry off any water that may escape from the stuffing-box, and if the surface is not covered with dust and lint, which would cause water to creep up by capillary attraction, the insulating plate *D''* will remain dry. Holes are drilled through the plate *E* for the bolts *E'*, and other holes are drilled opposite the screws that hold the magnet cores to the base. If it is desired to remove the magnets it is important to be sure that the current is turned off, as otherwise, if there should be a leak between the magnet coil and the core, the screw-driver coming in contact with the sides of the hole, which it would be sure to do, would form a ground connection.

In the construction shown in Fig. 53 there is as little liability of injuring the insulation around the bolts as in Fig. 49, as the surface on which the stuffing-boxes are held is depressed.

PROTECTION FROM WATER.

The only way that water can get onto the magnets in either Fig. 53 or Fig. 49 is by leaking out of the stuffing-boxes in sufficient quantity to form a spray. If this should happen, the best thing to do would be to put up a board, or a sheet of heavy cardboard or tin, to keep the water off until the stuffing-box is repacked. It is not a bad precaution to have galvanized-iron boxes made to permanently cover the magnets, not only to protect the coils from unexpected sprays of water from the stuffing-boxes, but also to keep out dust.

The drain-pipe that connects with the depression in the valve plate should be kept clear all the time, because the stuffing-boxes are sure to leak occasionally, and if the water cannot run off it will fill up the depression and reach the insulation.

The strainers through which passes the water that enters the pilot valve should be cleaned out as often as may be necessary to insure free circulation of the water. If this is not done the main valve will not open as rapidly as it should, and the elevator car will not start up quickly enough. No definite directions can be given as to the frequency with which the strainers should be cleaned, as this depends entirely upon the amount of dirt in the water, and while once a week may be often enough in one case, once a day may be too seldom in others. About the best guide is in the operation of the elevator. If, within a certain time after the strainer has been cleaned, it is found that the starts are too sluggish on the down trip, the idle strainer should be tried and the effect noted. If this improves the starting, it indicates that the first strainer requires cleaning. After a few trials the length of time the strainers can be used without cleaning will be readily apparent. On the up trips of the car the condition of the strainers will not have any effect on the starting, because the pilot valve is then lifted, and the water above the motor piston at the top of the main-valve chamber escapes into the discharge pipe, the rapidity of discharge being controlled entirely by the opening through the pilot valve.

FOR REMOVING THE MAGNETS.

If at any time it should be desired to remove the magnets it can be done without disturbing the base plate or the frame in Fig. 53, and it is advisable not to remove the latter if the insulation is sound, because there is always a certain amount of uncertainty about securing perfect insulation when the plate is replaced. The magnet coils and plungers may be removed by unscrewing the bolts that hold them to the frame, and the latter need not be removed unless it is desired to take out one of the valves *D*. In Fig. 49 the magnets can also be removed by taking out the screws through the holes in the valve plate *E*. The lever *A* can be removed by withdrawing the pin upon which it rocks, and this can be removed by loosening the set-screw shown in the drawing.

No part of the magnet apparatus shown requires lubricating, because the total movement of the levers is so small that the parts can run dry practically as well as if lubricated. If it is felt that some lubrication is necessary, it will be sufficient to remove the pins and rub their surface well with a soft-lead pencil. It is not advisable to use oil or grease on these joints, because such lubricants spread over the surface and catch dust, increasing the danger of injuring the insulation. Oil itself, if clean, is a good insulator, but the particles of dust that stick to it are likely to be good conductors. All the bearings of the floor indicator should be properly lubricated, but it is not necessary to oil the rollers *b b* or strips *a a'*.

PROPER TREATMENT OF CABLES

The stationary cables *B, E, F,* Fig. 96, require no especial attention, as there is no reason why they should deteriorate, but they should be examined to ascertain if they are exposed where they are liable to be injured. If they are found to be improperly protected at any point they should be shielded without delay. The flexible cable *H,* from the junction *G* to the car, requires frequent inspection because the continuous bending will cause it to give out at some point sooner or later. If it is found to be chafed anywhere a thorough investigation will probably show that it rubs against the car or the wall of the elevator well, and an effort should at once be made to remedy the defect, either by changing the end fastenings so the cable may run free, or by covering any rough surfaces against which it strikes, or by both means. If the cable is improperly hung or protected, the insulating covering will soon be destroyed, and the wires may be broken so that the circuit connection with some of the floors cannot be completed, in which case the car will fail to respond to the push-buttons on these floors; or, two or more wires may come together and form a short-circuit which would put the elevator entirely out of service.

HOW TO LOCATE THE PILOT VALVE.

The proper adjustment of the pilot valve as regards location is of decided importance, as upon it will depend the position of the main valve when the elevator is at rest, and also the velocity of the car. In Fig. 53 the motor cylinder *Z* is so arranged that the ports at both ends are connected with the discharge pipe when the elevator is not running, hence the spring draws the piston-rod to the central position when it forces the collars *Q Q′* against the collars *N N′*. From this it can be seen that if the sleeves *P P′* are raised, the motor piston *M* of the main valve will rest in a lower position when the pilot valve is central, because wherever the sleeves *P P′* may be set, they will be drawn to the central position, with reference to *N N′*, by the spring *S′*, and as the elevator will not stop until the pilot valve is central, the shifting out of position due to the depression of the piston in *Z* must be in the main valve. If the sleeves *P P′* are lowered, the displacement of the piston *M* will be in the opposite direction, that is, upward. The movement of *P P′* up or down from the position in which they are shown causes the piston in the cylinder *Z* to stop either higher or lower than shown, but it will be noticed that the cylinder is much longer than the stroke, to provide room for adjustment. The stroke of the piston in the cylinder *Z* is equal to the difference between the lengths *P P′* and *N N′*. In Fig. 49 the adjustment of the position of the pilot valve is effected by means of the centering screws *C C′*, and as will be seen the result is the same as in Fig. 53; that is, if the screws are adjusted so as to lift the joint *E″*, the main valve will be raised, and if the adjustment depresses *E″*, the piston *V′* will be depressed, since the pilot valve must come to the central position to stop the elevator.

ADJUSTMENT TO REGULATE SPEED.

The effect produced by adjusting the pilot valve in the manner above explained can be understood readily from Fig. 49. It shows the lower valve *V* in the stop position, and it will be noticed that the cup packings lap over the ports considerably at both ends—from *c* to *d* at the upper end; and from *e* to *g* at bottom—hence the valve can be moved a distance equal to $c\,d+c\,g$, without setting the car in motion in either direction. This simply means that the motor piston V' can be in any position between the dotted lines a and b when the car is stopped.

When the elevator runs down no water is used; the weight of the car furnishes the moving force, and the water is forced out of the upper end of the cylinder, down the circulating pipe, through the valve chamber in the direction indicated by the arrows $s\,s\,s$, through as much of the port P' as may be uncovered by the valve V, and into the lower end of the cylinder. The flow of water is impeded slightly by the frictional resistance in passing through the circulating pipe, the valve chamber and other passages, but this is not enough to prevent it from running through quickly enough to cause the car to descend at a very high velocity, especially if heavily loaded. Furthermore, whatever this resistance may be, it will remain about the same for all trips. To permit the regulation of the speed of the car, by compensating for different loads, the valve V is opened more or less so that the main impediment to the free circulation of the water is the passage through the holes in the brass lining of the valve chamber, opposite the port P'. When the car runs up, the water in the lower end of the cylinder passes out through the port P' into the discharge pipe D_s, and the rapidity with which it can escape will depend on the amount of port uncovered by the valve V. From all this it is evident that if the car does not run fast enough on the down trip, its velocity can be increased by adjusting the pilot valve so the motor piston V' will rest lower when in the stop position, which means that the end of the lever (A, Fig. 49) must be depressed, and the sleeves $P\,P'$, Fig. 53, must be raised. If this is done, the valve V can be moved farther down when running, so as to open the port P' wider, and thus permit the water to circulate more rapidly.

In the same way, if the car ascends too slowly the speed may be increased by adjusting the pilot valve in the opposite direction.

This explanation of the adjustment of the pilot valve applies to car-lever control as well as to electromagnetic control. It may be well to mention, however, that the required increase in speed cannot always be obtained by pilot-valve adjustment. Thus, if the up speed is too slow and the down speed just right, then increasing the up speed will reduce the down speed. If the down speed is too slow, increasing it will reduce the up speed. In cases of this kind pilot-valve adjustment will do no good, as the cause of insufficient speed is lack of power, and the proper remedy is to increase the pressure in the pressure tank slightly. It is only when the car runs faster in one direction, as well as slower than is desirable in the other direction, that anything can be done by pilot-valve adjustment.

CHAPTER XXI

HORIZONTAL HYDRAULIC ELEVATORS

DESCRIPTION OF THE "PUSHING" TYPE OF HORIZONTAL ELEVATORS, DETAILS OF OPERATION AND CONSTRUCTION OF VALVES AND PARTS

The main difference between vertical and horizontal hydraulic elevators is that the first-named have vertical lifting cylinders, while the latter have horizontal cylinders. The valve mechanism of a vertical machine can be used with perfect success on a horizontal machine, and in fact it has been so used in several instances, but the best type of horizontal machines have a valve mechanism of distinctive design; although the principle of operation is the same as in the vertical machine valve-gears described in previous chapters.

Horizontal hydraulic elevators are divided into two classes, known as the pushing and pulling types. In this chapter will be described the Crane pushing-type elevator, built in Chicago by the Crane Company, now a part of the Otis Elevator Company. This is one of the most extensively used elevators of this character. The general arrangement is clearly shown in Fig. 98. It will be seen that the rear end of the cylinder is adjacent to the elevator well, and that the pressure water forces the piston toward the opposite end, which results in pushing the traveling sheaves away from the stationary sheaves, thereby drawing the lifting ropes downward and lifting the car. Therefore, the "pushing" type of elevator derives its name from the fact that the cylinder is placed between the stationary and traveling sheaves and "pushes" them apart in order to raise the car. The steam pump draws water from the open tank shown and forces it into the pressure tank, whence it passes to the rear end of the cylinder, between the piston and the back head, and forces the piston forward. On the return stroke the water in the cylinder is discharged into the open tank through the discharge pipe shown entering at one side of the tank. The main valve through which the water enters and leaves the cylinder is located at *K*, and is automatically operated by the movement of a pilot valve located at *L*, the latter being actuated by the rocking of the shaft *M* through the movement of the rods *m m*, which are connected with a running-rope system operated by the car lever. In design the main valve and the pilot valve differ from those described in

FIG. 98

GENERAL ARRANGEMENT OF THE VARIOUS PARTS OF A HORIZONTAL HYDRAULIC ELEVATOR treating of vertical elevators, but the principle of operation is identical. An automatic stop valve is located at *R*, between the main valve and the cylinder, as is the case in vertical-elevator construction. It is actuated by the mechanism shown at *N*, which in turn is set in motion by the movement of the traveling-sheave crosshead.

THE OPERATION IN DETAIL.

The operation in detail can be more clearly explained in connection with Figs. 99, 100 and 101, the first being a side elevation, the second a plan view, and the third a vertical cross-section of the cylinder, a piston, sheaves and connecting parts. In Fig. 99 it will be seen that if the car lever *S* is moved in either direction the rods *m m* will rock the shaft *M*

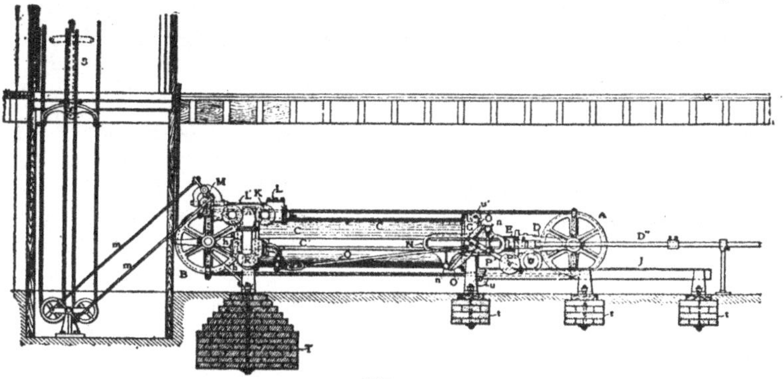

FIG. 99

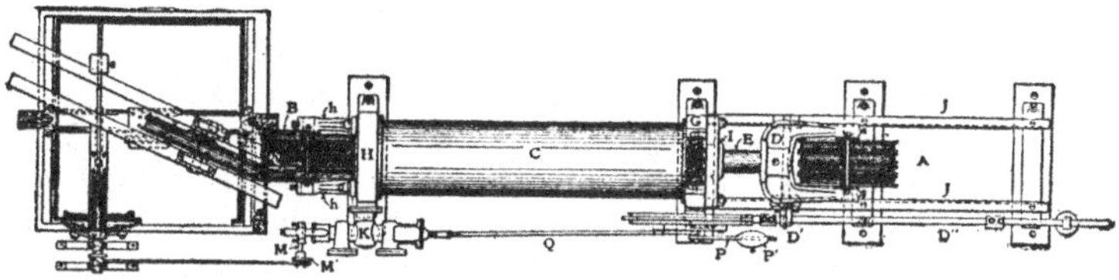

FIG. 100

and actuate the pilot valve through the medium of the valve-rod *L'*. The movement of the pilot valve opens or closes the main valve *K*, so as to let water in or out of the cylinder, according to the direction in which the lever *S* is moved. When the car reaches the upper end of the well, if the operator does not return the lever *S* to the stop position the frame *N* will be carried to the right by the movement of the cross-head, and the projecting arm *D'*, Fig. 100, will strike a stop mounted on rod *D"*, which is connected with the right-hand end of the frame. This movement of the frame *N* will cause a roller *n'* to strike lever *O'*, which will move to the right and pull rod *Q* in the same direction; and this action will close the stop-valve *R* and stop the flow of water into the cylinder, which, of course, will stop the movement of the piston, and of the car. If the car is coming down the piston will be moving to the left, and if the operator does not return the lever *S* to the stop position when the

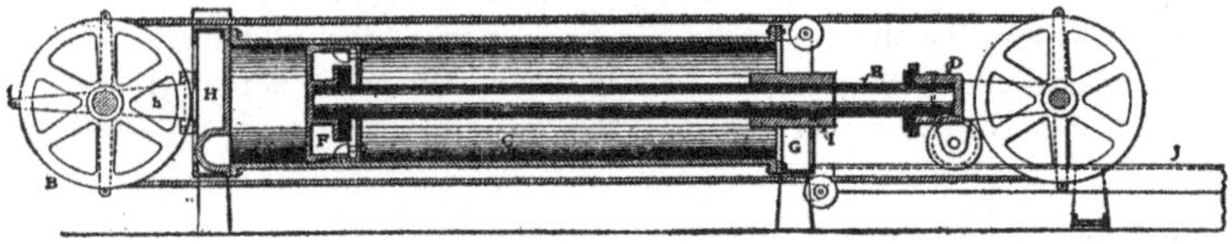

FIG. 101

lower floor is reached, the arm *D'* on the crosshead D will strike the stop on rod *D''* near the frame *N* and carry the frame to the left, so that arm *O* will be moved by the roller *n*, and, as before, rod *Q* will actuate the stop valve and prevent the flow of water from the cylinder. The stop valve *R* can be adjusted to act at any time desired, by means of the stops on the rod *D''*.

Referring to Fig. 101 it will be seen that there is a rubber ring around the piston end of the plunger *E* and a similar ring in the crosshead D, while attached to the front cylinder-head *G* is a strong buffer frame *I*. If the car overruns its normal travel at either end, the rubber ring at that end will strike the buffer *I* and prevent the car going farther in that direction. The adjustment of these parts should be such that the car and counterbalance weights cannot strike the overhead beams, in case the automatic stop gets out of adjustment and fails to act in time. With proper adjustment there will be no contact with the buffer *I* at all, but the stretching of the ropes may disturb the adjustment, when it should be restored to normal condition by the means to be explained. The buffer *I* and the back cylinder-head *H* are tied together by means of four strong bolts which extend the length of the cylinder, there being two on each side. The front bolts are shown at *C' C'*. The traveling-sheave crosshead is provided with rollers which run on the tracks *J J*, their purpose being to keep the piston plunger in line with the cylinder. The stationary sheaves at the back of the cylinder are mounted upon and revolve on a shaft held in position by the frames *h h;* and to increase the strength of the support these frames are braced by wrought-iron rods *h'*. The rear end of the cylinder is secured to a heavy foundation *T*, the weight of which is sufficient to overbalance the maximum pull on the lifting ropes. The foundation piers at *t, t, t* are not as large, being intended simply to insure alinement.

THE MAIN AND PILOT VALVES.

The operation of the main valve and the pilot valve may be understood from Figs. 102 and 103, the first a vertical elevation in section and the second a plan view. In Fig. 102 *I* and *J* are cup packings. The pressure water enters the valve chamber through the port *I'*, and if there is no pressure back of the piston *G* the valve will move to the left, because

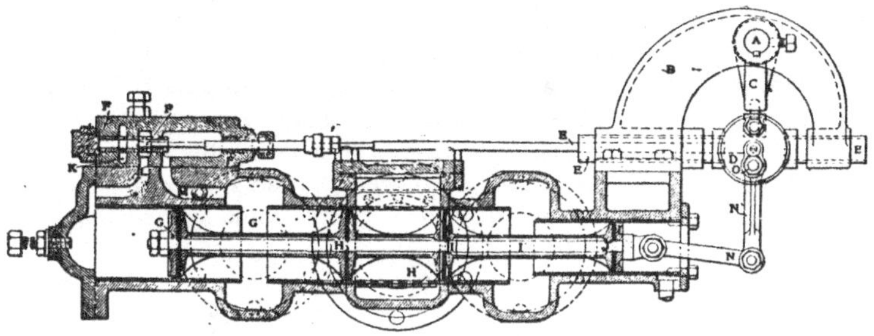

FIG. 102

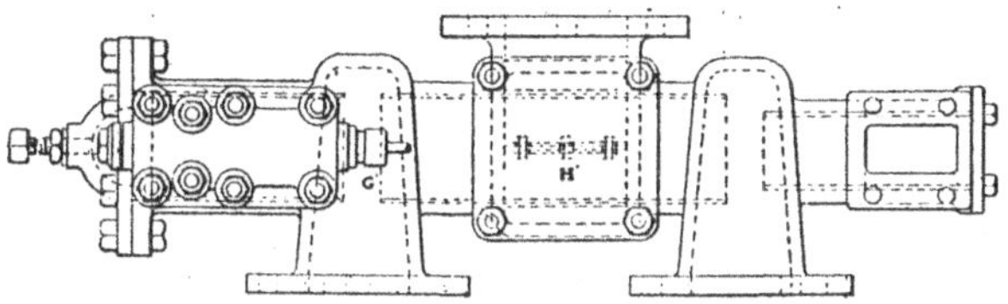

FIG. 103

the diameter of *I* is larger than that of *J*. On the other hand, if pressure water is admitted between the piston *G* and the cylinder-head the valve will move to the right, because the pressure on *G* will counterbalance that on *I*, and the pressure on *J* will force the latter to the right. In Fig. 98 it will be seen that the pipe *Z* connects the pressure pipe with the pilot-valve chamber. This pipe connects with the port *K*, Fig. 102; therefore, if the rocking lever *M'*, Fig. 99, is actuated so as to move the lever *C*, Fig. 102, to the right, the pilot valve will be shifted to permit the pressure water in the port *K* to pass through port *L* to the back of piston *G*. This will force the main valve to the right, connecting the center port *H* with the discharge port *G'*. In this manner the water will be discharged from the cylinder, the piston will move toward the rear and the car will descend. If the lever *C* is moved to the left, the pilot valve *F* will be shifted to allow the water behind piston *G* to run out through port *L* into and through port *M* to the discharge pipe.

This movement of the main valve will carry piston *I* to the left so that pressure water from *I'* will pass to and through *H* to the cylinder, forcing the piston forward and lifting the car. The lever *N'* is pivoted at the point *O*, in the circular frame *D*, and its upper end is connected with the pilot-valve stem *E'*; consequently, when the main valve moves to the right it carries the pilot valve to the left. When the pilot valve is moved to the right by the car lever, the main valve is also moved to the right through the pressure water passing from port *K* through port *L* to the rear end of the valve chamber; so, as soon as the main valve begins to move it shifts the pilot valve back to the stop position, the action being precisely the same as in all the pilot-valve gears previously described. In Figs. 102 and 103, for the purpose of simplifying the drawings, the flanged outlets of ports *G'*, *H'* and *I* have not been shown in their true positions. The outlets *G'* and *I'* may be on the same side of the valve

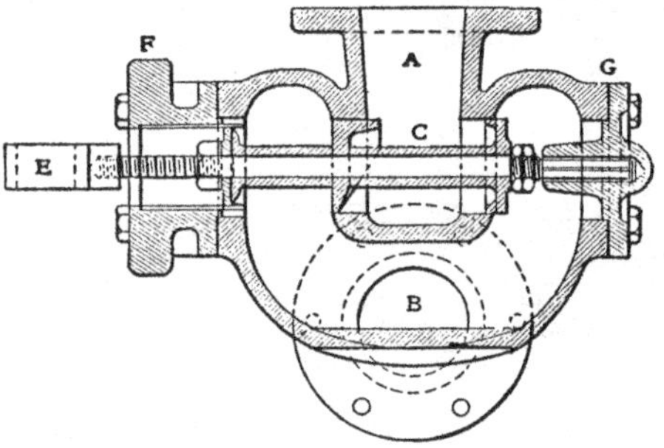

FIG. 104

chamber, or on opposite sides, according to which affords the best piping design. The center outlet *H'*, however, is always placed on the under side, so it can be run down to the top of the stop-motion valve, as will be readily understood from reference to Fig. 99.

THE AUTOMATIC STOP VALVE.

The operation of the automatic stop valve *R* is made clear in Fig. 104, which is a vertical section through the valve and casing on a line parallel with the axis of the cylinder. The construction of the valve-rod is shown in Figs. 105 and 106. The end of the lever *a'*, Fig. 99, is located within the square opening shown in the end of the valve-rod, which allows freedom of movement for shifting the valve in either direction.

The weight P' on the lever P, Fig. 99, is the medium by which the valve is pulled open (or on the left, in Fig. 104), and the rocking of the lever $O\ O'$ by the action of the frame N forces the valve to the closed position, or to the right. The piston of the valve C, which closes the passage, is tapered so as to gradually stop the flow of water and prevent the sudden stoppage of the car. Referring back to Fig. 99, it will be noted that when the elevator is at either end of the well the valve R is closed; therefore, unless its construction permits water to leak by, the car cannot be started. As a matter of fact, the valve does not fit perfectly, but leaks just enough to permit the piston to move slowly in starting from either

FIG. 105

FIG. 106

end. This not only prevents making a too sudden start, but it also prevents sudden stopping, because even after the valve has been closed some water can pass through it; but not enough to permit the elevator to run at anything like full speed.

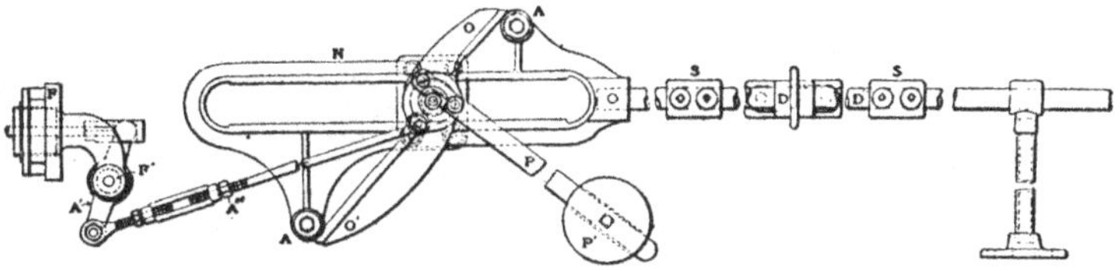

FIG. 107

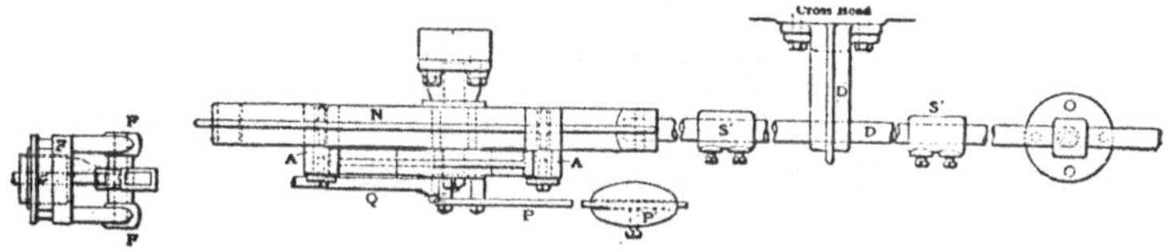

FIG. 108

The construction of the automatic stop-valve shifting mechanism is shown in Figs. 107 and 108, the former being a side elevation corresponding to Fig. 99 and the latter a top view, corresponding with Fig.

100. In these drawings it can be seen that *O* and *O'* are ends of the lever which operates the stop valve. This lever is mounted on a stud carried by a frame bolted to the side of the front cylinder-head, the frame also serving as a guide for the sliding frame *N* on which are carried the rollers *a* which actuate the lever. The front head of the stop valve is shown at *F*. The stops *S' S'* are adjusted on the rod *D''* so that they will be struck by the arm *D'*, projecting from the side of the traveling-sheave crosshead, at the right instant to bring the car to a stop just beyond the top or bottom floor of the building.

CONSTRUCTION OF OTHER PARTS.

The traveling-sheave crosshead is shown in Figs. 109, 110 and 111, which are side, top and end views, respectively. The piston plunger is secured in *A*, and the shaft on which the sheaves revolve is held in the ends *B B* by means of set-screws. The lugs *C C* are provided to hold a cable guard. The office of this guard is to hold the cables against the

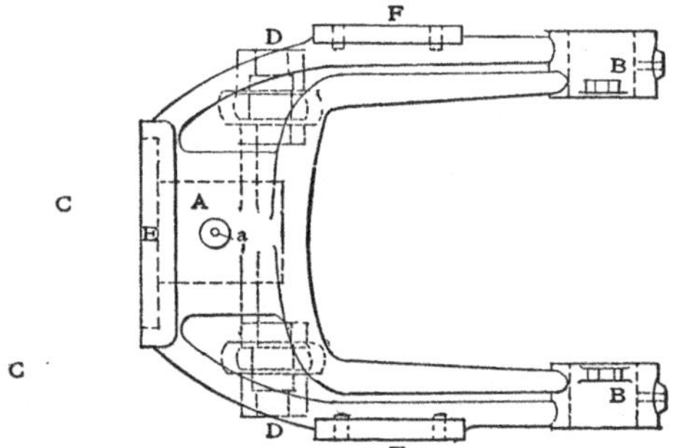

FIG. 110 FIG. 109

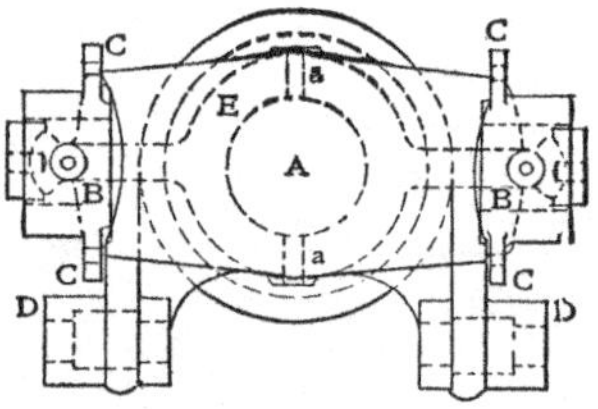

FIG. 111

sheave surface, in event of their becoming slack from any cause. The rollers upon which the crosshead moves are mounted on studs fastened in the lugs *D D*. The rubber buffer already mentioned is located in the recess *E*. To prevent the piston plunger from working loose a pin is inserted in the holes *a a*. The operating arm of the stop-motion mechanism is bolted to the seat *F*. There are two seats, one on each side, to permit placing the stop-motion to suit the convenience.

The piston is shown in Fig. 112, which is a section parallel with the axis. As indicated, the packing is compressed by means of a ring *A*, by

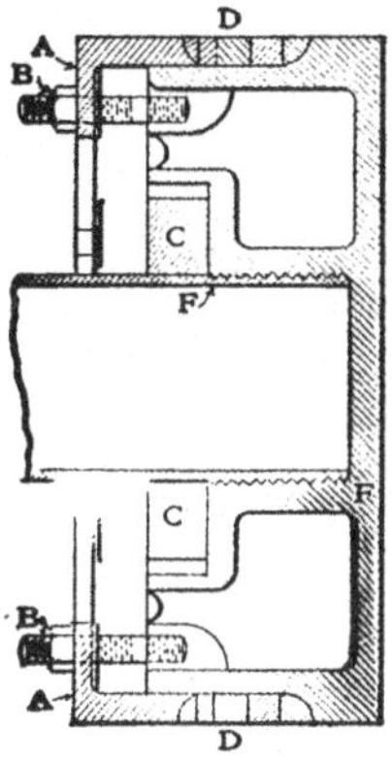

FIG. 112

tightening the bolts *B B*. The buffer ring is held in a circular depression surrounding the plunger, at *C*, the depression being somewhat larger in diameter than the ring, to give the rubber expansion room when it is acted upon by the buffer.

The construction of the front cylinder-head is shown in Figs. 113 and 114, the first being an end view and the second a side elevation. The head is secured to the cylinder flange by bolts passing through the holes *a a*. The supporting frame of the stop-motion mechanism is bolted to one of the faces *B B*. The guides upon which the crosshead rollers run are fastened to the lugs *C C*. These guides are made of heavy-section angle iron. The lugs *D D*, located just below the lugs *C C*, support a shaft upon which are mounted the shell sheaves for the lifting ropes running under the cylinder. The lugs *E E* on the top of the head hold a shaft on which are mounted the small sheaves which support the lifting ropes running over the top of the cylinder-head. The large holes *A A* are for the long bolts that hold the buffer against the head, and, as is apparent in Fig. 114, the part through which the holes extend being turned to form a raised seat to hold the buffer square with the cylinder, and prevent cramping the piston plunger.

The back cylinder-head is shown in Figs. 115 and 116. This head is

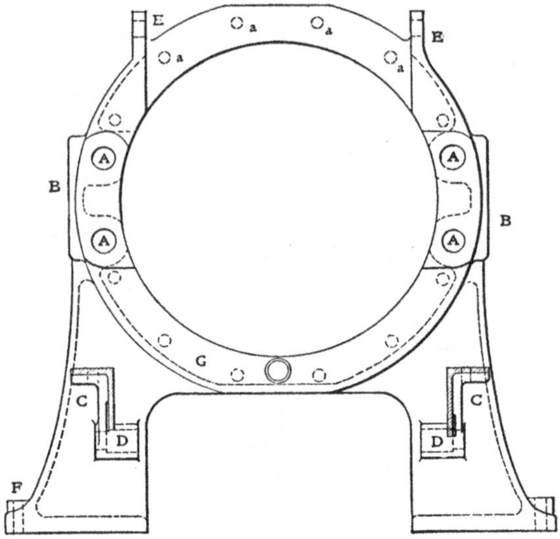

FIG. 113

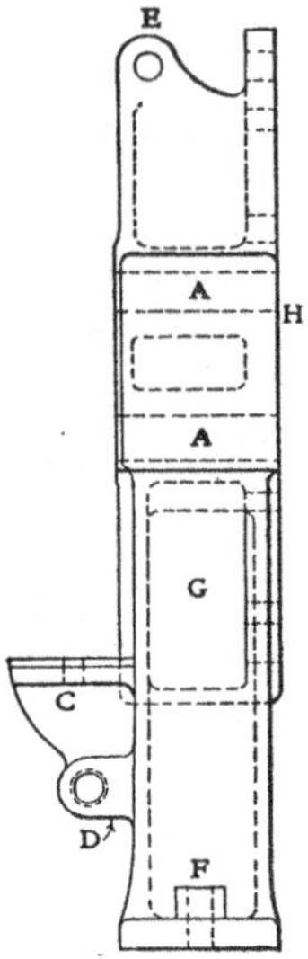

FIG. 114

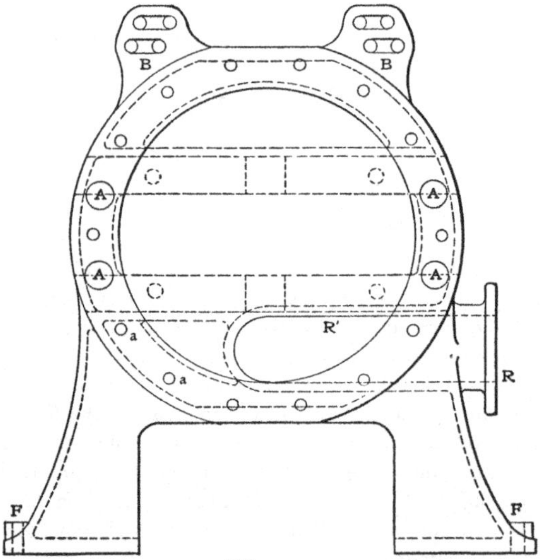

FIG. 115

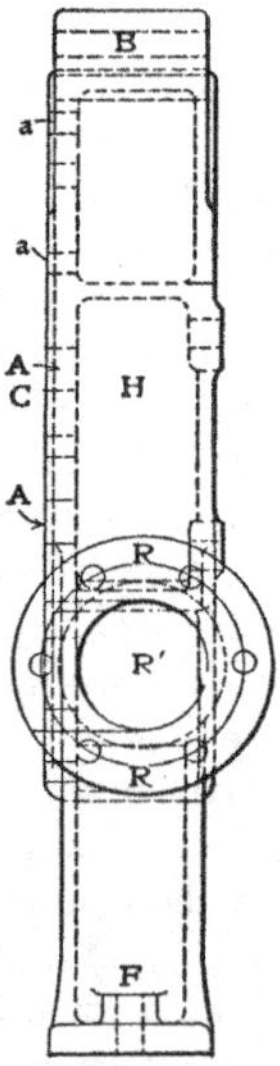

FIG. 116

with four large holes *A* to receive the long bolts that clamp the front cylinder-head. From this construction it is evident r the piston strikes the buffer, the strain is not impressed nder, but is carried by the bolts. The stop-motion valve is inlet *R*, a port *R'* being located in the body of the head. are anchorages to hold the ends of the lifting ropes, two ovided so that the ropes may be attached at whichever side

CHAPTER XXII

DESCRIPTION OF THE "PULLING" TYPE OF HORIZONTAL ELEVATOR, SHOWING THE OPERATING PRINCIPLE OF THE WHITTIER MACHINE

The "pulling" type of horizontal hydraulic elevator differs considerably from the "pushing" type. The general appearance of the pulling type as made by the Whittier Machine Company is illustrated in Fig. 117. The main operating valve is at *G*, and the pilot valve is placed directly above it, at *J*. The automatic stop valve is at *H* and is actuated by means of stop-balls *N* mounted on the rope *L*. These stop-balls are moved by contact with an arm attached to the crosshead which carries the traveling

FIG. 117

WHITTIER HORIZONTAL ELEVATOR, PULLING TYPE

sheaves *D*, and shoes *R* on the crosshead slide within the side guides. The weight *P*, suspended from a chain which travels between the two small guide-sheaves located just below the valve casing, is for the purpose of bringing the automatic stop valve to the central position as soon as the piston moves away from either end of the cylinder.

The pilot valve is moved by the lever *K*, to the ends of which the operating ropes that connect with the car lever are attached. The automatic stop valve is located under the main valve, at *H*, and is actuated by the movement of rope *L*, there being stop-balls *N* on the rope that are carried along by an arm that projects from the crosshead and surrounds the rope, as is clearly shown in the illustration.

If the sheaves shown in Fig. 117 are counted, it will be found that there are six stationary sheaves *E* and five traveling sheaves *D*. If there were one traveling sheave and two stationary sheaves, the rope starting from the hitching point *A* would run forward around one of the stationary sheaves, thence backward over the top and around the traveling sheave, then forward again along the under side to the secondary stationary sheave, taking a quarter of a turn around it, thence upward to the top of the building. From this it will be seen that there is one set of ropes passing from the top of the first stationary sheave to the top of the traveling sheave, and one set of ropes on the under side passing from the traveling sheave to the second stationary sheave; and according to the rule already given for determining the ratio of gearing, this would be a two-to-one machine. In this type of machine, then, the gearing ratio can be determined by counting the sheaves, as it is equal to the total number of sheaves less one. Therefore, if there are six stationary and five traveling sheaves the gearing ratio is ten-to-one.

Usually in the pulling type of elevator the ropes are secured near the cylinder, but they can also be built with the ropes hitched at the forward end above the stationary sheaves *E*. This construction is not as desirable, however, because it necessitates building an extension upward from the ends of the guide frames strong enough to afford a safe anchorage for the ends of the ropes. Another objection is that as the traveling sheave moves forward it carries along with it the twist in the ends of the ropes, thus developing a strong twisting strain in the shackle-bolts. With the ends attached near the cylinder there is no such strain, because the lengths of the ropes running from the hitching point to the stationary sheaves never change.

The traveling-sheave shaft is inclined so that each of the sheaves *D* may be in line with corresponding sheaves *E* on the under and upper sides. If the sheaves were not so set, there would be danger of the ropes running off when the traveling sheaves are close to the stationary sheaves. The crosshead is designed so as to hold the traveling-sheave shaft in the proper position, shown in Fig. 118. The trunnions *F F* carry the guide-shoes *R*, Fig. 117, and are in line with the holes *B′ B′* into which the piston-rods are secured, while the side rods *C′ C″* fit into the holes *A A′* set at an angle to the trunnions *F F* and the holes *B′ B′*. The bar *C* and the ends *C′* of the side bars are larger than the bar *E* and the ends *C″*, because they have practically to take care of all the tension, while the forward parts have simply to hold the crosshead in position and carry the forward guide-shoes *R*. All the parts of the crosshead are made of wrought metal, either steel or iron.

The main and the pilot valves of the Whittier machine are shown in detail in Figs. 119 and 120, the first being a plan view and the second a sectional side elevation. Looking at Fig. 119, it will be seen that the operating lever *K* is pivoted at the point *F*, so that when actuated by the operating ropes *A A'* it imparts an end movement to the pilot-valve rod *C*.

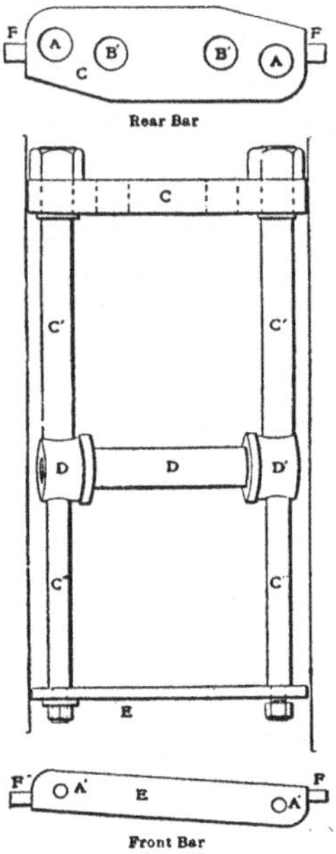

FIG. 118

The ropes *A A'* are connected with the operating lever in the car by either a running- or a standing-rope attachment identical with those used for vertical-cylinder elevators.

CONSTRUCTION AND OPERATION OF VALVES.

In Fig. 120 the pilot-valve rod *C* is connected with the upper end of the lever *D*, the latter being pivoted at *G*. The part *B* which holds the pivot *G* is actuated by the lever *K*. The supply pipe is connected with the right-hand end of the pilot-valve chamber through the pipe *E*. If the rod *C* is moved to the left, high-pressure water will pass through the pilot valve to the end *I* of the main valve and force the latter to the left, thereby connecting the cylinder with the discharge pipe, when the water will run out and the elevator car descend. The forward movement of the main valve will carry the lower end of the lever *D* to the left and the upper end to the right, until the pilot valve is returned to the closed position. If the pilot-valve rod *C* is moved to the right, the end *I* of the main valve will be connected with the discharge and the water will escape, then the pressure acting on the piston *L* will force the valves to the right and connect the supply pipe with the cylinder, which will fill with water

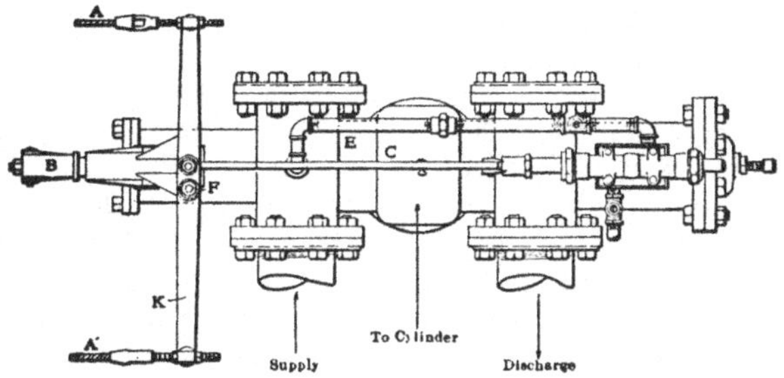

FIG. 119

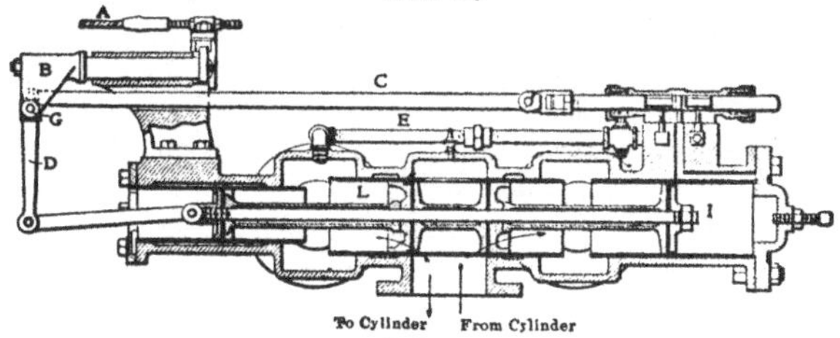

FIG. 120

from the pressure tank and the car will be forced upward. The movement of the main valve to the right will carry the lower end of the lever *D* in the same direction and the upper end to the left, and return the pilot valve to the central position. It is to be noted that the action of the valves is the same as in all pilot-valve devices previously described.

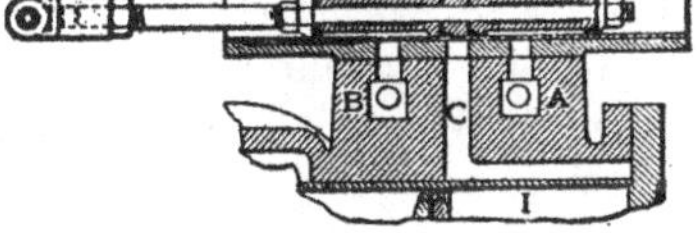

FIG. 121

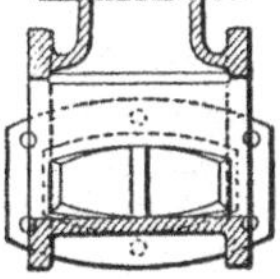

FIG. 122

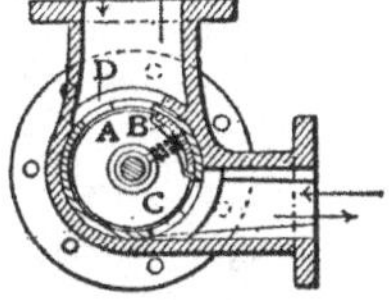

FIG. 123

The pilot valve shown in Fig. 120 is provided with stuffing-boxes at each end to insure tight joints with the valve-rod, but this construction is not used in all the Whittier elevators; in some of them the pilot valve is made as shown in Fig. 121, where the escape of water at the ends is prevented by the use of cup packings. The pressure water enters through the port *A*, the discharge being through the port *B*; consequently, the cups are set so as to oppose the pressure which is exerted in both directions from the port *A*.

The construction of the automatic stop valve of the type shown in Fig. 117 is clearly presented in Figs. 122 and 123, the former being a section at right angles to the axis of the valve, that is, parallel with the axis of the lifting cylinder, and the latter a section in line with the axis. Water flows through the valve into the cylinder from top to bottom, as indicated by the arrows, and flows out in the opposite direction. The valve *B* is rotated by a carrier *A*, and if water is flowing into the cylinder, that is, if the car is running upward, the carrier will rotate the valve in a clockwise direction so as to cover the port *C*. If water is running out of the cylinder (the car running downward) the carrier *A* will rotate the valve counter-clockwise so as to cover the port *D*; so that, in whichever direction the water may be passing through the valve chamber the valve will be moved over the port through which the flow is outward, and the pressure will force the valve *B* against its seat.

The valve *B* is held against the seat normally by the spring shown, which has sufficient tension for this purpose, but not enough to withstand the pressure of the water when acting to push the valve *B* toward the center. This construction is the same in principle as that used in the Otis vertical-cylinder elevators, and its object is to render it possible to start the elevator on the return trip at a fair rate of speed, notwithstanding that the automatic stop valve is in the closed position. When the elevator is started on the return trip the water coming to the valve chamber from the opposite direction forces the valve *B* away from the seat sufficiently to start the piston, and as soon as the piston moves, the valve *B*, under the influence of the weight *P* (see Fig. 117) is drawn away from the port to the central position in which it is shown in Fig. 123, when the water has an unobstructed passage through the valve chamber.

Referring to Fig. 118 it will be seen that there are two piston-rods, *b b*. This construction serves to strengthen the crosshead bar *C* by applying the strain at points nearer the side bars, and in addition it prevents the rotation of the piston, which would be undesirable, as it would increase the danger of pulling the rod out of the crosshead.

CHAPTER XXIII

THE MORSE-WILLIAMS "PULLING" MACHINE; CONSTRUCTION AND OPERATION OF THE VALVES

Another design of the pulling-type elevator is presented in Figs. 124, 125 and 126. This is called a "double-decked" machine, and is made by Morse, Williams & Co., of Philadelphia. Why it is called double-decked can be understood from the first illustration, which is a side elevation and shows two machines placed one over the other. In buildings where floor space is limited this construction is often adopted, in some cases three and four machines being installed one over another. Fig. 125 is a top view of Fig. 124, and Fig. 126 is an end view seen from the right

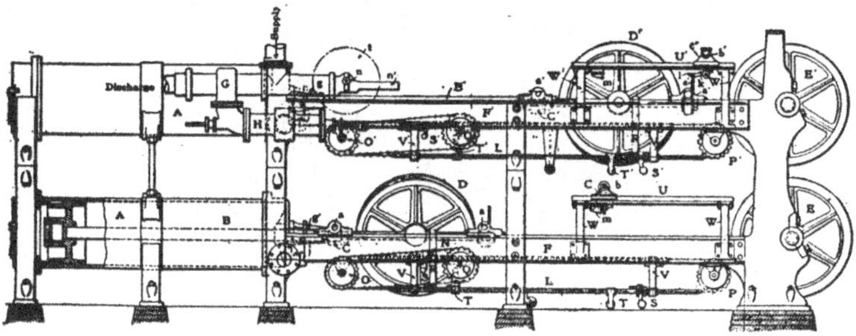

FIG. 124

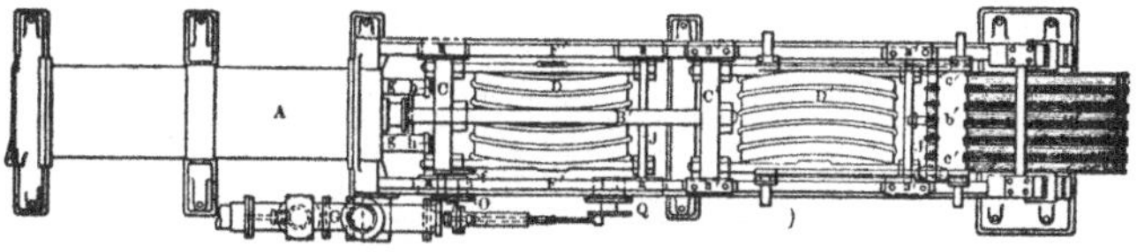

FIG. 125

side. In these machines there is but one piston-rod, as at *B*. The crosshead is similar to that in the Whittier machine, except that the sides of the end bars are square with the side frames, instead of in line with the traveling-sheave shaft, as shown at *J*. The guides *F* are set so that the crosshead shoes *a* slide on top of the upper flange, not between the flanges.

At the stationary-sheave end of the guides there are shorter guides *U* which carry a shaft provided with small rollers *b*, the function of which is to support the ropes running over the upper sides of the sheaves. The upper machine is shown with the traveling sheaves close to the stationary sheaves, caused by the car being at the lower floor of the building. In this machine the supporting rollers *b'* are at the extreme right-hand end of the guides *U'*. In the lower machine sheaves *D* are close to the cylinder, as they will be when the elevator car is at the top floor. In this case the supporting rollers *b* are at the extreme left-hand end of guides *U* and midway between the sheaves *D* and *E*, the better to support

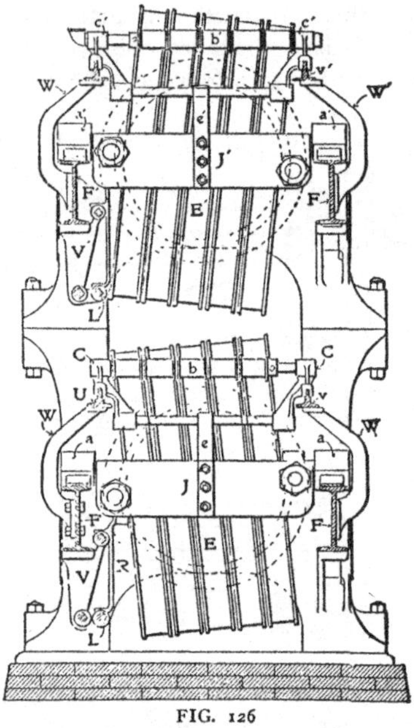

FIG. 126

the ropes at the central point. On the upper machine a hook *l* mounted on a shaft carried by the guide shoes *c'* engages a piece *e* secured to the part *J'*, as shown in Fig. 126, at the center. At one end of the shaft which carries hook *l* there is a lever *k*. When the sheaves *D'* move toward the cylinder, the hook *l* being engaged with lever *e*, the supporting rollers *b'* are carried along with the hook *l* until lever *k* reaches an inclined plane *m*, up which the rollers roll, causing the shaft to be rotated and hook *l* to be pulled up out of the way of the lever *e*, the rollers being left in the position of those shown on the lower machine. The supporting roller shaft is kept in line, notwithstanding that it is carried along by the part *e* acting at the central point, by reason of the guide-shoes *C* being provided with grooves that fit over the guides *U*; as clearly shown in Fig. 126. When the traveling sheaves move forward, the piece *e* engages hook *l* when the latter is reached, and the roller shaft is carried forward to the end of the guides, as shown at *b*. These supporting rollers relieve the ropes of considerable strain when the stroke is long and the traveling sheaves are near the cylinder, but they are of little service in short-stroke machines. The movement of the roller shaft is equal to one-half the stroke of the machine.

THE STOP AND MAIN VALVES.

In a machine of the pulling type the piston is forced toward the back end of the cylinder on the upward motion of the car. If the automatic stop valve is properly adjusted it will begin to close at the right time to stop the car even with the upper floor; but if it is improperly adjusted, the car is likely to run into the overhead beams, therefore buffers *g g*, faced with rubber cushions *h h*, are provided. In this machine the automatic stop valve does not fit perfectly, and if the main valve is not closed when the car reaches the upper floor, the car will not stop, but will slowly move upward until the crosshead brings up against the buffer cushions *h h*. On the downward trip, if the main valve is not closed when the car reaches the lower floor, the car will settle gradually until it rests on the bumpers, or the piston strikes the front cylinder-head.

The main valve is located at *G* and is actuated by a pinion at *n* which meshes with a rack in the neck-bearing *n'*. The automatic stop valve is contained within the casing *H* and is actuated by a rod connecting with a crank-pin on a crank-disk mounted on the shaft with the sprocket-wheel *Q*. This sprocket-wheel is rotated by means of a sprocket *O* mounted on the shaft with sprocket *f*, which latter is operated by a chain, the ends of which are affixed to the ends of two square rods, the lower of which is shown at *L*. Another chain around the sprocket *P* is connected with the opposite ends of these two rods. To stop the movement of the piston, the stop valve is actuated to the left. If the traveling sheave is moving toward the cylinder the actuating bar *R* attached to the crosshead will strike the stop *N* and move it to the left, which will set up a counter-clockwise rotation of the sheaves *O* and *Q*, and this will move the crank-pin and the stop valve to the left. If the traveling sheave is moving away from the cylinder, the lower end of bar *R* will strike the stop *N* on the square rod *L* and, by carrying the latter to the right, rotate sheaves *O* and *Q* counter-clockwise in the same direction. The stops *N* are hook-shaped; they slide over the side projections on bar *R*, Fig. 126, and lock with it, with the result that when the elevator is started on the return trip the movement of the crosshead carries the stop *N* with it and the automatic stop valve *H* is pulled open. When the elevator is started, it moves very slowly for a few inches, as only the water that leaks by the automatic stop valve is available to move it, but as the movement of the crosshead also operates the valve, the opening of the latter is rapidly increased and the car speed correspondingly accelerated. When the bar *R* has carried the stop *N* as far as the stop *T* the releasing lever *S* strikes the latter, and the hook on the stop *N* is raised so that the bar *R* may slide by and leave the stop *N* adjoining the stop *T*, ready to be struck by the bar *R* on the next stroke. The actuating stops *T* are not held on the rod *L*, but on a rod directly in front of it (see Fig. 126), and this rod is secured, so it will not move endwise, in the frame *V*.

CONSTRUCTION OF THE PISTON.

The construction of the piston is shown in Fig. 124, at the back end of the lower cylinder. Some of the other details are shown in Figs. 127

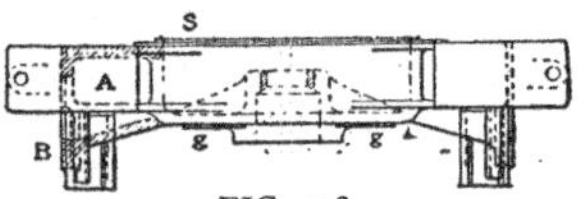

FIG. 128

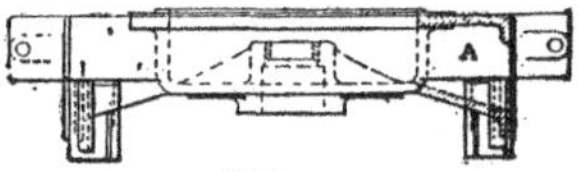

FIG. 130

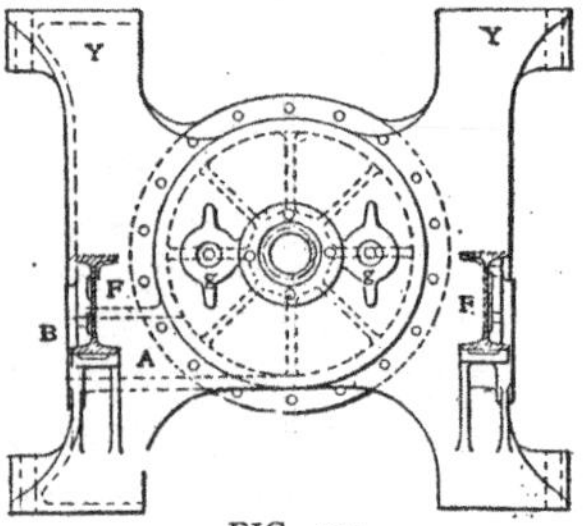

FIG. 127

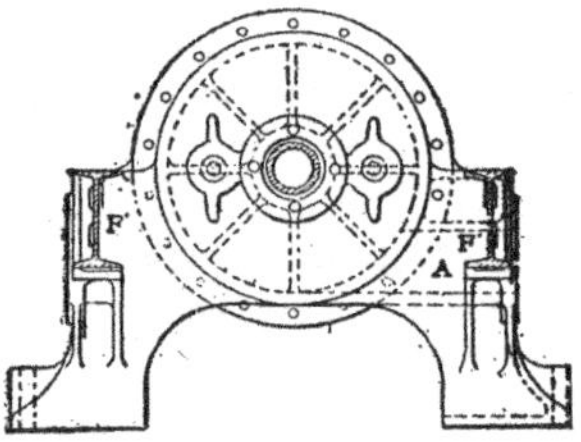

FIG. 129

to 135, inclusive. Figs. 127 and 128 show the front cylinder-head for the lower machine, the first being a front view, and the other a plan view. The automatic stop valve is attached to the face *B* surrounding the inlet port *A* through which the water passes in and out of the cylinder. The cylinder is a straight piece of pipe with suitable end flanges and is bolted to the back of the head, against the seat *S*. The manner in which the guides *F* are held is clearly shown; the buffer stands are fastened to the surfaces *g g*.

Figs. 129 and 130 show the top front cylinder-head, which is substantially the same as the lower head, the only difference being that the extensions *Y Y* of the latter are cut away. Figs. 131 and 132 show the top back cylinder-head, which is materially different from the front heads. This head is provided with a hinged door made to fit tightly enough to keep out dust, which is all that it is intended to accomplish. If there were no door to close the end of the cylinder the machine would run perfectly, but the inevitable accumulation of grit would cause the lower side to wear away faster than if protected. This head also serves as a drainage well to receive any water that may leak by the piston, and thus keep the floor clean. The dotted line *a a* marks the bottom of this well, and the circular line *b b* is the seat against which the door closes, the latter being hinged at *c c*.

Fig. 133 shows the supporting frame seen at the center of the guide tracks in Fig. 124. This is the frame for the lower machine; that for the

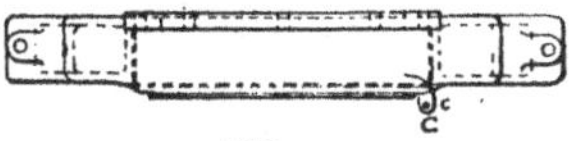

FIG. 132

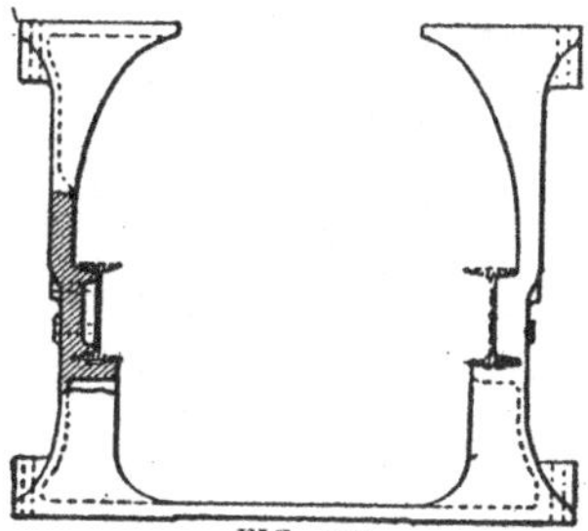

FIG. 133

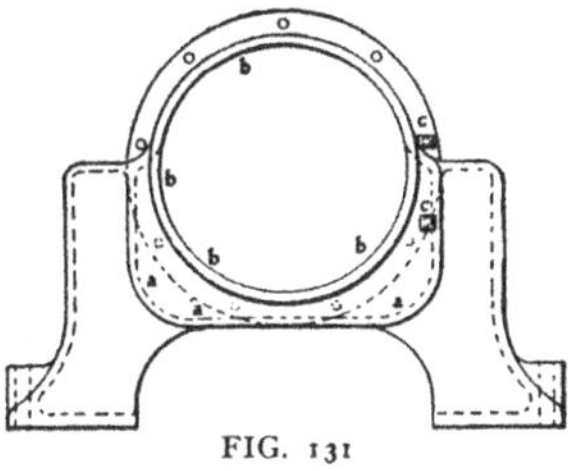

FIG. 131

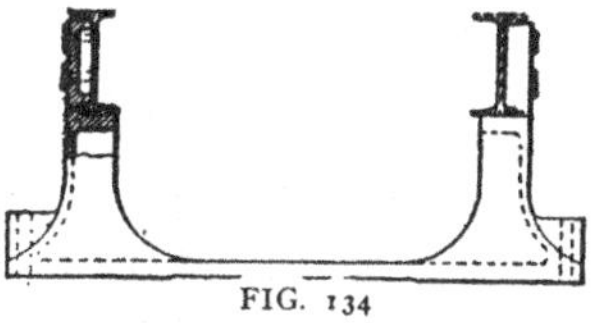

FIG. 134

upper machine is shown in Fig. 134. The construction of the stop valve *H* is shown in Fig. 135. The valve *D* is made hollow, so that there is no end pressure on it, and it acts by covering the port holes in the brass lining of the valve casing opposite the port *C* that connects with the cylinder. The main valve is connected with the stop valve at the port *B*.

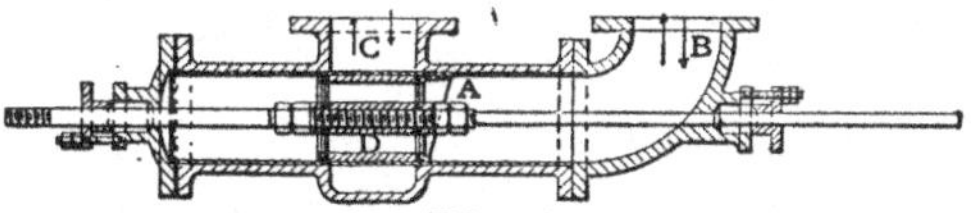

FIG. 135

As will be seen, the forward end *A* of the valve *D* is cut tapering so as to stop the flow of water gradually when it is closed.

The construction of the main valve is very simple. It consists of two pistons placed far enough apart to close the inlet port, and a third piston located at the end next the piston, which acts to balance the end pressure, the three pistons being of the same size, and also to prevent water from escaping through the front of the valve casing. This valve is moved toward the right to permit the water in the cylinder to escape into the discharge pipe, and to the left to permit water to flow into the cylinder

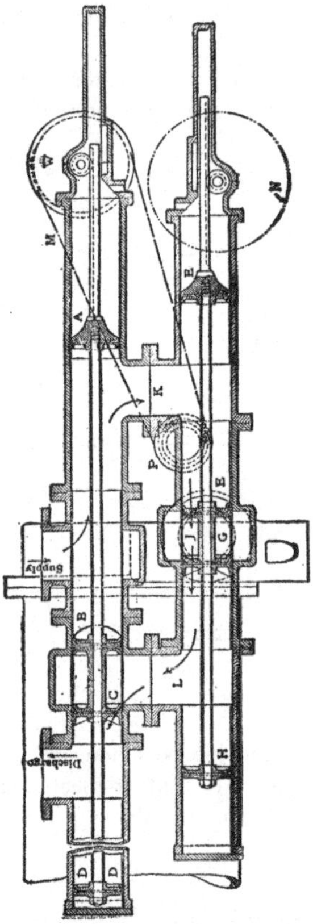

FIG. 136

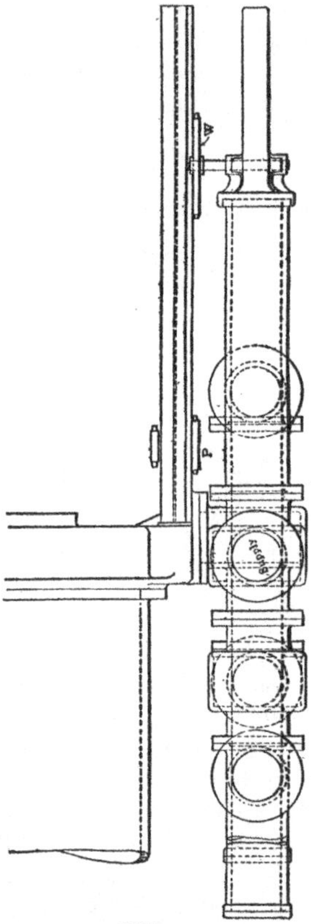

FIG. 137

from the supply pipe. More light may be thrown on its construction by the aid of Fig. 136, although this is not the same construction in every particular, it being a design of more recent date. In this figure, if the piston *D* is removed, it will leave a valve like the main valve of Fig. 124. Then if the port *K* is closed and the pistons *B* and *C* are moved to the right, water can flow up through the port *L* and to the discharge pipe, while if *B* and *C* are moved to the left, water from the supply pipe can pass down through *L*. If the pistons *F* and *G* are regarded as the stop valve *D* of Fig. 135, it will be seen that if this is open when the top valve is moved to the right the water will flow out of the cylinder and when the top valve is moved to the left the water will flow into the cylinder.

OPERATION OF THE VALVES.

The actual operation of the valves shown in Fig. 136 is decidedly different from this explanation, which is used to show the construction and operation of the main valve of Fig. 124. In Fig. 136 the lower valves form the main valve and are moved by a hand-rope which works on a

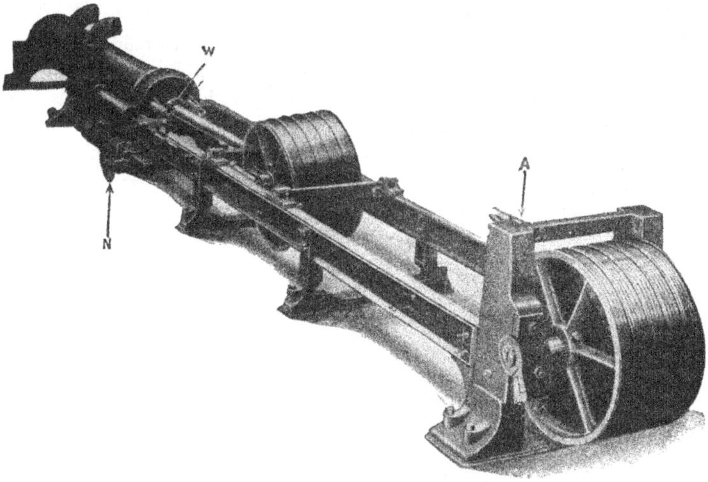

FIG. 138
HORIZONTAL ELEVATOR, PULLING TYPE

hand-rope sheave indicated by the circle *N*. The top valves constitute the automatic stop valve and are actuated by mechanism similar to that shown in Fig. 124, the sprocket-wheel *W* being rotated by the chain *M* which runs around another sprocket *P* mounted on the same shaft as the sprocket *f*, of Fig. 124, as is more clearly shown in the top view of the valves, Fig. 137. If the pistons *B* and *C* are in the position shown, and the main valve is moved to the left, water from the supply pipe can flow through *K*, as indicated by the arrows, and into the cylinder. If the pistons *B* and *C* are shifted to the right, so as to uncover the port, and the main valve is also shifted to the right, then the water in the piston can escape through *L* to the discharge pipe. The automatic stop valve, which is at the top, is in the position to which it is shifted to stop the car when reaching the lower floor on the down trip, and it does this by preventing the discharge water in *L* from escaping even if the main valve is open. To stop the car on the up trip, the pistons *B* and *C* are moved over to the right until they cover the port connecting with the supply pipe, and then the car will stop, even if the main valve is open, because the water cannot pass beyond the piston *B*, which as will be noticed is provided with double-cup packings so as to hold pressure from either side. From this it will be realized that the mechanism that operates this stop valve has to be modified so as to move the valve in one direction, i.e., to the right, to stop on the up trip, and in the opposite direction, or to the left, to stop on the down trip.

A half-tone view of a Morse & Williams single machine, equipped with valves similar to those in Fig. 136, is shown in Fig. 138. The sprocket wheel on the end of the pinion shaft of the main valve is seen at *N*, and the sprocket on the automatic stop-valve pinion shaft is at *W*. The chain can be seen running down to the sprocket *P* shown in Fig. 136, which is just below the side frame and mounted on one end of a short shaft that carries the sprocket *f*, shown in Fig. 137, on its inner end. Fig. 138 also shows the way in which the front guide-supporting frame is modified when the lifting ropes are hitched at the front end; the shackle-bolts being shown at *A*. It will be noticed that in this machine there are only five stationary sheaves, yet the machine is geared ten to one. A little reflection will show why one sheave can be dispensed with. The fact that one sheave less is required in a large measure offsets the objections to the front-rope hitch.

CHAPTER XXIV

CRANE HORIZONTAL "PUSHING" MACHINE; HOW THE STOP-MOTION GETS OUT OF ADJUSTMENT; THE CYLINDERS AND OTHER PARTS

The automatic stop-valve mechanism, above all other parts, should be kept in perfect working order in all types of elevator, for if this is not done, and some other part should be disarranged, the car is likely to "run away" and strike the bumpers violently. Even if it should not run away, it will strike the bumpers hard enough to cause the passengers discomfort and alarm if the operator neglects to move the lever to the stop position at the proper time. It is very common practice for an operator to depend upon the automatic stops to bring the car to a standstill at each end of its travel, consequently if the adjustment should be imperfect the car would probably strike the bumpers.

The proper treatment of the automatic-stop mechanism of the Crane horizontal pushing machine can be fully understood by the aid of Figs. 107 and 108. It should be understood that the frame *N* is moved by the motion of the traveling-sheave crosshead, the arm *D′*, on the latter, striking against stops fastened to the operating bar attached to the right-hand end of the frame *N*. As the movement of the frame, therefore, depends on the position of the stops on the operating bar, it follows that these must be in such position that when the car is moving with normal velocity the automatic stop will cause it to come to a state of rest even with the top or bottom floor, as the case may be. The location of the stops is determined by actual trial, usually by the elevator erectors, so that as a rule all that is necessary afterward is to keep an eye on the stops and make sure that they do not shift and are not likely to.

If the automatic-stop mechanism works hard there will be more danger of the stops shifting than if it works freely, therefore every part that is liable to stick should be examined and tested frequently. If the faces of lever *O, O′* become rusty, or even covered with dirt, they will offer more resistance to the rollers *A, A* than if bright and smooth, and this is also true of the inner faces of the frame *N*, particularly the upper one. The pin upon which the lever *O, O′* rocks carries a roller that runs within the groove of the frame *N*, and the weight of the latter, in addition to the pull of the valve connecting-rod *A″*, has to be carried by this roller whenever the frame is moved by the stops on the operating rod to close the valve. When the elevator starts from either the top or bottom floor the frame *N* is moved by the weight *P'*, when the pressure on the roller on the center pin becomes somewhat less than the weight, because the force required to move the valve to the open position acts to lift the frame *N*, the rod *A"* having to push against the lever *A'* in order to pull the valve-rod to the right, the direction in which it should be moved to open the valve. In this way more wear and strain are brought to bear on the top inner face of the frame *N* than on the lower inner face. The rollers *A, A* should be kept clean and well lubricated, otherwise they are liable to slide instead of roll. The joints of the valve lever *A'* should also be kept clean and free, and the upper end should be protected so grit will not get into it and cut the surfaces. Each of these moving surfaces considered by itself will not appreciably affect the power required to move the stop mechanism if it should get rough or sticky and run a little hard, but several of them together will make a decided difference.

HOW THE STOP-MOTION GETS OUT OF ADJUSTMENT.

Notwithstanding that the stop-motion is properly adjusted by the elevator erectors at the time of installation, it can get out of adjustment thereafter through the wearing away or displacement of a part. If the car stops a short distance beyond the top and bottom floors, it is a sign that the stop valve has become worn along the edges that shut off the flow of water, and to restore the adjustment so that the car will stop even with the floors all that is necessary is to shorten the rod *A"* by means of the right-and-left coupling. If the car stops short of the top floor and runs below the bottom landing, it indicates that the lifting ropes have stretched, and, while the proper way to restore the adjustment is by shortening the ropes, a slight overrunning can be remedied by shifting the stops on the operating rod to the right, so that the crosshead will have to move farther from the cylinder to close the stop valve when the car is going upward, and less near to the cylinder when the car is descending. This last expedient cannot be resorted to in case much adjustment is required, however, because it will cause the piston to move out of the cylinder too far and strike the buffers, which would cause the elevator to stop even if the stop valve were not entirely closed. In fact, this adjustment cannot be made unless when the car stops at the top floor the piston buffer is not in contact with the stationary buffer; then the stops may be moved a distance less than the clearance between the buffers. It is possible for the adjustment to become changed so that the car will overrun the mark at the bottom floor only. This may be the case if the wear of the stop valve and the stretch of the ropes just offset each other. Such an occurrence is likely to be remote, because the stretch of the ropes is almost sure to be greater or less. In any case, if the car does not stop about the same distance below both landings, it is because, in addition to a stretch in the ropes, there is a leak over the front edge of the stop valve, and to correct the adjustment the ropes and the rod *A″* must be shortened. Unless the car runs far enough beyond the top and bottom landings to inconvenience the passengers, however, it is better not to make any adjustment.

The roller *C′* on which the frame *N* of Fig. 107 rides, together with the pin *C* on which it is mounted, are shown in Fig. 139. The pin is

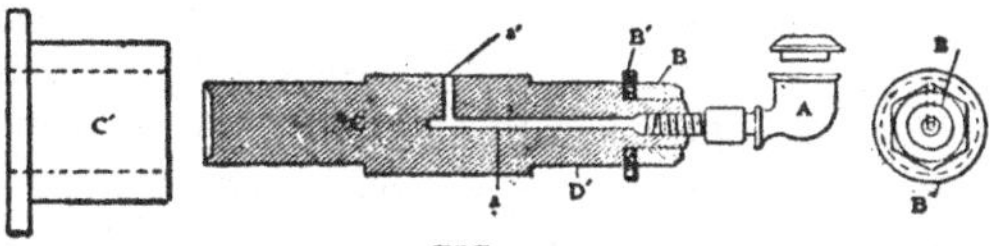

FIG. 139

provided with an oil cup *A* and oil holes *a a′*, through which the surface with which the roller *C′* contacts is lubricated. The difference between the diameter of the roller *C′* and the stud *C* is not very great; therefore, if the surface is permitted to run dry it will be very liable to cause the roller to stick, and then the frame will slide over it, eventually wearing a flat spot. This should be guarded against carefully, because if once a flat surface is worn on the roller it will be next to impossible to keep it

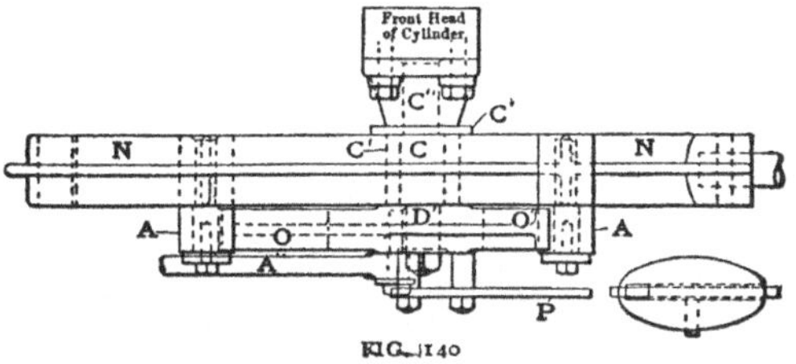

FIG. 140

revolving. Owing to this fact it is advisable not to lubricate the upper surface of the frame very often, so that the sliding friction at this point may be sufficient to cause the roller *C′* to revolve, even if it is not as well lubricated as it should be. The weight lever *P*, Fig. 107, is bolted to the side of the cam lever *O, O′*, and the latter swings on the portion *D′* of the stud *C*. This stud is riveted into a stand, bolted to the side of the front cylinder-head, in the manner clearly shown in Fig. 140. This drawing also shows the position of the flange of the roller *C′*, which is between the frame *N* and the stand *C″*. The cam lever *O, O′*, being placed outside of the frame *N*, serves to retain the latter in place. The cam lever is held in position by means of the washer *B'* and the nut *B*, as shown in Fig. 139. The weight *P'* on the lever *P* returns the stop valve to the open position each trip as the car moves away from the landing. By varying the position of the weight on the lever the rotative force can be varied. To be in proper adjustment, the cam lever should follow up the retreating rollers *A A* on the frame *N*, so that the stop valve may be drawn out of the way fast enough to permit the car to pick up the requisite speed in starting. If the cam *O*, *O'* does not follow up the rollers the weight *P'* must be set farther from the center, but it must not be set any farther than is necessary, because by so doing more effort will have to be exerted by the mechanism when the valve is moved to the stop position. From this it will be evident that if the cam lever fails to follow up the rollers it will not do to shift the weight *P'* out to the end of the lever to make sure of having enough moving force, but it must be shifted a little at a time, until the cam lever follows the roller throughout the whole travel; and then shifted just a trifle more so as to be on the safe side.

CHAPTER XXV

CRANE HORIZONTAL "PUSHING" MACHINE; DESCRIPTION OF AUTOMATIC STOP VALVE; PACKING THE MAIN PISTON AND CYLINDERS

The automatic stop valve of the Crane machine is shown in Fig. 141, the valve casing being in the center, the valve on the right-hand side, and the back head on the left, at *D*. The valve has only one tight-fitting piston, which is the smallest of the three seen at *E*. This piston has a cup packing that fits into a brass lining fastened into the front head of the valve casing. To remove the valve from the casing the valve lever *A'*, Fig. 107, is taken out by removing the center pin, then the back head *D* is taken off and the valve, with the valve rod, drawn out. The valve casing is placed on the side of the back cylinder-head in the position

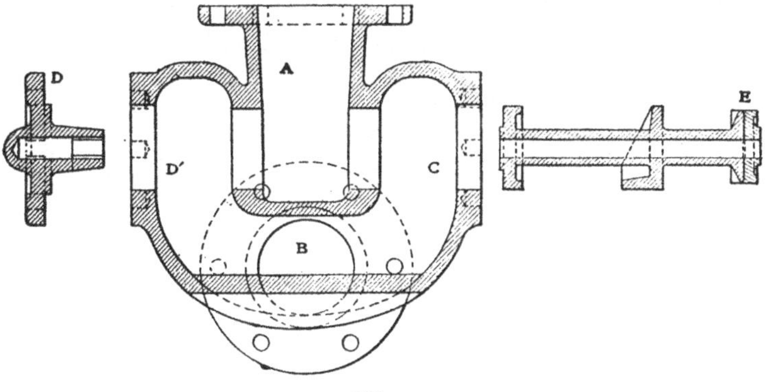

FIG. 141

shown in Fig. 141, the flange around the outlet *B* being bolted against the cylinder-head. The flange around the inlet *A* is bolted up against the under side of the main-valve casing.

Whenever it is desirable to repack the main valve, the latter is removed by taking off the back-casing head and disconnecting the connecting lever at the front end of the valve-stem. In the back head is a stop-screw (see Fig. 102) to prevent the valve from moving too far backward if for any reason the pilot valve does not stop the flow of water at the proper time, as for instance if it does not fit tightly, which is often the case with pilot valves without cup packings. This stop-bolt is adjusted to permit the main valve to open just enough to lift the car with the proper velocity at full load, and even if the operator should try to run it faster he could not. If this stop were not provided the car speed in most cases would be too great, because as a rule an elevator machine is made somewhat larger than is actually necessary to enable it to meet fully the requirements of the contract as to maximum lifting

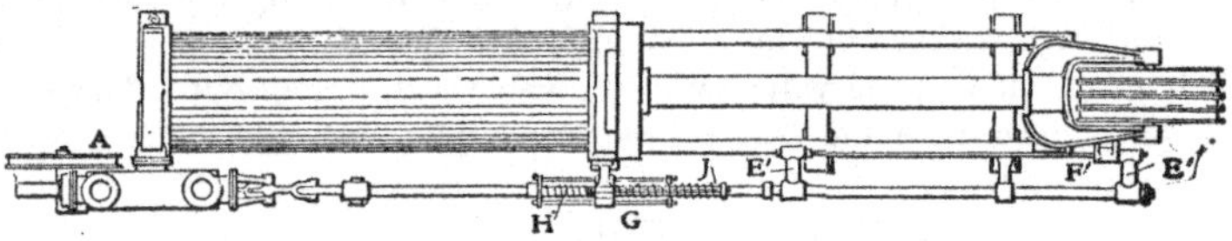

FIG. 142

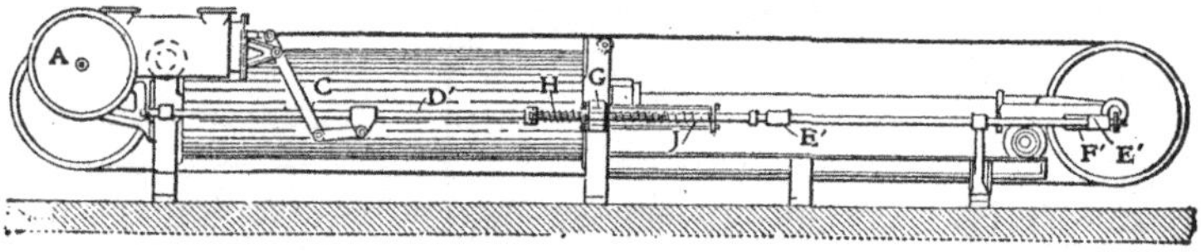

FIG. 143

capacity and speed. When the machine is installed the stop is set, by actual trial, to give the proper car speed, and generally the car lever is provided with stops also set for this velocity. The stop-bolt does not govern the opening of the main valve for the downward motion of the car, this being controlled by a stop on the car lever.

The valve gear of the Crane machine described in the foregoing is of the pilot-valve type, but many of these machines are provided with simple valve gear, and the operator in the car, by means of a hand rope or hand-wheel, moves the main valve directly. A machine of this type is shown in Figs. 142 and 143, the first being a plan view, and the second a side elevation. The valve proper is shown on a larger scale in Fig. 144, which shows a section through the axis of two valve cylinders, together with an end view. The main valve is located in the lower cylinder, the top valve being the automatic stop-valve. Looking at the left-hand end of the lower valve rod it will be seen that it terminates in a rack into which meshes a pinion mounted on a shaft carrying a hand-rope sheave on its outer end, as clearly shown in the end view at *A*. The pressure water enters through the inlet *B* and follows the path indicated by the arrows to the outlet *C*, passing to the latter when the piston *D* is moved far enough to the left. The outlet *C* leads into the cylinder, therefore if the main valve is moved to the left the car will ascend. If the main valve is moved to the right far enough for the piston *E* to uncover the outlet *C* the flow of water from the inlet *B* will be shut off, the water in the cylinder will escape to the outlet *F* and the car will run down. Referring to Fig. 143, it will be seen that the top valve, which is the automatic stop, is controlled by the movement of the rod *D'*, which is shifted by the stop-balls *E'* and the shifting-arm *F'* attached to the traveling-sheave crosshead. If the rod *D'* is shifted to the right, which is the case when the car is running upward, the upper valve pistons will be shifted to the left, and from Fig. 144 it will be seen that this movement will carry the piston *G* over the inlet port *B* and thus stop the flow

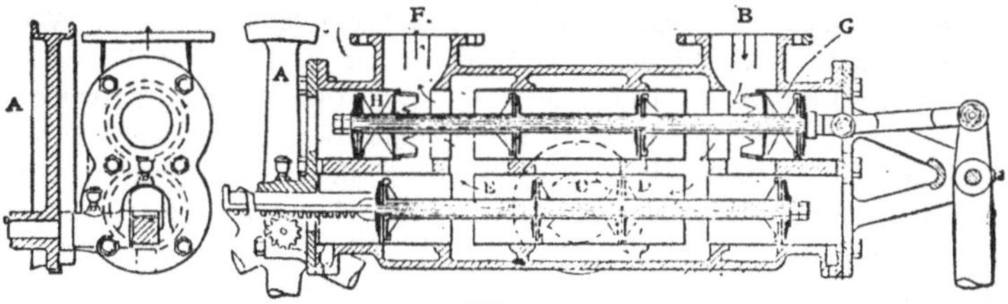

FIG. 144

of pressure water, even if the main valve is not closed. If the rod *D'* is shifted to the left, as it will be when the car is running downward, the top valve will be shifted to the right, and then the piston *H* will close the outlet *F*, so that water cannot flow out of the cylinder, and the car will stop even if the main valve is left open. These valves can be removed from either end of the valve cylinders, but generally it will be found more convenient to remove them through the right-hand side, that is, on the side where the automatic-stop lever is connected.

HOW MAIN PISTON IS PACKED.

The main piston of the horizontal pushing machine is packed by running the piston out to the end of the cylinder, where it can be easily reached. From this it will be seen that the packing is done while the car is at the top of the elevator well, therefore the latter must be firmly tied up to the overhead beams. To pack the piston, the first thing to do after securing the car to the overhead beams is to remove all the water from the cylinder, and from such piping as may extend higher than the cylinder, then close the valve in the pressure-water pipe, and in the discharge as well, if the discharge tank is higher than the cylinder. The first step when the valves are to be taken apart is to run the car to the bottom floor, remove all the water as just stated, and then open the casings and remove the valves.

In those Crane machines which are provided with pilot-valve gear, a strainer is provided to clean the water that passes through the pilot valve. The construction of this strainer is shown in Fig. 145. Its body is a short cylinder, closed at one end and provided with a flange at the other. The actual shape can be understood from the section at *A*; the cover of the chamber is shown at *B* in the side elevation and in the plan. The small disk *E* forms the strainer proper and is secured in the cylinder *A*, at the location indicated by *E'*, being forced into place. The two pipes *C* and *D*, shown in the plan view of the head, are for the water to flow through, the inlet being the pipe *D*, which does not project through the strainer disk. The outlet pipe *C* projects far enough through the head to pass through the large hole shown in the strainer disk *E*. Wire gauze is placed over the disk *E*. As the water enters above the disk *E* and passes through the strainer before it can reach the pipe *C*, when it reaches the pilot valve it is clean, all the impurities being retained in the cylinder *A* above the strainer *E*; and to clean the device all that is necessary is to remove the cover *B*. A valve placed in the pipe *D* will stop the flow of water from the pressure pipe while the strainer is being cleaned.

The rollers which support the lifting ropes revolve on a pipe, capped at one end and having an upturned elbow at the other, as shown in Fig. 146. Above the elbow is an oil cup *D*. There are perforations in the pipe under the rollers, as at *E*. The construction of the rollers is shown at *F* and *F'*, the first being a two-rope and the other a four-rope roller. It will be seen that the bore is enlarged at the center, so as to catch the oil. The oil cup *D* can be made to swallow sufficient oil to flood the pipe *A*, and if this is done the front cylinder-head and the floor under it will soon be anything but neat. By using a little judgment these rollers can be kept properly oiled without dripping. If a rope roller appears to run hard it will probably be due to the clogging of the oil hole *E*, Fig. 146, and by removing the pipe and cleaning out the oil hole the trouble may be overcome.

Each set of sheaves is provided with rope guards to hold the lifting ropes against the sheaves in case the ropes become loosened. The guards for the traveling sheaves support the ropes at the top and bottom, and those for the stationary sheaves support them at the side as well as at the top and bottom. These guards consist of side frames and rods running over the faces of the sheaves from one side of the set to the other. The side frames are secured to the crosshead in one case, and to the side frames that hold the sheaves in the other case. The guard rods that pass over the faces of the sheaves consist of long bolts, which extend through end holes, and pieces of gas pipe large enough to slip over the bolts. The pipes are of the proper length to fit between the two frames, the bolts being longer than the pipes; tightening the nuts causes both bolts and pipes to be held firmly in place.

THE CYLINDERS AND THE PACKING.

The cylinders of horizontal pushing machines are open at the front end and are, therefore, liable to become filled with dust and grit, which

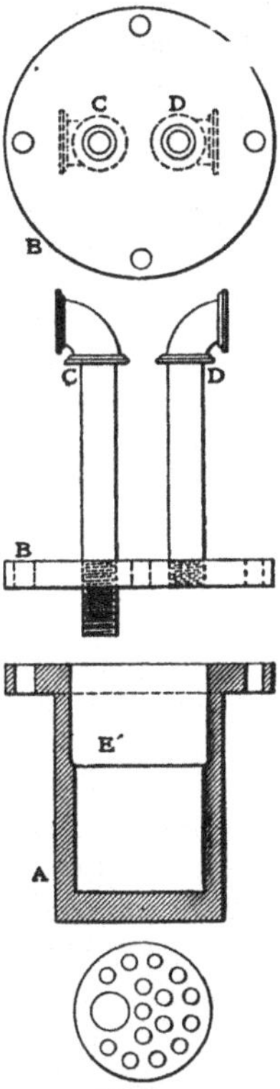

FIG. 145

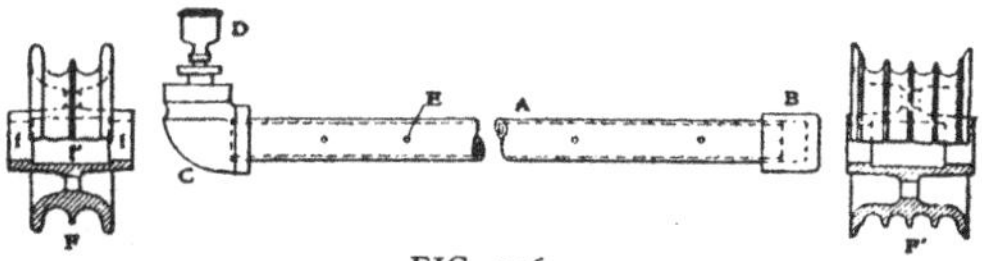

FIG. 146

should not be permitted to accumulate because it will increase the wear on the lower side of the cylinder and the piston. The packing will act as a cleaning agent and keep the surface apparently clean, but some of the grit will embed itself in the packing and cause it gradually to grind the metal away. If the machines are located where there is much dust, a cloth cover should be provided to keep the dust out, while occasionally the cylinder should be swabbed out. This is not as easy as might be supposed, if the cylinder is long, as the cleaning must be done when the piston is as far back in the cylinder as it will go, which brings the sheave crosshead so close to the open end that there is not much room to work in. By the use of patience and with a little ingenuity, however, the work can be done. A telescopic fishing rod will greatly facilitate the work; in its absence a bamboo rod can be made effective.

The same kind of packing is used in the pistons of horizontal machines as in vertical cylinders, that is, a soft square packing. It should be soaked before being put in, if it is likely to swell with water, and it should not be screwed up too tightly. As much care, if not more, is required in this respect with horizontal cylinders, because they are of larger diameter and therefore subject to greater strain from the water pressure and from the tightly compressed packing. These cylinders are made thicker on account of their greater diameter, but their strength is certainly no greater in proportion to their size than that of the smaller vertical cylinders. There is no excuse for getting the piston packing of a horizontal machine too tight, because the side of the piston from which the packing is done is always accessible. This being the case, the proper procedure when new packing is put in is to tighten it slightly at first, then put on the water pressure and make one trip, if the packing does not leak too much; then tighten the follower a little and make another trip; if still leaky, tighten it more, and keep on in this manner until it is perfectly tight. This cannot be done with a vertical cylinder. With the latter, as it is considerable trouble to take the cylinder-head off to tighten up the packing, it is necessary to depend upon judgment and try to make the packing tight enough, but not too tight, the first time. With both vertical and horizontal cylinders it is necessary to be careful not to get the follower ring screwed up farther on one side than on the other, for if this is done it will cramp in the cylinder. To guard against this all that is necessary is to measure from some finished surface square with the axis at the end of the cylinder and from a finished surface on the follower ring.

CHAPTER XXVI

CONSTRUCTION DETAILS OF THE WHITTIER MACHINE WITH SPECIAL REFERENCE TO THE STOP-VALVE MECHANISM

The Whittier main operating valve with pilot-valve control is very similar to that of the Crane machine; the slight difference between them is in the shape of some of the minor parts. In the Whittier valve there is an adjusting screw in the center of the back head, by means of which the car speed on the up trip is regulated so as to not exceed a certain amount, just as in the Crane elevator. In both machines the valves are removed from the valve casing through the back end, after unshipping the connecting-rod that links the valve-rod to the lever which moves the

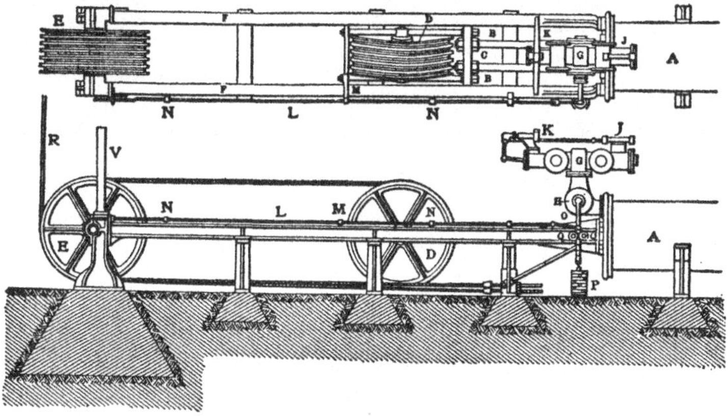

FIG. 147

pilot valve back to the stop position. Between the pilot valves there is practically no difference, so that all that has been said about the Crane valve and pilot will apply equally well to the Whittier apparatus.

In the automatic stop valve and its operating mechanism there is a considerable difference between the two machines. One type of Whittier automatic stop-valve mechanism is shown in Fig. 147. In this drawing the stop valve is at *H*, and the mechanism for moving it consists of the rod *L* upon which are fastened the stops *N* and *N*, and a projecting arm *M* attached to the front end of the traveling-sheave crosshead. The position of the arm *M* is more clearly shown in the plan view. By means of a short link the rod *L* is connected with the crank *O* on the shaft of the stop valve *H*, so that when the traveling sheaves are moving toward the cylinder (the car going up) the crank *O* is pushed to the right; and when the sheaves move to the left, the crank *O* is drawn to the left. In both cases the weight *P* is lifted, the chain to which it is attached being hung from the end of the crank. As soon as the car starts to move from either end of its travel the arm *M* moves away from the stop *N* against which it may be pressing, and then the weight *P* will pull the crank *O* to the central position; the chain attached to the crank runs between guide wheels *Q Q*, in order to avoid oscillations of the crank before coming to rest. To secure satisfactory operation of this mechanism it is necessary that weight *P* be heavy enough to swing the crank *O* around to the stop position just as fast as the arm *M* moves, so that the latter may not run ahead of the stop *N*. At the same time, however, it is desirable to make the weight only as heavy as is necessary to cause the stop *N* to follow up *M* closely, because the heavier the weight, the greater the effort required to move the stops *N N*, and the greater the liability of their slipping out of place on the rod *L*. The friction of this mechanism is very small, so that there is very little danger of its becoming sufficient to cause the weight to stop, in its descent, if it is made only a trifle heavier than is actually necessary to do the work.

STOP VALVE NOT ADJUSTABLE.

The stop valve of this machine is not adjustable in any way, so that any adjustment that may be required must be done on the outside apparatus. If the car stops below the upper floor, it can be made to run higher by moving the stop next to the cylinder a little to the right. If the car stops short of the lower floor it can be made to run lower by moving the stop next to the stationary sheaves a trifle to the left. It must be remembered, however, that only a small adjustment can be made in this way, as it throws the piston out of position, causing it to run up against the front cylinder-head if the stop near the stationary sheaves is moved too far to the left, and against the back head if the stop near the cylinder is moved too far to the right. When the car is installed, the apparatus is properly adjusted, and about the only way in which it can get out of adjustment is by the stretching of the lifting ropes, which is sure to occur. The stretching of the ropes will cause the car to run below the lower floor, and stop short of the top floor, and the proper way to restore the adjustment is by taking up the ropes, but sometimes it is not convenient to do this, and then a temporary adjustment can be made by shifting both stops *N* to the right, that is, toward the cylinder. To determine how much to shift them ascertain the distance by which the car falls short of the landing, and then divide this by the gear of the machine, and the quotient will be the proper distance. Thus, if the car falls short of the upper landing ten inches, and runs below the bottom floor a like amount, and the gear is ten to one, the proper distance to shift the stops is one inch.

Another arrangement of stop-motion gearing used on the Whittier machines is shown in Fig. 148. The difference between this and the one shown in Fig. 147 is that instead of the rod *L* of the latter, there is

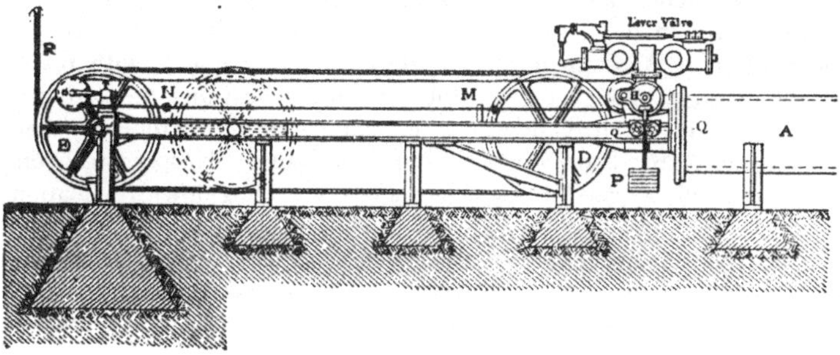

FIG. 148

substituted a rope running over a sheave *S* at the end of the machine, and another sheave attached to the casing of the stop valve *H*. This latter sheave is mounted upon a shaft which is geared to the valve shaft, so that when the rope is moved by the arm *M* striking either one of the stops *N N*, the stop valve is rotated precisely the same as in Fig. 147. The only advantage of the rope driving gear is that it cannot buckle, while the rod *L* of Fig. 147 can under certain conditions; for instance, if it is bent slightly in some accidental manner so as to cramp in its guide bearings, and therefore refuses to slide through.

In addition to what has been said in relation to the stop-gear of Fig. 147, all that is necessary with reference to Fig. 148 is to call attention to the fact that as the sheave *S* is held in a frame that is adjusted to the proper position by means of the screw *T*, it can work loose if not properly looked after. It may not get loose enough to wabble out of place, but enough so to twist around, and interfere with the proper shifting of the stop valve.

MUST BE ANCHORED TO FOUNDATION.

Horizontal elevator machines, like the vertical machines, must be anchored down to the foundation or otherwise held in order that they may not be lifted by the weight of the loaded car. The tendency to lift the machine, however, is not so great in horizontal as in vertical elevators, because the lifting force is simply the weight of the car and its load, less the weight of the counterbalance. In vertical machines, the lifting force is this same weight multiplied by the gear of the machine, so that if the latter is four to one the lifting force will be four times as great as in a horizontal machine. The bolts that hold the lifting end of the machine down to the foundation should be looked after from time to time to see that they are tight, but the other foundation bolts require little attention. The difference between the foundations under the lifting end of the machine and the other supports is shown in a somewhat exaggerated form in Figs. 147 and 148. In some buildings of steel-frame construction it is common for the elevators to be put in with foundations only large enough to hold the machine in line, and the lifting end of the machine is held down by bracing from the stationary sheave bearings up to the floor beams of the floor above. This construction is simple and fully as good as the use of a heavy holding-down foundation.

HAND ROPE ALSO USED.

The Whittier machines are also made so as to be operated by a main valve moved by the direct pull of a hand rope, instead of by the action of a pilot rope, being moved either by hand-wheel in the car, or by being

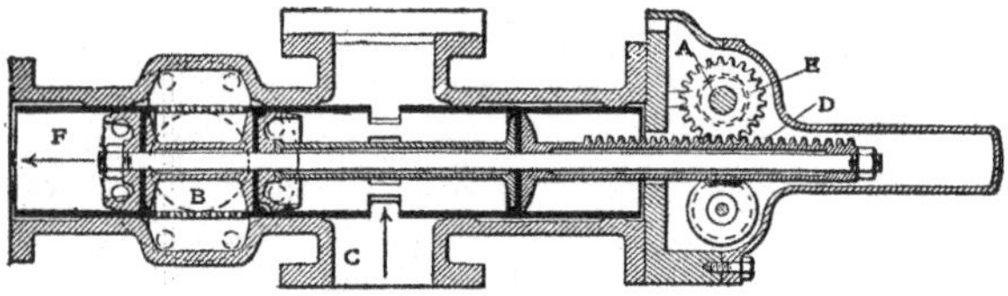

FIG. 149

pulled directly by the operator. The type of valve used for this type of operation is shown in Fig. 149, which is a section parallel with the valve rod. The hand-rope sheave is mounted on the shaft *A* of the pinion *E* and the latter meshes into the rack *D* attached to the end of the valve rod. Water from the pressure tank enters through the inlet *C*, and if the valve is moved to the left, it passes to the chamber *B* and to the cylinder. If the valve is returned to the position in which it is drawn the flow of water will be stopped, and the car will come to rest. If the valve is now moved farther to the right, the water in the cylinder will rush out through the chamber *B* and the end *F* of the valve casing, and thus to the discharge tank. This valve, it will be seen, is substantially the same as the same type of valve used with vertical cylinders, and it is taken apart, cleaned or packed in the same way.

THE PISTON, SHEAVES AND BEARINGS OF THE WHITTIER PULLING MACHINE.

The pistons of pulling machines are slightly different in design from those used with pushing machines. The difference is only that made necessary by the direction in which the force is applied. In pushing machines the piston pushes the piston-rod or plunger ahead of it, but in

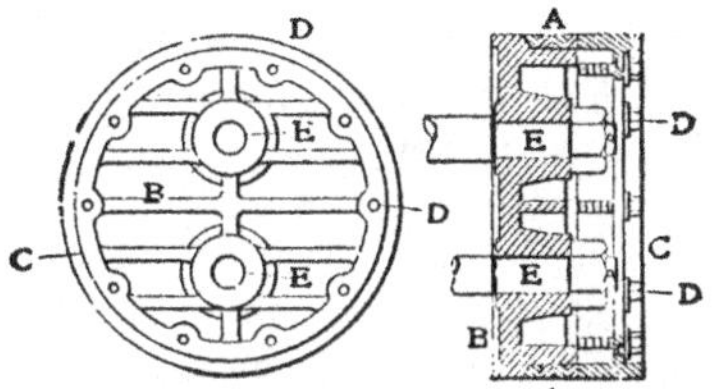

FIG. 150

pulling machines the piston pulls the piston-rod along. The piston used in the Whittier pulling machine is shown in Fig. 150, which gives two views, the one a face view and the other a section through the center, in line with the piston-rods. These pistons are easier to pack than those of pushing machines, because they are moved to the back end of the cylinder, where they are wholly exposed, instead of being partially obstructed by a large piston plunger. The same kind of packing is used in them as in other types of elevators already explained, and the process of packing is the same; that is, the piston is forced as far back as it will go when the elevator car is at the top of the building. The car is then securely fastened to the overhead beams, and all the water is drawn from the cylinder and pipes, after which the packing is done. As the

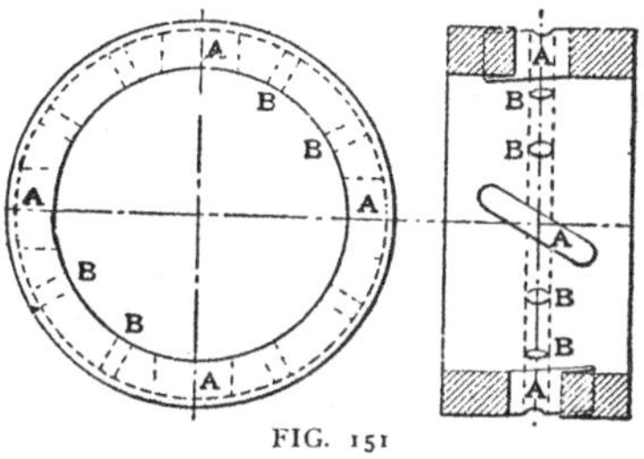

FIG. 151

piston is so accessible the packing should not be made very tight at first, but should be tightened a little at a time, until it stops leaking, so as to run no risk of bursting the cylinder. Dry packing that will swell when it gets soaked with water should not be used; if it is there will probably be trouble.

The sheaves of horizontal elevator machines are not easily reached for lubricating purposes, and on that account are commonly made with pockets which hold a considerable supply of lubricating grease. The bushing of the Whittier sheave, in which are provided such spaces, is shown in Fig. 151, which shows a side view of the bushing and a section parallel with the bore. It will be seen that there are four diagonal openings, cut through the bushing at *A A A A*. In addition to these there are several holes *B B*, which are filled with a lubricating grease by means of grease cups set over each pocket.

All sheaves are not made with brass bushings, and the manufacturers who use such bushings do not always provide grease pockets. In most machines grease cups alone are used, there being three or four on a sheave, so that one of them may easily be reached at any time. Screw-cap cups are preferred, and with these the cap is screwed down a half turn or so each morning before starting up, which is sufficient to keep the bearing well lubricated.

CHAPTER XXVII

OVERHEAD SHEAVES AND BEARINGS; KIND OF LUBRICANT BEST SUITED FOR THEM; DISCHARGE PIPE RELIEF VALVES; THE USE OF STRAINERS; WHEN REQUIRED

The overhead sheaves and their bearings are made practically the same for all types of elevators. The general practice is to mount the sheave on a short shaft that projects five or six inches on each side, and these ends run in bearings that are lined with babbitt, and are made in some cases solid, in some cases with a cap, and in others in a more elaborate manner. The shaft could be made fast in the side bearings, with the sheave to revolve on it, as is done in the traveling sheaves, but this construction is not so desirable, because the sheave if loose on the

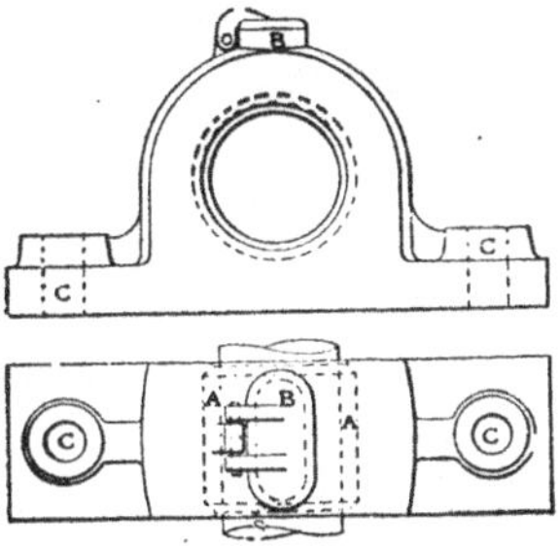

FIG. 152

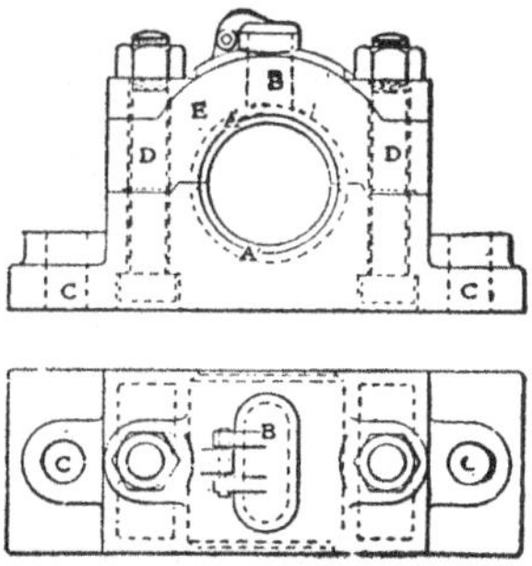

FIG. 153

stud could in time wear the center hole out of true so as to wabble badly. In traveling sheaves of horizontal machines this construction is unavoidable, and in vertical machines it is used because it is necessary to make the sheave and frame as narrow as possible so as to be able to run in a small space; with the shaft fast in the side frames the structure can be made of considerably less width than if side bearings were provided with the sheave fast on the shaft. A solid overhead-sheave bearing is illustrated in Fig. 152, which shows the side elevation and plan. The babbitted lining of the bearing is at *A*. On the top of the bearing is a grease chamber *B*, covered with a hinged cap. The bearing is shown with two holes, *C C*, for holding-down bolts, this construction being used when the bearings are supported on wooden beams. If the supporting beams are of steel, the bearing is made with four holes *C* foı the holding-down bolts.

A bearing provided with a cap is shown in Fig. 153, which shows the side elevation and plan view. In this case the two halves can be babbitted at *A* and *A'*, but as all the pressure is on the lower half there is no occasion to babbitt the top half, and it is better to leave a space there for lubricating grease. The advantage of the split bearing over the solid form is that if at any time it becomes necessary to remove the sheave, all that is necessary is to remove the caps and lift out the sheave.

The overhead beams on which bearings are held can spring out of line, or get displaced through the settling of the walls of the building.

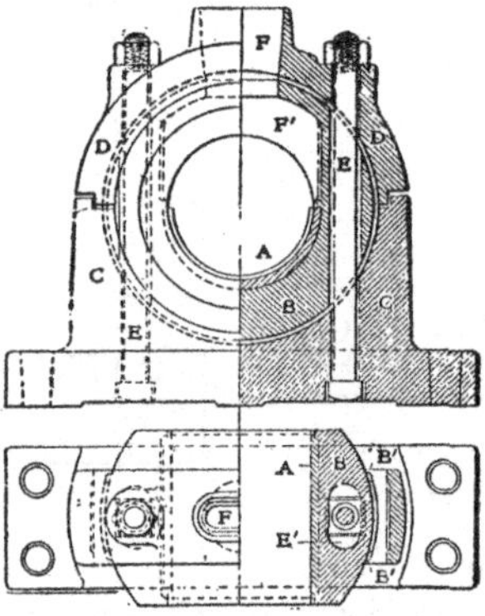

FIG. 154

If in any way the beams get displaced, the sheave bearings are almost sure to be thrown out of line, with the result that they will bind at one side or the other and thus run hard, and possibly heat enough to cut. For the purpose of preventing trouble in this way the Otis company makes the ball and socket bearings shown in Fig. 154. The sleeve *B* is turned to spherical form and sits in a spherical seat that bears at the points *B' B'*. With this construction it makes no difference how much the supporting beams get out of position, because the center part *B* will swing around on the surfaces *B' B'* until the bearings are in perfect alinement with the shaft. In this bearing only the lower half is babbitted, as indicated at *A*. The upper half is cored out to hold the lubricant. The clamping bolts *E* pass through openings *E'* in the two halves of the sleeve *B*, that are large enough to permit the bearing to swing more than it is ever likely to get out of line in actual service.

GREASE PREFERABLE FOR LUBRICATION.

Grease is used in preference to oil for lubricating the overhead sheaves, because as the velocity of rotation is very low, a hard lubricant answers every purpose, and it is cleaner, there being much less danger of the drip dropping down on the elevator car; nevertheless, drip pans should always be provided.

DISCHARGE PIPE RELIEF VALVES.

In every form of elevator thus far described, and, in fact, in all types of hydraulic elevators, if they run at high speed it often happens that when the car is stopped suddenly on the down trip, the water flowing out of the valve chamber into the discharge pipe will not reduce its velocity as fast as the valve is closed, the result being that the water is drawn away from the end of the valve, forming a vacuum at that point. The vacuum, together with the friction of the water flowing through the pipe, soon brings the column of water to a state of rest, and then the vacuum draws the water back, and in rushing in to fill the vacant space it strikes a violent blow against the under side of the valve, which does the latter no good, and in addition produces objectionable noise. To prevent this water hammer, all high-speed elevators and some of not very high speed are provided with a relief valve connected with the discharge pipe just beyond the end of the main valve casing. A valve of this type is shown in Fig. 155. The body of the valve is shown at *A*, and *B* is the cap, a top view of which is shown at *C*. This cap has openings *E*, on each side of the center, and in the center it is bored at *D* to fit the stem of the valve, the latter being shown at *F*. To assemble the valve the head *F′* is removed from the valve stem, and then the latter is slipped through the center holes in the cap *B*; then a helical spring is slipped over the valve stem and compressed in order to replace the head *F′*. Next the cap *B*, with the valve in place, is screwed into the top of the casing *A*, the lower end of which is screwed into the discharge pipe. The action of this valve is as follows:

If when the car is coming down at a rapid rate, the main valve is suddenly closed, the water flowing out through the discharge pipe will form a vacuum in the latter, and then the valve will be drawn in and thus opened so that the air will rush in and reduce the vacuum. When the water stops flowing and is drawn back by the slight vacuum still remaining, it will not strike the end of the main valve, because the air drawn in through the relief valve will act as a cushion, and as soon as it is compressed to atmospheric pressure will act to hold the water back.

WHERE STRAINERS ARE REQUIRED.

When hydraulic elevators are installed in a building where water under sufficient pressure to operate them can be drawn from the street

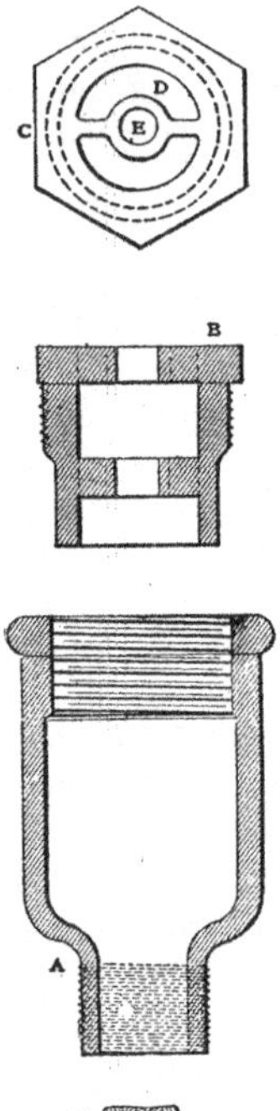

FIG. 155

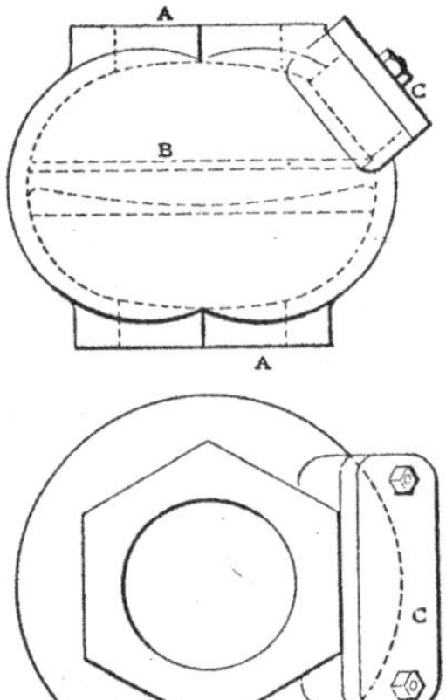

FIG. 156

mains or some other source, a fresh supply of water enters the cylinder at each stroke, and, as it is liable to contain many impurities that could clog up the valves and prevent their proper action, it is necessary to provide strainers. One of the many designs of strainers in use is shown in Fig. 156. This is used with the Crane machines, and is made so as to be inserted in the supply pipe, preferably with the strainer grating *B* in a horizontal position. Above the grating *B* a piece of wire gauze is placed. Through the opening *C* the strainer is cleaned out whenever required.

Very few elevator installations are operated with water drawn from the street mains nowadays, as it is not possible to obtain sufficient water for the purpose except in small places where there is a large supply at high pressure, and only a few elevators. In all large cities a pump is installed in the building and used to force water into a pressure tank from which the elevators are supplied. In such installations the same water is used over and over again, the cylinders discharging into an open tank from which the pump transfers the water to the pressure tank, the latter supplying the cylinders. In such cases it is evident that if the water when drawn into the system is clean, it can be kept so; hence, a strainer is not required, and it is not desirable because it only adds resistance to the flow of water. The general practice in such installations is for the makers of the elevators to place strainers in the supply pipe between the cylinder and the pressure tank, when the plant is first started, and to keep them in place for several weeks. This is done in order to get out any impurities that may be lodged in any part of the system. These strainers sometimes bring to light many things besides dirt and chips; nails, screws, bolts and pocket knives are quite common. After the elevators have been running three or four weeks the strainers are removed. These strainers generally consist of a sheet of perforated brass that is clamped in one of the joints in the supply pipes, so that its removal does not require any reconstruction of the piping. As water has to be supplied to the system from time to time to make up for the inevitable loss by leakage, it is advisable to put a strainer in the pipe through which this water is admitted, although it is not always done.

CHAPTER XXVIII

DISCUSSION RELATIVE TO THE DETAILS OF THE MORSE AND WILLIAMS PULLING MACHINES

In the Morse and Williams pulling machine, as in all other forms of elevator, the automatic-stop mechanism should be most thoroughly understood, and always kept in perfect order. This part is described in connection with Figs. 124 and 125 in Chapter XXIII. These drawings, however, do not show the construction and operation of the mechanism as well as might be desired, and on that account the writer presents here a simplified diagram which illustrates the action of the several parts more clearly than is possible with actual working drawings of the apparatus. In Fig. 157 a side elevation of the vital parts of the stop-motion mechanism is shown at *A*; a plan view of *A* is shown at *B*, and an end view at *C*. The several parts in these diagrammatic illustrations are lettered the same as in Figs. 124 to 126, in order to facilitate com-

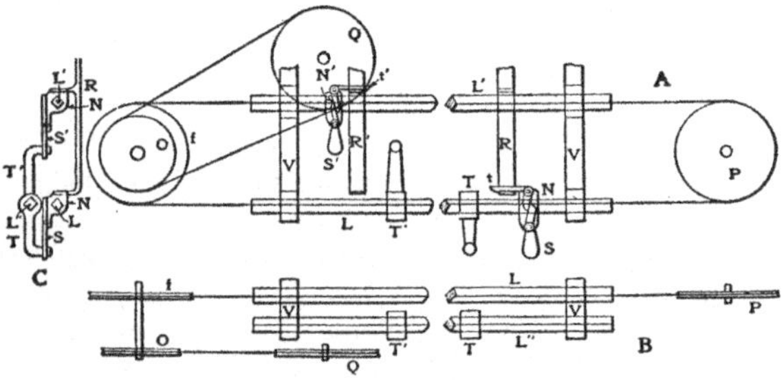

FIG. 157

parative references. The rods *L, L'* and *L''* are of square-section, as clearly shown at *C*, and to their ends are attached sprocket chains that run around the sprocket wheels *P* and *f*. The frames *V* and *V* support the rods *L, L'* and also the rod *L''*, which is directly in front of *L*, as seen at *B* and also at *C*. The rod *L''* is held firmly in the frames *V, V*, and its function is to hold the arms *T, T'*, which are fastened upon it by means of suitable screws. The stops *N* and *N'* are mounted up *L* and *L'*, respectively. The bar *R* is fastened to the crosshead that carries the traveling sheaves, as shown in Fig. 124. The latch lever *t* is pivoted upon the stop *N*, and is arranged to slide over a lateral projection from *R*, as shown at *C*, in Fig. 157. The lever *S* is also pivoted upon the stop *N*, and its upper end is extended so as to engage with the lower end of the latch *t* and be able to move the latter so as to lift it out of the path of the lateral extension on the bar *R*.

The operation of this stop-motion mechanism is as follows: When the elevator car is at the lower landing, the bar *R* is at the right-hand end of the rods *L, L'* and engaged with the latch of the lever *t*. To move the car upward, the traveling sheaves have to move to the left, and the bar *R* carries the stop *N* with it, but when the lever *S* reaches the stationary arm *T* it is prevented from going farther, and then its upper end swings over to the left and thereby lifts the latch *t* and disengages it from the bar *R*. The object of this operation is to open the stop valve quickly when the car is started from the lower landing. As will be seen by looking at Fig. 124, the stop valve is connected with the stop mechanism through a connecting-rod that extends from the end of the valve stem to a crank-pin on a disk mounted on the same shaft with the sprocket *Q* (see Fig. 125); hence, the movement of the block *N* by its engagement with the bar *R* acts to move the stop valve from the closed to the open position. If the bar *R* did not carry the block *N* forward, the stop valve would remain closed and on that account the elevator would only move as fast as the water could pass by it, no matter how much the main valve might be opened.

Inasmuch as it is necessary for the sprocket *O* to rotate a certain angular distance in order to move the stop valve from the closed to the wide-open position, it follows that the amount of opening of this valve can be controlled by the position of the stationary stop *T*. Thus the position of the stop *T* with reference to the block *N* will determine the amount of opening of the stop valve, and thereby the maximum speed at which the car can run; since this speed is dependent upon the volume of water that can pass through the stop valve, that is, providing the main valve is opened wide enough to permit more water to pass through it than can pass through the stop valve. The stop valve is set so as to permit the elevator to run at the highest velocity desired, so as to make it impossible for the operator to run it at a faster speed. If the stop valve were not adjusted so as to control the velocity, it would be possible for the operator to impart a very high velocity to the car if it were empty or nearly so, and this velocity might be high enough to be dangerons. The position of the stop valve only limits the speed of the car to the pre-determined maximum, but does not in any way interfere with running at any desired lower speed. Any velocity less than the maximum can be obtained by the operator by moving the car lever so as to open the main valve to the proper point.

HOW THE STOP VALVE OPERATES WHEN STOPPING.

In the foregoing I have shown the operation of the stop-valve mechanism in the act of starting; the following refers to the operation in the act of stopping. Assuming the car to be on the upward trip, with the bar *R* moving to the left, the block *N* will be left resting against the right side of stationary stop *T*, and the bar *R* will continue moving to the left as the car runs up the elevator well. When the car reaches a point near the top floor, the bar *R* will be in the position *R'* and will be carrying the stop *N'* to the left with it. This movement of *N'* will rotate the sprocket *Q* in a direction opposite to that in which it was moved when the bar *R* pulled the block *N* along with it toward the stop *T*; hence, the stop valve will be carried back to the stop position, and the elevator will be stopped regardless of whether the operator moves the main valve or not. Considering this action of the bar *R* on the stop valve through the movement of the stop *N'* it is easy to see that if the stop is secured to rod *L'* too far to the right, the car will be stopped before it reaches the top floor, while if it is set too far to the left, the car will run above the top floor before it is stopped by the action of the automatic stop.

TO SET THE STOP-VALVE MECHANISM.

From the foregoing it can be seen that to set the automatic stop-valve mechanism so that the car cannot run beyond a certain speed, it is necessary to shift the stationary stops *T T'* mounted on the rod *L"*. If these stops are moved toward each other, the maximum car speed will be increased, and if moved away from each other it will be decreased. If for any reason it is desired to run faster in one direction than the other, say on the up trip, this can be done by simply setting the stops *T T'* in the proper positions. To run faster on the up trips, the stop *T'* is set farther away from *V*, that is, farther to the left.

By properly setting the stops *N* and *N'* on the rods *L* and *L'* the car can be made to stop automatically and accurately at the top and bottom floors. If the car runs beyond the top floor, set *N'* farther to the right, and in the opposite direction if the car stops before reaching the floor. If the car runs below the lower floor set the stop *N* to the left, or to the right if the car stops short of the floor.

In order that this mechanism shall operate properly it is necessary that the several stops remain in correct adjustment. It can also be seen that as the bar *R* is fastened to the crosshead of the machine, if it is not released by the latches *t t'* at the proper time, it will continue to pull

N or *N'* along, and thus also pull *T* or *T'* unless these are secured firmly enough to withstand the pull. In that case something else will have to give way. If the car is running upward and the latch *t* fails to release the bar *R*, and the pull of the latter is great enough to draw the stop *T* along with it, the result will be that the sprocket *Q* will be rotated until the rods *L L'* run into the sprockets *f* and *P*, and if this is sufficient to carry *Q* through about one-half of a revolution, the stop valve will be returned to the stop position and the car will stop. In practice, however, the general result if the catch *T* fails to release the bar *R* is that something about the mechanism will break; therefore, it is necessary to keep these catches in proper working order. It is also necessary to keep the stops firmly secured so that the mechanism may not get out of adjustment, and thus fail to stop the car at the top and bottom floors.

To insure that the stop-valve mechanism shall work well it is necessary that all the moving parts be kept clean and well oiled; this is particularly the case with the latch *t* and dog *S*. The oil used for these parts, as well as for the bearings of rods *L* and *L'* through the frames *V, V*, should be of a kind that will not gum. If the latch *t* and the dog *S* should become clogged in any way, they should be cleaned immediately with kerosene so as to work freely. It is necessary to keep all the parts in good condition so that there may not be too great a strain on the sprocket chains that pass around the sprocket wheels *P* and *f*. If one of these chains should break, the apparatus would become useless, and as the automatic stop valve and its mechanism constitute the most valuable and reliable safety device attached to an elevator, they should never be allowed to run if not in perfect condition at every point.

THE CARRIER ROLLERS FOR THE ROPE SHEAVES.

Another part of the mechanism of the Morse and Williams pulling machine that was not shown as clearly as might be desired in Figs. 124 to 126 is the set of carrier rollers provided to hold up the ropes between the stationary and the traveling sheaves. These rollers are mounted in a carriage *c*, Fig. 158, that slides along on short guides, *V*. The construction is such that when the car runs down to the lower floor the crosshead catches the roller carriage, and on the next upward trip carries it along with it until the piston has traveled about one-half its stroke. At this point the carriage is released, and the rollers remain in that position, so that when the piston has moved the full stroke the rollers are about midway between the two sets of sheaves. This construction is only used with very long-stroke machines, but as these are common in high buildings, it is necessary to understand the construction and operation of the mechanism.

The upper sketch, Fig. 158, shows the position of the stationary sheaves *E*, and the traveling sheaves *D*, when the car is at the lower floor; and the lower drawing shows the position of the traveling sheaves when the carriage supporting the guide rollers is about to be released, and their position when the car is at the top of the building. The solid-line circle *D* shows the first-mentioned position of the traveling sheaves, and the broken-line circle the position when the car is at the top floor. The flange *e* is fastened to the traveling-sheave crosshead frame, and when the car is at the lower floor it engages with the latch *l*, on the end of the lever *k*. When the car runs upward, the traveling sheaves move to the left, and the finger *e* pulls the carriage *c* along with it by means of the latch. At the left-hand end of the guides *V* are located two trip-blocks *m* having inclined surfaces in the path of projections extending from the side of the latch *l*, and when the traveling sheave has

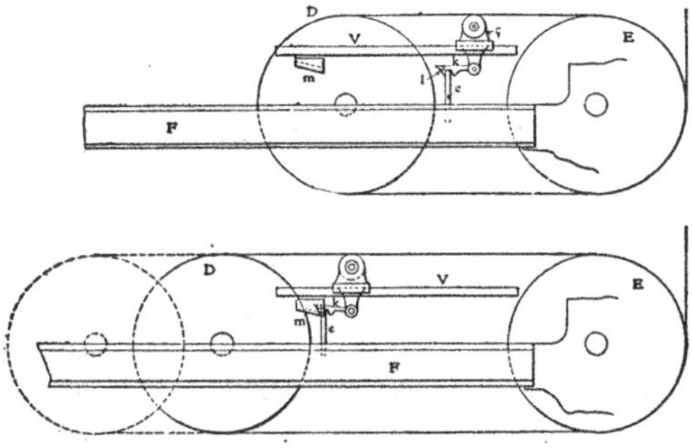

FIG. 158

reached the position shown in the lower view, these projections begin to slide upward on the inclined ribs of the block *m* and thus lift the latch *l* out of engagement with the finger *e*. As soon as the latch is free from the finger, the movement of carriage *c* stops in a position just a trifle in advance of that shown in the lower view, and the traveling sheave continues moving to the end of the stroke.

It must be understood that Fig. 158 does not show the actual construction of these parts in the Morse and Williams machine; these are made as shown in Fig. 124, but Fig. 158 illustrates the mechanical effects obtained. The reason for modifying the construction is that the parts as actually made, when presented on paper, do not reveal the operation as clearly as those shown in the simple diagrammatic representation. From an inspection of this diagram it can be seen that to keep the apparatus in proper running order it is necessary not to permit the latch mechanism to get out of adjustment so as to not release the carriage at the proper time; if this should occur, the rollers and their carriage would be carried along with the traveling sheave and do considerable damage. If all the parts are well lubricated with an oil that does not gum, and all joints are prevented from working loose, there will be no trouble.

The axle that carries the rope-supporting rollers is made of extra strong piping, and is provided with an oil cup at one end through which it can be filled so as to lubricate the rollers. Care should be taken to keep these well oiled so that they may not stick and let the ropes be dragged over their stationary edges. The guides *V* should also be looked at frequently and be kept free from dust or grit of any kind, and well enough lubricated to prevent undue friction. The finger *e* is located midway between the two guides, so that if the friction of one of the shoes of the carriage *c* is greater than that of the other, the carriage is liable to be twisted out of line, and possibly enough so to make it catch on the guides.

The carriage *c*, with the rope-supporting rollers, is moved back from the position shown in the lower drawing to that shown in the upper one at each downward trip of the car, because the finger *e*, when it reaches the latch *l* strikes the projection *k* back of the hook end, and thus pushes the carriage back.

THE PISTON OF THE MORSE AND WILLIAMS MACHINE.

The piston of the Morse and Williams pulling machine is packed in practically the same way as that of the Whittier machine, so that what was said on this subject in connection with the latter machine applies to the former. The construction of the piston is fully shown in Fig. 159, in which a section of the body of the piston parallel with the axis is shown at *A*, and a top view at *B*. In the latter the positions of the bolts for pressing down the stuffing-box gland are shown. A section of the gland is shown at *C*. A cap is bolted against the seat surrounding the opening *H*, and is held central by fitting into this opening. This cap is for the purpose of preventing leakage of any water that may escape through the hole into which the piston-rod fits. The rod is made with an enlarged head that fits against the beveled seat at the end of the hole in *F*. The packing is done by running the car to the top of the building so as to bring the piston out to the back end of the cylinder, the car is securely fastened to the overhead beams, and the water is drawn from the cylinder and pipes in the manner previously explained, and then the packing is done.

As it is so easy to get at the tightening bolts with this type of machine it is advisable to not screw up the gland ring any more than is necessary to keep the water from leaking, then make a trip, and if the piston leaks tighten up a trifle more. It will not be much trouble if you have to stop the car several times before you get the packing tight enough, and if this course is pursued the machine will run much better, and if the pressure used is barely sufficient to give the required speed with a full load, the operation of the elevator will be far more satisfactory.

LUBRICATION.

The cylinder of a pulling machine can be easily lubricated through the rear end, and it can also be cleaned out if at any time it should look as if there was grit in it that might cut away the surface of the cylinder. Elevators of all kinds are generally made with ample provision for oiling

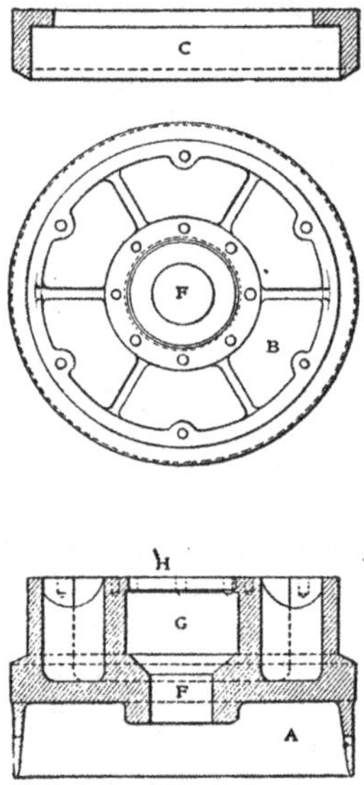

FIG. 159

the cylinder, the valves and other parts inside of the valve chamber, but at the present time the most generally approved method is by using lubricating compounds that have come into use that are dissolved in the water. These compounds are dropped into the open tank and become mixed with the water, and as the latter circulates through the system they are deposited on all the surfaces, and thus lubricate all the valves, pistons, rods, etc., not only of the elevator machine, but also of the pumps. These compounds cannot be used except in plants where the same water is used over and over, but there are practically no elevators at the present time in which this is not done, at least not where there are two or more elevators in use.

The slides on which the crosshead shoes run should not only be kept well oiled, but in addition should be examined frequently and cleaned off so as to remove any grit that may settle on them. The sheaves are as a

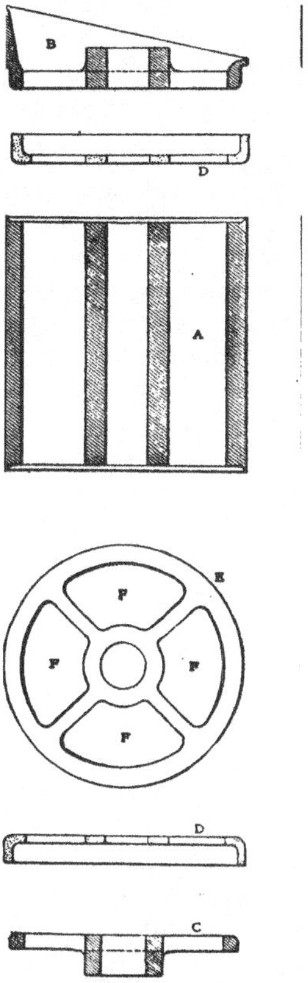

FIG. 160

rule lubricated with grease fed from cups provided with screw caps. Most sheaves have a number of cups so that one of them may easily be reached in whatever position the sheaves may be. Every morning the cap should be given about a half turn, and this will force out enough grease to last all day. All the sheaves should be attended to, and as there is danger of having the arm caught while reaching in to the center sheaves, if the car should be started, the car should be run down to the lower floor and the hand valve in the supply pipe closed before doing this work. If the valve is located where it cannot be seen by the man working at the sheaves, it is advisable to provide means to lock it in the closed position so that meddlesome persons may not be able to open it and start the machine without his knowledge. All this may appear to be unnecessarily cautious, but it only takes a little time to do these things, and it is better to spend this time every day of one's life than to have an arm mangled just once.

THE VALVE PISTONS.

The valve pistons of the Morse and Williams pulling machine are constructed substantially the same as those used in the other machines which have been fully explained in previous chapters. The only one that requires notice here is the stop valve used in the type of machine shown in Fig. 124. The construction of the several parts of this valve is clearly shown in Fig. 160. A sectional view of the body of the valve is shown at *A;* the clamping head for the forward end is shown in section at *B;* the back head also in section is shown at *C,* and the leather cup packings are shown at *D, D.* At *E* is given an end view that serves for the body and the two heads, these being made of the same section. The cup packings can also be made of this form if desired, with the openings *F* just as large as those through the metal parts, but it is better simply to punch four holes, say one-half inch in diameter, as by doing so more leather is left to hold the cup in shape, and the opening through these holes is all that is required. The object of these openings is to permit the water to circulate freely through the valve so as to prevent end pressure, and this freedom of circulation can be fully obtained through four holes of the size named.

WHY THERE IS NO PILOT VALVE.

It will be noticed that in the Morse and Williams pulling machines that have been shown there is no pilot valve. The reason why this valve is not used is that a specially constructed wheel-operating device is provided in the elevator car, by means of which it is possible for the operator to move the main valve with little effort; therefore, no pilot valve is required. The only reason why a pilot valve is used with any type of hydraulic elevator is that on account of its small size it can be moved with so much less effort than the main valve that there is no difficulty in moving it by means of a lever in the car, the end of which need travel through a distance of only about one foot. If it required no more effort to move the main valve than the pilot, the latter would not be used. In the Morse and Williams arrangement a hand-wheel about fifteen inches in diameter is placed in the car, and this is so geared with the main valve that it has to be rotated through nearly one revolution to open the valve fully; therefore the effort required to turn the wheel is greatly reduced, the distance through which the hand is moved being correspondingly increased. The operation of this device may be understood by reference to Fig. 161, which is a diagrammatic representation of the system. The

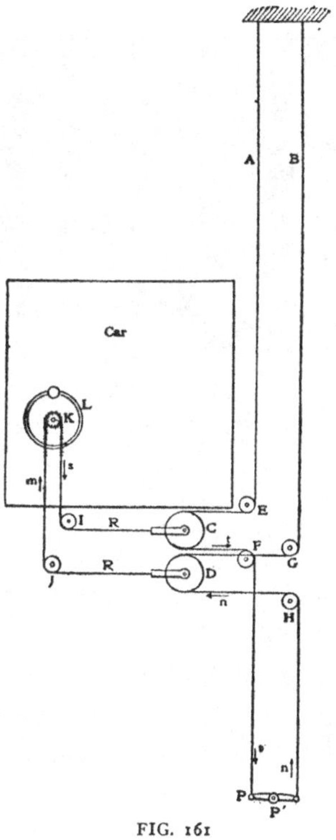

FIG. 161

two stand ropes *A, B* are fastened at the top of the elevator well and pass around the sheaves *C, D,* under the car, and thence down to the ends of the lever *P,* which is fixed on a rock-shaft connected with the main valve. The hand-wheel carries on an extension of its shaft a small sprocket wheel *K,* and by rotating the hand-wheel the ropes *R* and *R′* are drawn in or paid out, and thus the positions of *C* and *D* are varied. If the wheel is turned clockwise, the movement of the ropes *R, R′* will be as indicated by the arrows *m* and *s,* and the sheave *D* will be drawn to the left and the sheave *C* to the right, the rope *A* being pulled down, as indicated by arrows *t, t;* by the drawing up of the rope *B*, as indicated by arrows *n, n'*. This will cause lever *P* to turn its shaft counter-clockwise, and by the connection with the main valve the latter is moved. This arrangement cannot be used with a lever in the car because it would require too great an effort to move the lever. The hand-wheel moves

FIG. 162

with a small effort because of the great leverage over the valve mechanism; the hand of the operator travels through many times the distance traversed by the lever *P*.

In the diagram all the sheaves and guide wheels are drawn so as to show their sides, for the purpose of making the operation perfectly clear, but their true position can be better judged from Fig. 162, in which the sheaves *C, D, J* and *I* are clearly shown. The guide sheaves *E, F, G, H*, are not shown in Fig. 162, but they are also mounted upon the car. In fact, all the parts indicated in Fig. 161 are mounted upon the car and move with it except the standing ropes *A, B* and the lever *P*.

CHAPTER XXIX

HIGH PRESSURE HYDRAULIC ELEVATORS

CONSTRUCTION AND OPERATION OF THE OTIS VERTICAL MACHINE; THE FUNCTIONS OF THE PUMPS AND THE ACCUMULATOR

All the elevators discussed in former chapters belong to the low-pressure class, being operated by water under pressures not over 200 pounds per square inch; the average pressure is probably about 100 pounds.

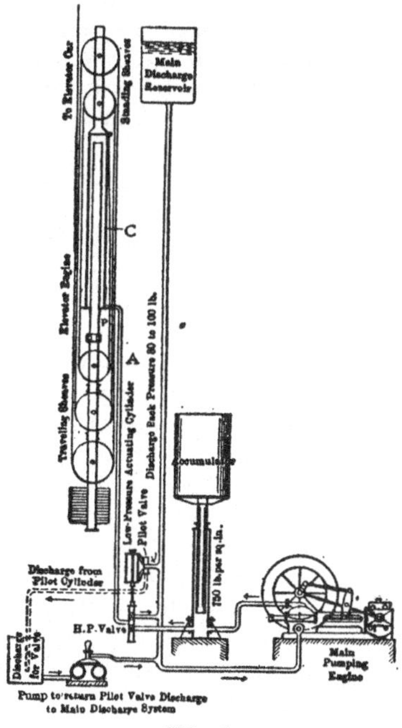

FIG. 163

Such machines are very well suited to comparatively low buildings, small capacities and moderate car speeds. In very high buildings, where high car velocity is desired they are unsuitable on account of the room required for the machinery and pipe connections. One advantage of low-pressure machines is that there is very little difficulty in keeping the piston, valves and other sliding joints tight—so little, in fact, that ordinary packings meet all the requrements except for the valves, and in these cup packings are necessary simply because the packing has to slide over the port openings.

It is evident that if the pressure of the water be increased the diameter of the lifting piston can be reduced and therefore the machine can be made more compact. If, however, the pressure is carried above 200 pounds, the ordinary type of packing will not be satisfactory; it will be necessary to substitute cup packings for it at practically every point, and so long as it is necessary to make such a change, it is advisable to make a decided increase in the pressure so as to gain in a high degree the advantages derived from high pressure. On this account high-pressure elevators are operated with pressures that average about seven times the average pressure with low-pressure systems. Generally the high-pressure machines are operated with a pressure of 750 pounds per square inch. The reduction in the size of the machine and piping that can be effected by using this pressure is much greater than would be supposed by those who have not investigated the subject. To give a general idea of how great the reduction actually is, suppose a low-pressure elevator has a cylinder 16 inches in diameter and works with a pressure of 100 pounds. For such a machine the supply pipe would probably be not less than 6 inches in diameter. Substitute for this a high-pressure machine working with 800 pounds pressure per square inch; then, if everything else remains unchanged, the area of the cylinder will be reduced to one-eighth, and this will make the diameter a trifle under 5¾ inches, as compared with the low-pressure cylinder of 16 inches diameter. This is not all the gain that can be made; there can also be effected a great reduction in the size of the supply pipe, for as only one-eighth of the quantity of water is required the size of the pipe can be reduced to the same degree as that of the cylinder, provided the water is to run through it at the same velocity. This reduction would cut the pipe down from 6 inches to a trifle over 2 inches in diameter.

These reductions are not exactly what would be made in actual practice, because the frictional loss in the small high-pressure cylinder would not be as great as in the large low-pressure cylinder, and the velocity of the water through the supply pipe could be made greater for the same percentage of loss; this would permit a further reduction in the size of the pipe. In practice the gain in this direction is utilized in part to reduce the size of the apparatus and in part to reduce the loss of energy in forcing the water through the pipes. As a result, the loss of energy due to the friction of the water passing through the pipes, lifting cylinder and valves is reduced to about 5 or 6 per cent., whereas in low-pressure machines it runs from, say, 10 to 30 per cent. The change of pressure from 100 to 700 or 800 pounds brings about other changes in the construction of the machine and apparatus and also in the general arrangement of the system.

The arrangement of the various parts of a high-pressure system is indicated in the diagram Fig. 163. This diagram shows a machine geared six to one. The cylinder is shown at *C*, the plunger at *P* and the traveling

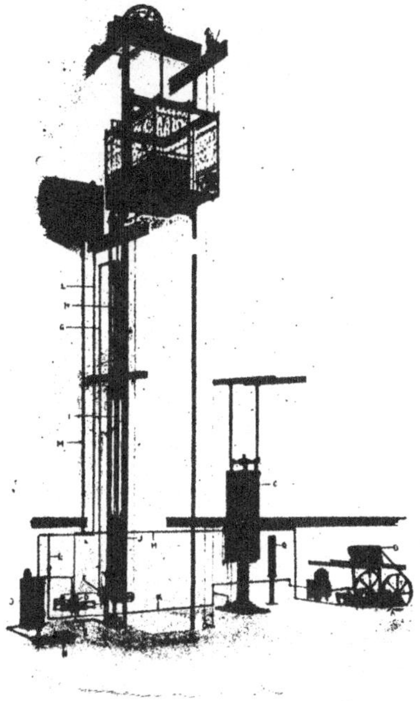

FIG. 164

sheaves below it; the cylinder is inverted, the plunger being forced downward by the pressure of the water. This construction is used because the small size of the cylinder makes it impracticable to use a piston and piston-rod, therefore a solid plunger is provided and the pressure acts to push it out of the cylinder. If the cylinder were set with the mouth up, the weight of the car would tend to pull the plunger out of the cylinder. Machines of this type have been made, but are not in use at the present time. In these machines the gear was reduced to two to one, so as to not have to add so much weight to the plunger; as the weight of the plunger has to lift the car it must be sufficient to lift the maximum load and in addition overcome all the friction.

In Fig. 163 the pump forces water into the lower end of an accumulator, from which a pipe runs to the main valve, through which it passes to the pipe *A* and thence to the lifting cylinder. On the return stroke the water passes out of the cylinder through the pipe *A* and through the upper end of the main valve to the discharge pipe, which runs up to a tank placed on or near the roof of the building. The object of this arrangement is to provide a low pressure to operate the pilot valve, which is shown in the diagram just above the main valve. In the first high-pressure elevators made, the pilot valve was operated with water at the same pressure that was used for the lifting cylinder, but these valves were not successful, owing to the fact that they had to be very small and the packings would not withstand the wear due to the pinhead jets of water striking them at terrific velocities; in addition, the small holes through which the water passed were soon enlarged so that the valve would not work satisfactorily. With the low-pressure pilot valve there is no trouble. A small tank is provided to receive the discharge from the pilot valve and its actuating cylinder, and this water is returned to the roof tank by means of a small pump as shown in the drawing.

THE ACCUMULATOR.

The accumulator takes the place of the pressure tank of the low-pressure system. A pressure tank cannot be used with the high-pressure system, owing to the fact that it is troublesome and expensive to pump air against a high pressure, and it is necessary to do this so as to replenish the air that gradually leaks out of the pressure tank. Even if there were no difficulty in pumping air into a high-pressure tank, the accumulator would be preferable, because with it the pressure depends upon the weight on top of the plunger, not on the hight of the water in the cylinder. With a pressure tank the pressure drops as soon as you begin to draw water out, and it runs up as soon as the outflow stops, consequently the pressure is continually varying.

The actual appearance and location of the apparatus of an Otis high-pressure vertical elevator are shown in Fig. 164. This illustration shows several parts not represented in the elementary diagram. The main pump is at *A* and at *B* is shown the prime mover, which in this case is an electric motor, although in practice steam power is almost always used. The accumulator is shown at *C* and the main valve and pilot valve are at *D*. From the main valve the water passes to the lifting cylinder through a

FIG. 165
HORIZONTAL HIGH-PRESSURE ELEVATOR

pipe *E*, passing first through an automatic stop valve *F*, thence through the pipe *G* to the cylinder *H*. The plunger is shown at *I*, and the traveling sheaves at *J*.

The high-pressure water from the accumulator reaches the main valve through the pipe *K* and is discharged from the valve through the pipe *L* which runs up to the tank at the top of the building. Through the pipe *M*, the water returns to the pump *A*. An air chamber is provided at *Q* to smooth out any pulsations of the pump that its own air chamber does not subdue. The small pump to return the water discharged from the pilot valve to the roof tank is shown at *N*. This drawing shows all the devices generally used, but several that are added for large plants where all the most approved refinements are desired are not shown here.

It will be noticed in Fig. 164 that the machine proper of a vertical high-pressure elevator is not very elaborate. This, however, is not true of the horizontal machines, on which most of the devices required for its operation are mounted, as can be seen in Fig. 165, which shows a "double-decker." The main and pilot valves in this illustration are shown at *A*, and at *B* is shown the lever with which the ropes that connect with the car lever are connected. At *C* is shown the automatic stop valve which is actuated by the rope *D*, that passes around a sheave *E*, located at the forward end of the cylinder. This rope D is clamped by the arm *E1* which is fastened to a rod *E2* that runs to the forward end of the machine and carries two stops *E E* and *E E*. These stops are struck by a projection carried on the traveling-sheave crosshead when the latter approaches either end of its travel; the action being the same as in the low-pressure horizontal machines previously described. The only object of using the rod *E2* is to reduce the length of the operating rope *D*. If the sheave *E* were placed at the extreme forward end of the machine, and the rope *D* were run over it, then stops could be placed on the rope *D* in the usual way adopted in other types of horizontal machines; but with this construction there would be more danger of getting the stop-motion out of order by reason of having so long and exposed an operating rope. The design of Fig. 165 increases the cost of construction but as an offset to the increased expense makes the machine more reliable in operation.

At *F* is the automatic stop valve, and at *R* is a speed regulator to prevent the car from running at an excessive velocity if the operator should move the lever far enough to open the valve wide when the car is lightly loaded. In many of the elevators heretofore described the maximum speed is determined by the set of the automatic stop valve, but in this machine, as well as in the vertical-cylinder high-pressure machines, the device *R* is used for this purpose; in fact all the valves used in the vertical and horizontal high-pressure machines shown in Figs. 164 and 165 are the same.

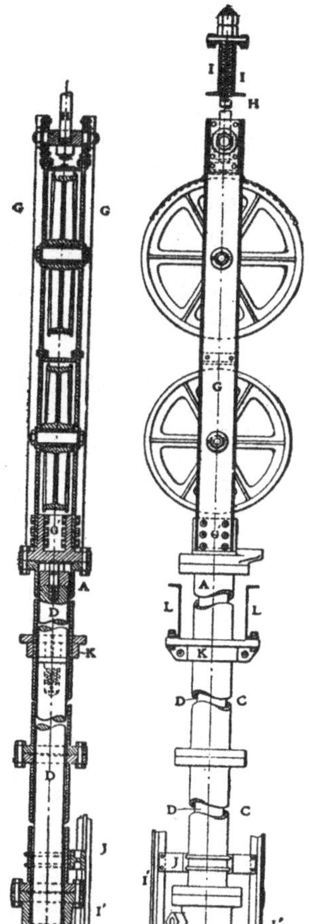

FIG. 166

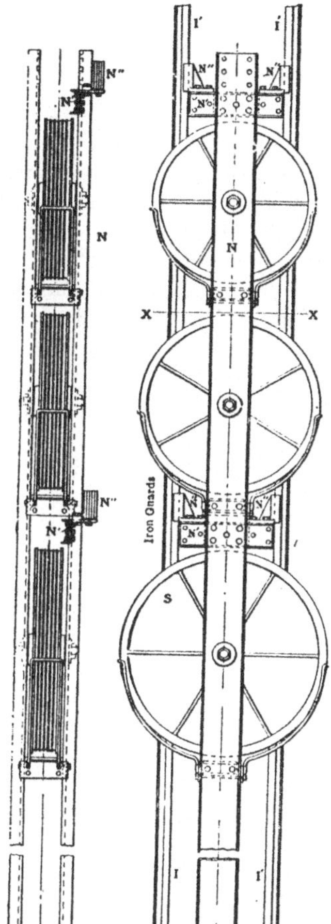

FIG. 168

The devices shown at *S* are strainers through which the water for the pilot valve flows.

CONSTRUCTION OF THE CYLINDER, PLUNGER AND SHEAVES

The construction of the cylinder, the plunger and the sheaves of the Otis high-pressure vertical machine is shown in Figs. 166 to 169. Fig. 166 gives external and sectional views of the cylinder, the upper end of which is seen at *A* and the lower end at *B*. To shorten up the drawing the cylinder is broken at *C C*. The plunger is indicated by *D*. Above the cylinder are shown the stationary sheaves held between side frames made of channel iron *G*, to the lower end of which the cylinder is bolted, as

FIG. 167

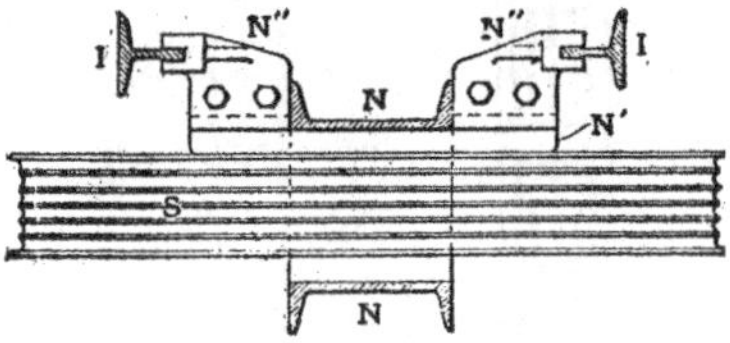

FIG. 169

shown at *G'*. The channel frames *G* are bolted to a rod *H* at the upper end, and this is held between beams *I* that are secured to the wall or floor framing of the building. The traveling sheaves are carried in a cross-head attached to the lower end *F* of the plunger.

The internal construction of the cylinder is shown in the vertical section on the left side of the drawing, which is taken at right angles to the exterior view. The upper end of this drawing shows the way in which the bearings of the stationary sheaves are held between the side frame channel beams *G G* and in like manner the lower end shows the construction of the cap *F* that forms the end of the plunger and the support for the traveling-sheave frame. This cap is constructed cup-shaped on its upper side to receive the drip from the cylinder. The plunger, it will be noticed, does not fit the cylinder throughout its entire length, but only for a short distance at the lower end, where the stuffing-box is located. The cylinder is held up by the rod *H*, and is sustained against side displacement by means of one or more rings *K* and the frame *J*, the construction of both of which can be readily understood from the drawings. An end view of the frame *J* is shown in Fig. 167, which is a section through the cylinder at a point above *J* in Fig. 166, viewed downward and showing a top view of the traveling sheaves, which in a complete view of the elevation would be seen below *F* in Fig. 166. The outlet *M* is the pipe connection through which the actuating water enters and passes out of the cylinder. The pipe itself is seen at *M'* in Fig. 167.

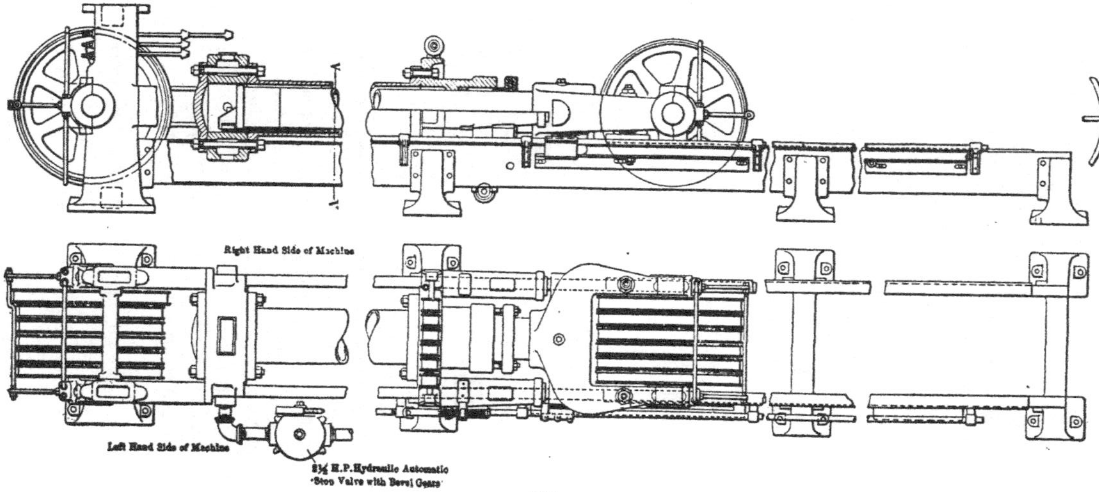
Right Hand Side of Machine
Left Hand Side of Machine
2½ H.P. Hydraulic Automatic
Stop Valve with Bevel Gears

FIG. 170

The T-bars I' I' to which the frame J is bolted form the guides for the crosshead of the traveling sheaves, and the cylinder is held true with these by means of the frame J so as to keep the plunger and the crosshead guides in line.

The traveling sheaves, the crosshead and the guides are shown in Figs. 168 and 169, the first being edge and side elevations, and the second a plan section taken on line X X of the side elevation. The cross-head is made up of side frames N consisting of channel iron, with cross connections between the sheaves and at the ends, and additional cross beams N' to which are fastened the shoes N'' that run on the guides I'. The construction of the crosshead is so clearly shown in the drawings that a detailed explanation is not necessary. To realize clearly the arrangement

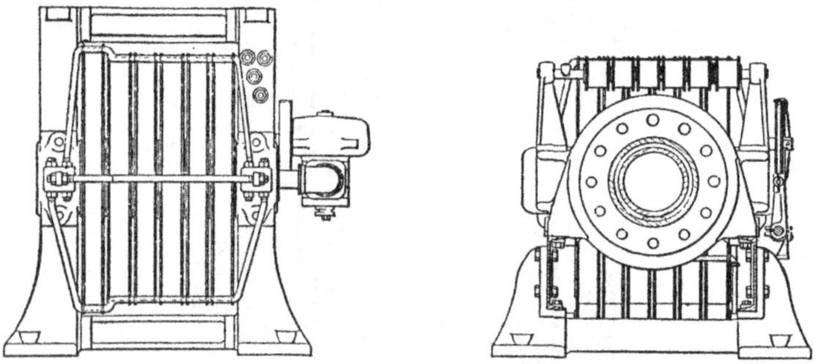

FIG. 171

of the complete cylinder and the sheaves, stationary as well as traveling, Fig. 168 should be seen below Fig. 166, of which it is a continuation. The construction is divided into two illustrations on account of the length; the details would be entirely too small to be shown clearly if the complete construction were shown in one drawing the length of these pages.

The construction of the cylinder and other parts of the horizontal machine can be understood from Figs. 170 and 171, the former showing a side elevation with the cylinder in section and a plan view of the whole machine, the latter a rear-end view and a section at right angles to the cylinder taken on line A A of Fig. 170. The cylinder barrel is made of wrought iron with the flanges welded on. To the back end is bolted a casting having an annular chamber from which a number of port holes open into the central bore. This casting is closed with a head held by the bolts that hold the casting to the cylinder, as shown in Fig. 170. At the front end of the cylinder is a casting that contains the stuffing-box; this casting is bored to fit the plunger. On the rear end of the plunger is bolted a shoe which slides on a brass strip along the lower side of the cylinder from one end to the other. The object of this strip is to hold the back end of the plunger in line with the front end and also to keep the plunger from dragging on the cylinder wall and the lower side of the stuffing-box. The position as well as sectional shape of the wearing strip is shown in Fig. 171.

In the plan view of the machine, Fig. 170, the automatic stop-valve is shown and also a top view of the mechanism by which it is moved. An end view of this valve is shown in Fig. 171.

Looking at Fig. 170 it will be noticed that one of the stationary sheaves at the back end of the cylinder is larger in diameter than the others: this construction is used so as to be able to stack machines one above the other without setting the top sheaves back of the lower ones, or making them of smaller diameter in order to afford room for the ropes from the lower machine to run up the elevator well. It is obvious that if the upper machine is made with the large sheave on the right side and the lower one has it on the left, the ropes from the lower machine can run up past the upper set of sheaves without striking them, as they are of smaller diameter on that side of the machine. This arrangement is clearly shown in Fig. 165.

CHAPTER XXX

OPERATION OF THE MAIN AND PILOT VALVES OF THE OTIS VERRTICAL ELEVATOR—THE ELECTRICAL CONTROL SOMETIMES EMPLOYED

The main valve shown at *A* in Fig. 165, and at *D* in Fig. 164, is shown in section and plan in Fig. 172. The pilot valve is at *A,* and the main valve at *C; B* is a motor cylinder, the piston of which moves the main valve. In this construction, the pilot valve is not much smaller in diameter than the main valve, and the motor piston is very much larger than the main valve. The difference in the proportions of these parts as compared with the valves described in connection with low-pressure machines is due to the fact that in the high-pressure system the motor piston *B,* is actuated by low-pressure water, so as to make it possible to use a pilot valve of large enough size to be durable. As is shown in Fig. 164, the tank into which the lifting cylinder discharges is placed high enough to give ample pressure to operate the motor piston, and from this tank water passes through the pilot valve *A* to the cylinder *B.* If the motor piston were operated by the high-pressure water, the pilot valve and its port holes would have to be so small that the parts could not be made sufficiently substantial. For this reason water at a pressure of about 80 pounds per square inch is used to operate the motor piston.

It might be thought that having to discharge the water in the lifting cylinder against a back pressure of 80 pounds would cause considerable loss and make the high-pressure system objectionable on the score of low efficiency, but this is not the case because the main pump draws water from the same discharge tank; therefore, the back pressure against the lifting cylinder acts to help the pump, so that in reality the work the pump has to do is to force water against a pressure equal to the difference between the pressure of the accumulator and that of the discharge tank. The net result is that if the accumulator pressure is 750 pounds, and that of the discharge tank is 80 pounds, the actual pressure against which the pump acts is $750 - 80 = 670$ pounds, and the pressure that acts in the lifting cylinder to raise the elevator car is 670 pounds, not taking into account the losses due to friction of the water through the pipes and valves on its way from the accumulator to the cylinder.

OPERATION OF MAIN AND PILOT VALVES.

The operation of the main and pilot valve of Fig. 172 is as follows: If the operator desires to run the car upward he moves the car lever so as to pull up the rope N' on the right side, thus tilting the rock lever N in a counter-clockwise direction. The levers N and L are secured to the shaft P; hence, the end of L will move down and through the connecting-rod L' will pull down the lever L''; and the latter, through M, will depress the pilot valve. The center pipe E is connected with the

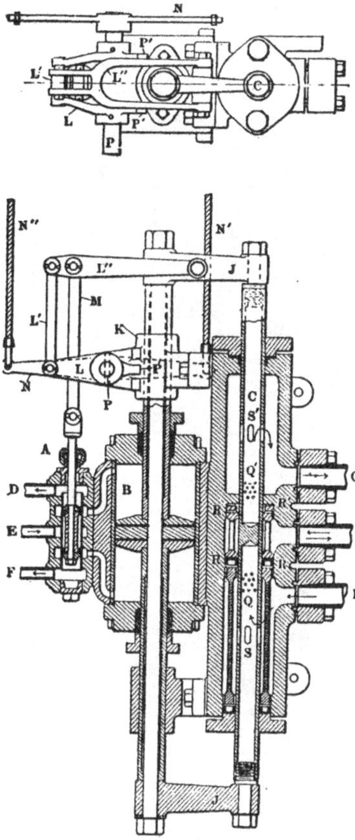

FIG. 172

upper discharge tank; hence, water will flow in and through the lower end of the pilot-valve chamber, pass to the lower end of the motor-piston cylinder B, and raise the piston, the water above the latter passing out into the pilot-valve chamber above the valve, and thence to the pipe D. As the motor piston-rod is connected at both ends by arms J J with the ends of the main valve *C*, the upward movement of the piston will lift the main valve, and then the water from the accumulator coming through the pipe *I* will pass into the center of the main valve through the port *S*. The ports *Q* will be above the packing *R*, so that the water will pass out into the central pipe *H* and thence to the lifting-cylinder, and by pushing the plunger out of the latter will lift the elevator car. If the

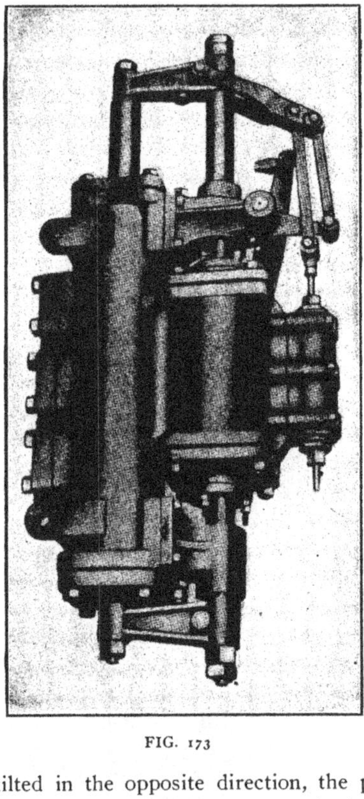

FIG. 173

rock lever *N* is tilted in the opposite direction, the pilot valve will be raised, and then water will pass to the upper end of the motor cylinder and depress the piston, thus moving the main valve down so that the water in the lifting cylinder may escape through the ports *Q'* into the upper end of the main valve and thence through the ports *S'* to the upper discharge pipe *G*; from there it passes to the discharge tank near the top of the building.

Upon examining this valve-gear it will be found that as soon as the main valve moves it acts to return the pilot valve to the central position; thus, if the pilot valve is raised, the main valve will immediately thereafter start to move upward, and then the lever *L″*, which is pivoted on the arm *J*, will start to draw up the link *M*, thereby moving the pilot valve toward the stop position. When it has lifted the pilot valve to

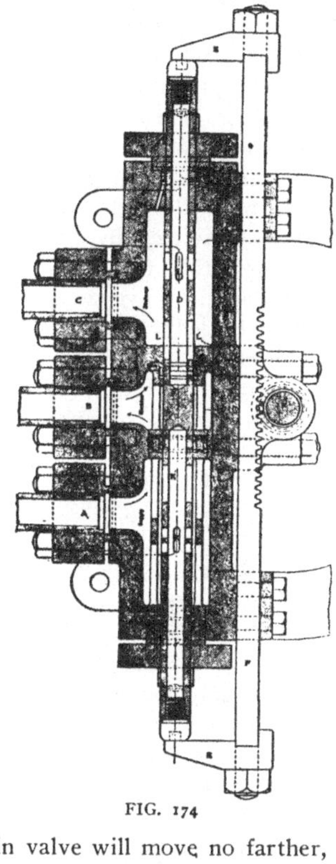

FIG. 174

this position, the main valve will move no farther, so that the extent of the opening of the main valve will depend upon the distance through which lever *N* is moved. This action, as will be noticed, is precisely the same as that of the pilot valves of the low-pressure systems, so that the principle of operation of the valve-gear of Fig. 172 is the same as that of those used in all other hydraulic elevators operated by pilot valves, the only differences being in the details of construction. These are greater than the difference between different designs of low-pressure valves owing to the fact that the motor piston and the pilot valve operate with one pressure of water, while the main valve controls the flow of water from another source, which is of much higher pressure. The location and relations of the levers *L, L″* and *N* can be more fully understood from an examination of the plan view at the top of Fig. 172. The external appearance of these valves is shown in the photographic view Fig. 173. Allowance must be made, however, for the fact that the latter is the reversed view, from the rear of Fig. 172.

In some machines the pilot valve is rendered unnecessary by the use of a hand-wheel in the car, the leverage of which makes it possible to move the main valve directly. When this arrangement is used with high-pressure machines, the valve is altered in construction in the manner clearly shown in Fig. 174. In this drawing the inlet and outlet connections are on the left instead of on the right, as in Fig. 172. This is due to the fact that the drawing shows the valve from the reverse side. The only actual difference between this valve-gear and the one shown in Fig. 172 is that in place of the pilot valve and the motor cylinder, a rack *F* and actuating pinion *F′* are used; the rack being connected with the ends of the main valve in the same manner as the motor piston-rod. The arms *E E* are the counterparts of the arms *J J*. In Fig. 174 the ports *L L* are shown as long slots, but the common construction is to make them in the form of numerous small holes, so as to prevent the water from drawing the edges of the cup packings into them.

MODIFICATIONS IN VALVES MADE TO BE OPERATED BY MAGNETS.

The valves of Fig. 172 are also made so as to be operated by magnets whenever it is desired to control the car by the movement of a small electric switch instead of the lever used with the running- and standing-rope gear. The advantages of the electric switch in the car are that less space is required for its manipulation, which is a matter of considerable importance if the car is small, and that all the more or less complicated rope connections between the car lever and the lever of the pilot valve are dispensed with; in their place are substantial small copper wires within a flexible cable extending from the under side of the car to a point about half the way up the elevator well.

The modifications made in the valve when arranged to be operated by magnets are shown in Figs. 175 and 176, the one being a sectional elevation through the main and the pilot valves, and the other an external view taken at right angles to the former. In Fig. 175 it will be noticed that the main valve and the motor cylinder and piston are the same as in Fig. 172, but there is some modification in the pilot valve. This consists in substituting a solid valve for the valve with cup packings, and this change is made in order to reduce the frictional resistance and make it possible to operate the valve with magnets of reasonable size. The solid pilot valve is made a perfect fit by grinding and with ordinary care will remain tight for a long time, especially if the pressure of the water used to operate the motor piston is low, and this can be made as low as desired by simply increasing the diameter of the motor piston. For the purpose of making the pilot valve move as easily as possible and with

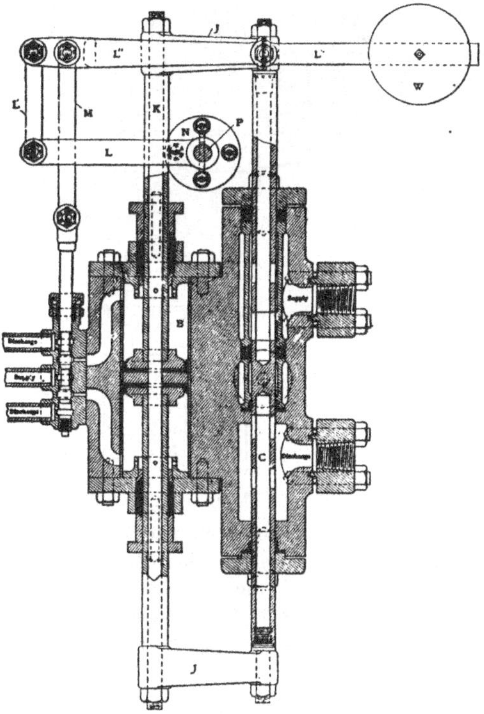

FIG. 175

the same effort whether lifted or depressed, the lever L'' is extended to the right, as shown, and carries at its outer end a weight that is carefully adjusted to balance the combined weight of the pilot valve, the lever L and the connecting-rods L' and M. The shaft P is provided with a coupling N that connects it mechanically with the rock-shaft of the magnets which operate the valve; but the connection is made through insulating material, as is indicated by the heavy black lines in Fig. 176, so as to prevent any electrical connection between the two parts.

The magnets are of the same type as those described in connection with vertical low-pressure elevators and act in the same manner. While

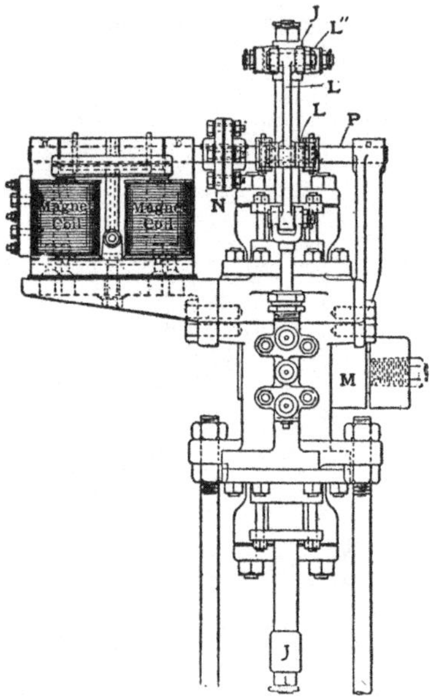

FIG. 176

magnet control operated by a small electric switch in the car is very desirable so far as compactness and simplicity of apparatus are concerned, it has the objection that with it the operator cannot vary the car speed in the same way as with the car lever and standing-rope rig, and for that reason the lever, although more cumbersome, is still used in nearly all cases.

CHAPTER XXXI

CONSTRUCTION AND OPERATION OF THE ACCUMULATORS AND THE AUTOMATIC VALVES USED WITH THEM

For high-pressure elevators accumulators are always used in place of pressure tanks. The latter are practically out of the question for such service, because no matter how perfect all the parts may be, the air in the pressure tank is sure to escape, so that from time to time it has to be renewed. With low-pressure systems this is easily done by drawing in air with the water, the pumps being made so as to draw in air whenever required, that is, the same pump can be used to pump the air as well as the water. With the high-pressure system this arrangement would not be desirable, and to provide a regular air pump to force air into the tank whenever necessary would be objectionable on account of the high pressure. To pump the air successfully would require two-stage compression to the required pressure. Even after doing this, the tank system would not be equal to the accumulator because with it the pressure cannot be kept constant, whereas with the accumulator it cannot vary, because it does not depend upon the amount of water within the accumulator, but upon the weight of the accumulator plunger.

In large elevator installations, where there are many elevators and a number of pumps to supply the water, one accumulator may supply several elevators, and these may be connected all in one system, so that water can be drawn from any one of the accumulators to operate any one of the elevators, or the plant may be divided into a number of independent sections. If there is only one elevator, there will be one accumulator, and its capacity will be very much larger in proportion to the elevator than where there are several elevators operating from one accumulator. Thus if there are two elevators, the accumulator will not be much larger than for one car, and for three cars it will not be much larger than for two. The accumulator in a plant where the elevators are in constant service is only of sufficient capacity to supply water during any short interval when the pumps are overtaxed, and this is not often.

CONSTRUCTION OF THE ACCUMULATOR.

The construction of an elevator accumulator made by the Otis Elevator Company is shown by Fig. 177, a sectional elevation and a plan view being shown. This is one of many designs used, but is the most common, because it is so constructed as to require the least amount of head room, and in practically every installation head room is limited. The accumulator cylinder is shown at *C*. At the upper end of the plunger *P* is mounted a crosshead *B* from which are suspended four strong rods *A*.

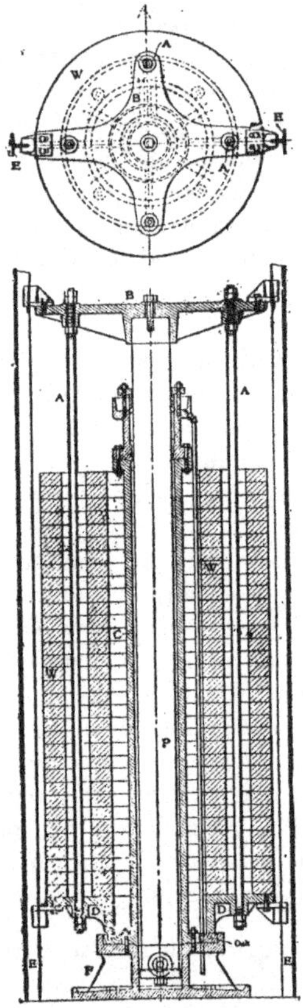

FIG. 177

These rods pass through a number of weights *W* and a lower retaining flange *D*. The number of weights depends upon the pressure required, and is made such that the combined weight of the plunger, the crosshead, the rods *A*, the flange *D* and the weights *W* produces the required pressure upon the end of the plunger. The weights have an opening through the center that is large enough to pass the weights over the cylinder and any projecting parts connected with it. In this way the hight is much less than it would be if the weights were placed directly on top of the plunger. When nearly all the water is drawn out of the accumulator, the plunger

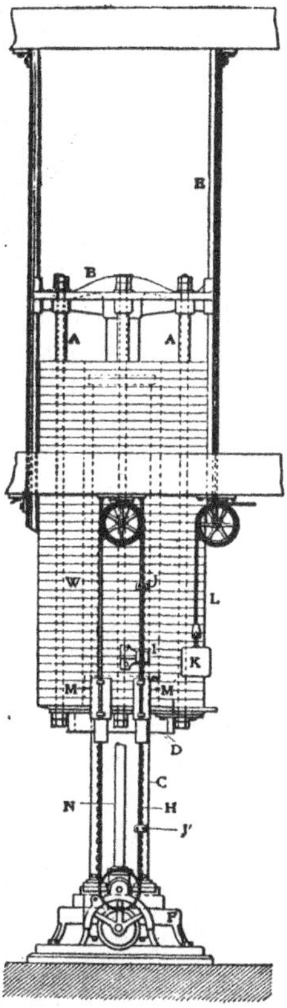

FIG. 178

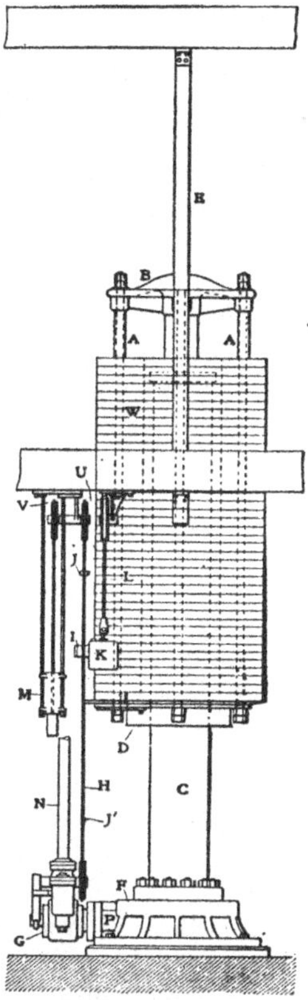

FIG. 179

descends to the position shown in the drawing and the lower part of the flange *D* strikes upon the buffer block of hard wood in the top of the base casting *F*. When the plunger is raised to its highest position it is prevented from being forced entirely out of the cylinder by the enlarged head, which is securely held in position by bolts. This head brings up against a shoulder at the top of the cylinder and prevents the plunger from rising any farther. The plunger is held in line with the cylinder by means of the guides *E*, as shown in the plan view. It is not desirable to allow the plunger to come down and strike the lower buffer, or to run up until it hits the upper stop; hence, means are provided to stop the plunger automatically before it runs either too far up or too far down. This is accomplished by stopping the pump when the plunger has reached the upper limit, and if there are several pumps and accumulators, by in addition closing the entrance so that no more water can be pumped in, although the water in the accumulator will be free to run out. If the water is drawn out so as to make the plunger descend to the lower safe limit, the opening into the accumulator is closed so that no more water can run out, but the closing does not prevent more water from passing into the cylinder.

The way in which all this is accomplished is shown in Figs. 178 and 179. These drawings show two elevations of an accumulator, at right angles to each other, equipped with an automatic stop-valve, and additional means for stopping the pump. The accumulator is shown at about the mid position. Should it rise, a plate secured to the lower weight supporting the flange *D* directly under the weight *K* will lift the latter. The rope *L* runs to the starting lever of the pump and when the weight *K* is hanging in the air, it holds up the lever and thereby keeps the pump running, but as soon as it is lifted by the plate on the accumulator, a counterweight or spring pulls the lever in the opposite direction so as to stop the pump. The automatic stop valve is located at *G* and is moved by the sprocket chain *H*, upon which are mounted stop balls *J* and *J'*. When the accumulator is full the stop *J* is struck by the arm *I*, attached to the weights *W*, and the valve *G* is rotated so as to prevent more water being forced into the cylinder. When the water is drawn out enough to cause the arm *I* to strike the lower stop *J'*, the stop valve *G* is moved in the opposite direction until it is closed, preventing further escape of water from the cylinder.

When the accumulator is in any intermediate position, the weights *M* and *M'* act to move the valve *G* into the central position, in which water can pass through it freely in either direction. These weights draw the valve into the central position because they are hung on a chain that passes over the sprocket *V* which is fixed on the same shaft as the sprocket *U*, over which the valve-moving chain *H* passes. The weights run on guides, as shown in Fig. 179, and when one of them is raised by the rotation of the sprocket *U* the other one does not go down, because it is held by the nuts on the lower end of the guide rods; hence, the lifted weight hangs on the wheel *V* and acts to rotate it back to the central position as soon as the weights *W* have moved far enough away to permit it to descend to the normal position.

CONSTRUCTION AND OPERATION OF THE AUTOMATIC STOP VALVE IN THE OTIS VERTICAL MACHINE.

The construction and operation of the stop valve *B* can be understood from Fig. 180, which shows vertical central sections at right angles to each other. In the cut it will be seen that there are two spring-retained check-valves *T, T'*, set so as to open in opposite directions. If the main valve *R* is in the central position, water can flow through from the pipe coming from the pumps into the accumulator, following the path of the

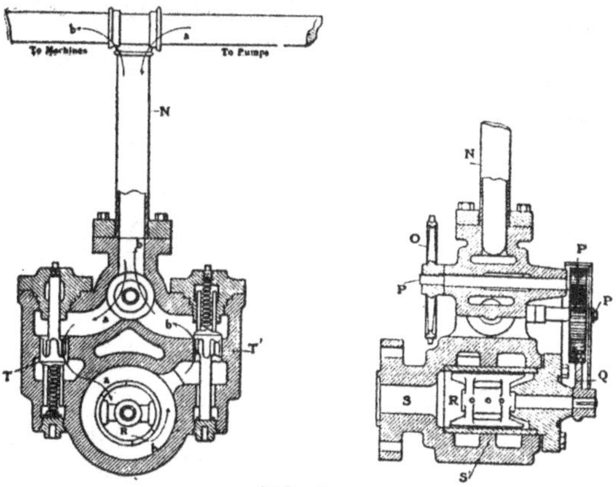

FIG. 180

arrows *a*. If the elevator cylinders draw more water than the pumps can supply, the deficiency will pass out of the accumulator along the path of the arrows *b*. If the accumulator is filled, the rotation of the sprocket wheel *O* on the shaft *P* will turn the pinion *P'* and through the intermediate wheel *P''* will rotate the segment *Q* and thereby turn the main valve *R* so as to close the inlet along the path of the arrows *a*. If the water in the accumulator runs to the lower limit, the sprocket *O* will be rotated in the opposite direction so as to close the outlet path of the arrows *b*. When the inlet is closed, the path of the arrows *b* remains open so that water may run out of the accumulator, and when the outlet path is closed, the inlet remains open so that water may flow in and fill the accumulator. To understand clearly this action it must be understood that the partition *S'* separates the ports in which the check-valves *T, T'* are located and that the openings in the valve *R* are so located that when the valve is in the central position both ports are open, but when it is turned in one direction it closes the port of the valve *T*, and when turned in the opposite direction it closes the port of the valve *T'*.

ELECTROMAGNETIC DEVICE USED FOR CONTROLLING THE PUMP.

The arrangement shown in Figs. 178 and 179 for stopping the pump when the accumulator is full and starting it when the water is nearly all drawn out is purely mechanical, and works perfectly in practice except in cases where the accumulator and pump are not very close to each

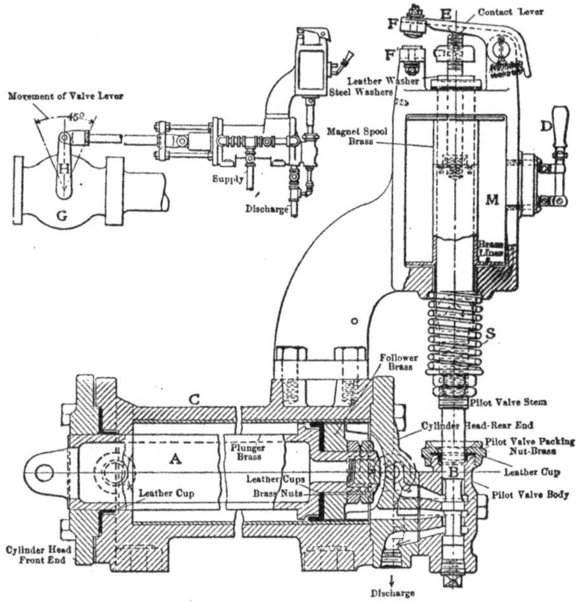

FIG. 181

other, or where the construction of the building makes it impracticable to obtain a direct rope connection between the pump and the weight *K* on the accumulator. In such cases an electromagnetic device can be used that will simplify the construction and also render its operation more certain. One of these is shown in Fig. 181, which gives a sectional elevation of the device and a diagram of its connection with the pump-starting valve. This type of apparatus is generally actuated by placing near the travel limits of the accumulator weights a pair of switch contacts that are controlled by an arm projecting from the weights, in precisely the same way that the arm *I* in Fig. 178 actuates the valve *G*.

Looking at both the illustrations it will be seen that when the differential plunger *A* moves, it either opens or closes the valve *G* and thereby either stops or starts the pump. Suppose that when the accumulator is in the highest position the lever *H* is over in the extreme right-hand position, then the plunger *A* will be in the position shown and the valve B will be drawn up so as to connect the right-hand end of the cylinder to discharge its water through the lower ports into the discharge pipe. For the valve *B* to be raised into this position, the magnet *M* must be energized so as to be able to compress the spring *S;* hence, the arm on the accumulator must close the switch contacts when the weights reach the upper position. This movement of the lever *H* will close the valve *G* and stop the pump, so that if water is drawn from the accumulator the plunger will descend, but the stationary switch contacts that were moved by the arm on the weights will remain closed, and the magnet *M* will continue energized. When the accumulator weight descends low enough to require the pump to be put into action, the stationary switch will be moved so as to open the circuit through the magnet *M* and then the spring *S* will push the valve *B* down so that water from the supply pipe will pass to the right-hand end of cylinder *C* and force the plunger *A* to the left, thereby moving the lever *H* of the valve *G* to the open position to start the pump, which will continue running until the accumulator is filled again to the upper lmit, when the former operation will be repeated, and the pump will be stopped.

The switch on top of the magnet in Fig. 181 is a type that is commonly used in connection with magnets that are intended to remain in service for a considerable length of time. When the magnet *M* draws up the plunger, the upper end lifts the switch lever *E*, thereby opening the contacts *F*, *F'* and putting a high resistance into the circuit of the magnet winding; this reduces the current to probably 10 or 15 per cent. of its initial strength, so that it can flow through the coil for a long time without heating it to a dangerous degree.

CHAPTER XXXII

AUTOMATIC STOP VALVES USED WITH OTIS HIGH-PRESSURE MACHINES

The construction and operation of the automatic stop valve for the elevator machine shown in Fig. 164 can be understood from an examination of Fig. 182, which is a sectional elevation. This valve is actuated by the lever *A*, and there are two of these levers, one on the valve proper, and the other at a point near the cylinder. The lower one is moved when the traveling sheaves reach the lowest position, and the upper one when they reach the highest position. The rod *L* extends to the upper limit of the sheave travel and there connects with a lever *C* that is mounted on a stud similar to *B* and connects with a lever like the lever *A* in the same way that these parts are connected in the drawing. The only difference is that the lower *A* lever stands normally about in the position of arrow *M* while the upper *A* lever stands normally in the position of arrow *N*. The lower lever is pushed to the position in which it is here shown by a projection on the traveling-sheave frame, and the upper lever is similarly pushed into the position of arrow *O*.

If the car is going up the traveling sheave will be coming down, and water will be running into the lifting cylinder through the valve from the lower supply pipe up into the machine pipe. Through the valve the path will be by way of the port *H* through the spring-supported check-valve *V'* into the port *I*. When the traveling sheave descends low enough to move the lever *A* to the position in which it is shown the valve *E* will be moved down opposite the port *H* and then the flow of water into the cylinder will be stopped, as the only other path is closed by the check-valve *K*. If now the operator sets the car lever so as to run down, the water in the cylinder will flow out and down the machine pipe into the port *I* and thence into the port *J* and through the valve *K* to the lower pipe. When the traveling-sheave frame reaches the upper limit of travel, the upper *A* lever will be moved into the position of arrow *O*, and the rod *L* will pull up the valve *E* so as to close the port *J* and then no more water will be able to flow out of the cylinder and the car will be stopped. When the car and traveling sheaves are in any intermediate position, the valve *E* is moved to the central space, *E'* by the action of the spring

S, which forces the upper head *F* up against the end *G* of the spring box, or the lower head *F'* against the lower end *G'*.

THE AUTOMATIC STOP VALVE OF THE OTIS HORIZONTAL MACHINE.

The automatic stop valve shown on the horizontal machine, Fig. 165, is very different from that above described. Its construction is

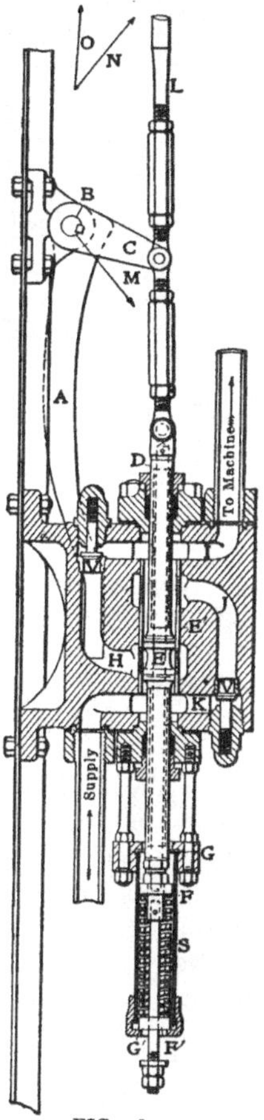

FIG. 182

clearly shown in Fig. 183. The actuating sprocket chain passes around the sprocket wheel *G* which carries a beveled pinion *G'* meshing with a segment *H*, mounted on the valve shaft *H'*. The body of the valve *F* carries a loose segment F' that fits against the valve chamber when forced outward, but can be moved away from the seat some distance when the force acts toward the center of the shaft. When the water flows in the direction indicated by D and E, the valve is turned counter-clockwise to stop the flow and, consequently, the elevator car. But when the car lever is reversed to run in the opposite direction, the flow of water through the valve will be reversed and the segment F' will be pushed away from

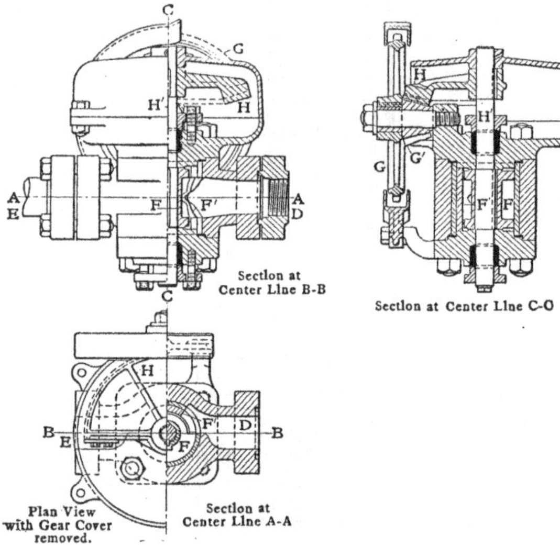

FIG. 183

the valve seat so that a sufficient amount of water may pass through to enable the car to start up at a moderate speed. The operation of this valve will be recognized as identical with that of the automatic stop valve used in the low-pressure vertical elevators described in former chapters. When the car moves away from the landing, the valve F, F' is returned to the central position, in which it is drawn by the action of a centering weight or spring. No device of this kind is shown in the drawings, but the valve is not used without it. The general appearance of this valve can be seen in the photographic views, Figs. 184, 185, 186.

Another design of automatic stop-valve used generally with vertical-cylinder high-pressure machines is shown in Fig. 187. The way in which it is actuated is clearly shown in Fig. 188. The operating chain passes around the sprocket G which is mounted on one end of the shaft I, Fig. 187. On the other end of this shaft there is a pinion that meshes into a gear *J* mounted on the valve shaft. The valve is constructed in the same manner as that of Fig. 183 and operates in the same way. On the shaft *I* is mounted a large wheel *L* and to this is fastened a chain that carries

FIG. 184 FIG. 185 FIG. 186

the centering weight, as shown in Fig. 188. This chain is held in the central position by the guide wheels *L' L'*. It will be noted that this part of the construction is the same as that of the stop-motion valve of the

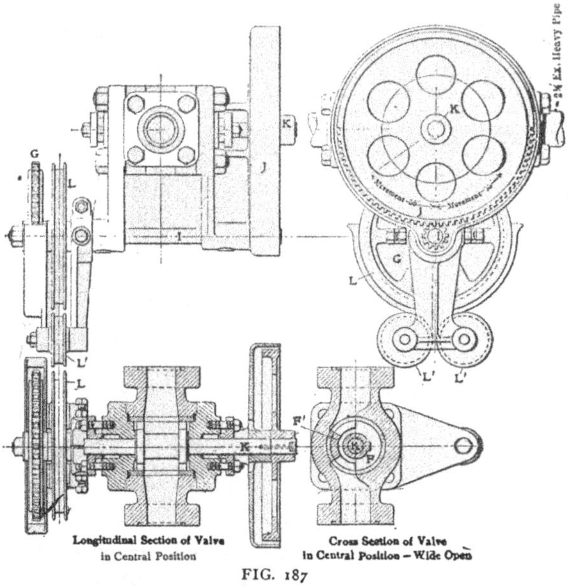

FIG. 187

Whittier horizontal pulling machine. The operating rope carries stop-balls that are struck by the arm projecting from the traveling-sheave frame. This rope runs the entire length of the guides in which the sheave frame travels. At the lower end a sprocket chain is connected with the ends of the rope and passes around sprocket wheel *G* so as to move the valve in exact time with the movement of the sheave frame.

OTHER DEVICES USED WITH HIGH-PRESSURE SYSTEMS.

The air chamber shown at *Q*, in Fig. 164, is constructed in the way shown in Fig. 189. The object of this air chamber is to smooth out any

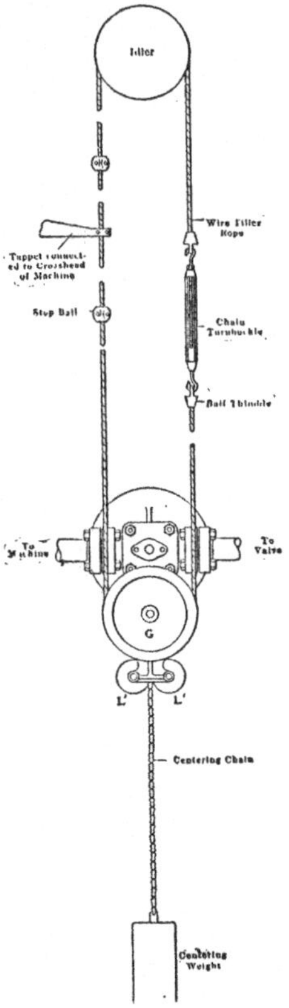

FIG. 188

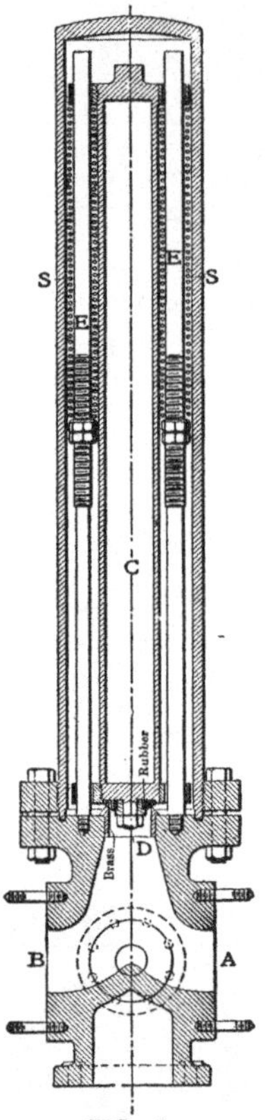

FIG. 189

slight pulsations in the water that may not be taken out by the air chambers attached to the pump. In the low-pressure systems such a device is not required because the pumps deliver into a pressure tank and the water flows from this into the elevator cylinders in an even stream. In high-pressure systems this is not the case; the pumps are continually forcing water into the system and the lifting cylinders are drawing it out, and unless some device is provided that can act like a cushion every pulsation of the pumps that is not subdued by the pump air chamber is sure to be transmitted to the elevator cylinders and thence to the elevator cars.

The difference between the air chamber in Fig. 189 and those commonly used for similar purposes is that it is provided with a check-valve to prevent the air from getting into the pipes. If the lower end of the chamber were not closed by the valve *D* the air could expand enough to force all the water and a part of the air out of the chamber and into the pipe line connecting with *A*, *B*, if for any reason there should be a momentary drop in pressure of, say, 25 or 30 per cent.; and although this is not very likely to occur, it can happen, and would cause trouble, as the air would eventually get into the lifting cylinders and cause the car to bounce when stopped at the floors. The valve *D* is carried on the lower end of a float *C* so that when the water rises in the chamber above a certain point the valve will float off its seat, but if the pressure in the pipe line drops below this point the float will not sustain the valve, so it will settle down on its seat and thus prevent the water and air from being forced out into the pipe line. The level of the water in the chamber is adjusted by so setting the nuts on the rods that hold the adjusting springs that more or less of the weight is supported. If the nuts are run down more weight will have to be lifted by the float and as a result the water level will be raised; in like manner running the nuts up on the rods will lower the water level.

THE SPEED CONTROLLER.

The speed controller shown at *R* in Fig. 187 is constructed as indicated in Fig. 190, which shows a side elevation and a section at right angles to this elevation. The external appearance is shown in the photographic view, Fig. 191. Looking at Fig. 190, it will be seen that the spring *K* acts through the levers *G*, *G* to press against the ends of the valve-stem *A*. These levers *G*, *G* are pivoted on the heads of the valve casing so that their tendency is to keep the valve *B* in the central position. The governor is connected in the pipe between the main valve and the lifting cylinder, and the water in entering or passing out of the cylinder flows through its valve chamber, from the side *C* to the side *D* or in the reverse direction. The water flowing through must pass through the openings *E*, *E* in the valve *B* before it can reach the outlet, no matter in which direction it passes through. When water passes through contracted openings it suffers a considerable loss of pressure so that if it has a pressure of, say, 100 pounds when it reaches the lower side of the valve *B*, coming in through the port *C*, it may not have more than 95 pounds pressure after passing through the openings *E*, *E*.

As the pressure on the forward side of the valve *B* is five pounds more than on the leaving side, there will be this pressure available to move the valve away from the central position against the tension of the spring *K*. This pressure will remain constant even if the pressure of the water that passes through the valve varies widely, because the loss of pressure sustained by the water in passing through the contracted openings *E, E*

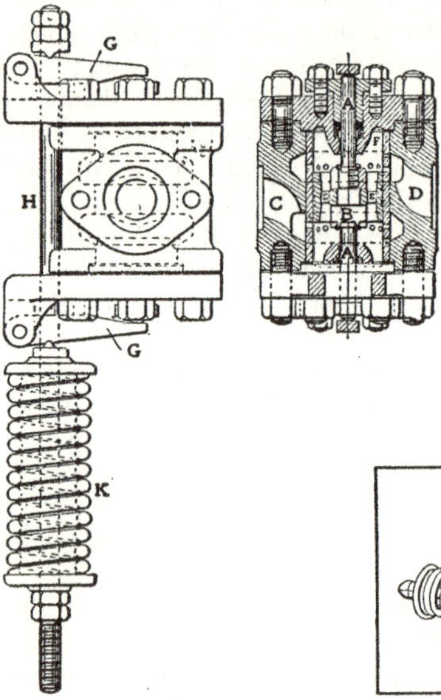

FIG. 190

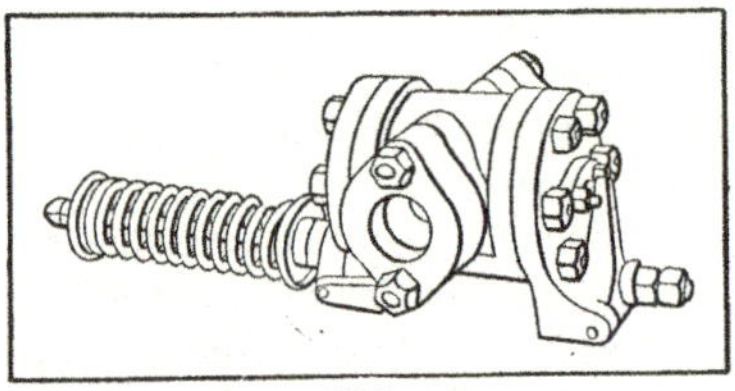

FIG. 191

depends entirely upon the velocity of flow and is not changed by variations in the pressure of the water. This being the case, the extent to which the valve *B* is carried beyond the central position by the difference in the water pressures on its two sides will depend upon the velocity of the stream flowing through the openings *E, E,* or upon the quantity of water that passes through in a unit of time.

The water that passes through the openings *E, E* gets into and passes out of the valve cylinder through a large number of holes that are drilled on spiral lines, and are in such a position that when the valve *B* moves slightly away from the central position it begins to cover these holes, and the farther it moves, the more holes it covers. Consequently, its movement away from the central position closes more and more of the openings through which the water passes out of the valve cylinder, and thereby reduces the quantity of water that flows through; and as the car speed depends upon the rapidity with which the water passes in or out of the lifting cylinder, the velocity cannot vary much from the standard for which the governor is adjusted.

If the car is lightly loaded and starts to run fast, the increased velocity of the water through the openings *E, E* will develop a greater difference in pressure between the two sides of the valve, and as a consequence the valve *B* will be carried farther away from the central position and will close up more outlet holes in the cylinder, reducing the quantity of water

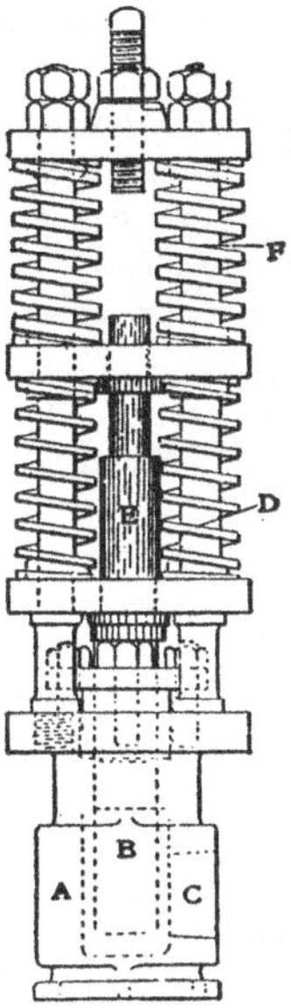

FIG. 192

passing through it and thereby preventing the car speed from increasing. This controller cannot maintain a perfectly constant speed, because it cannot act unless the velocity of the water changes to some extent just as an engine governor cannot act unless the speed changes slightly, but it can be proportioned so as to keep the speed very nearly constant. In practice the maximum speed variation is not usually more than 5 per cent.

As is well known, if when water is flowing through a long pipe at a high velocity its motion is suddenly stopped, the tendency of the stream is to keep on moving, and if there is no space into which it can move, a violent water-ram effect is produced. In high-pressure elevator systems this water-ram effect is much greater than in low-pressure systems, owing to the fact that the water flows through the pipes at a higher velocity. The loss of pressure by the flow of water through the pipes depends upon the velocity, and as in low-pressure systems the pipe loss must be kept much lower than in high-pressure systems, the velocity of flow through the pipes must be much lower. Suppose the system operates with a pressure of 100 pounds, and that the water flowing through the pipes at a certain velocity loses 10 pounds pressure by the time it reaches the cylinder; then if there is another system operating with a pressure of 750 pounds, a loss of 75 pounds in pipe friction can be allowed without giving any lower efficiency. In practice this is not done; the efficiency of the high-pressure systems is made greater, but the actual amount of loss in the pipes is several times as great, this increase being due principally to increasing the velocity of the water through the pipes. It is on this account that while in low-pressure systems the water hammer effect is not serious, in high-pressure systems it is sufficient to require the use of devices to subdue it if the pipes are unusually long. Such a device is shown in Fig. 192 and is simply what may be called a mechanical air chamber. If an ordinary air chamber were placed at the end of a long pipe line it would eliminate the water-ram effect, as the water would rush into it and compress the air, but if the air worked out, as it most likely would in a short time, the chamber would be useless until filled with air again. To arrange such an air chamber so as to replenish the air whenever necessary would involve considerable additional piping, which in addition to the expense would be objectionable as it would afford another way in which air could get into the lifting cylinder. When the device shown in Fig. 192 is connected to the end of a long pipe, the stream of water rushing into it when the flow through the regular channel is suddenly closed causes the plunger *E* to move up out of the cylinder and compress the springs *D,* and the stronger ones above them if necessary. Thus the force of the water ram is expended in compressing springs instead of air, and when the velocity of the water is exhausted the springs force the plunger *E* back into the cylinder, ready to receive the next blow struck by the water.

CHAPTER XXXIII

ADJUSTMENT AND CARE OF AUTOMATIC STOP VALVES AND MECHANISM OF HIGH-PRESSURE ELEVATORS; HOW TO PACK THE DIFFERENT PARTS; KINDS OF PACKING USED

The automatic stop valves of high-pressure elevators and their actuating mechanism require the same attention as those of low-pressure machines, and keeping them in perfect adjustment and running order at all times is just as necessary, because the safety of the elevator depends on them in the same degree. Several designs of automatic stop valve are shown in the description of high-pressure machines and there are several others that were not shown. All those described, with the exception of Fig. 182, are substantially the same as those shown in connection with low-pressure machines. Stop valves of this kind, in almost every case, are operated by means of an actuating rope upon which stop balls are secured at the proper points to bring the car to a state of rest level with the top and bottom floors; and what has been said in previous chapters relative to the care and adjustment of such mechanism applies equally to the stop valves and actuating devices used in high-pressure systems.

The valve in Fig. 182, however, requires some additional consideration owing to the fact that it can be got so far out of adjustment as to cause considerable damage. The illustration shows that the connecting-rods that connect the crank lever *C* with the valve stem and with the rod *L* that runs up to the top stop lever are provided with right- and left-hand screw couplings by means of which their length can be varied. It also shows that the valve *E* is held normally in the central position by the spring *S*, which forces the upper head *F* against the upper end of the spring casing, and the lower head *F'* against the lower end *G'*. This being the case, any change made in the length of the connections, whether with the valve or with the rod *L*, will have the effect of varying the position of the lever *C* and likewise that of the corresponding lever at the upper end of the rod *L*. All this will appear perfectly clear by the aid of the diagram, Fig. 193, where *E'* represents the position in which the valve *E* is normally held, while line *A* represents the normal position of lever *A* and *a*, the position into which it is pushed by the lower roller *R* mounted on the traveling-sheave frame *T*, *b* represents the normal position of the upper *A* lever and *B* and *B'* are the centers around which these levers move.

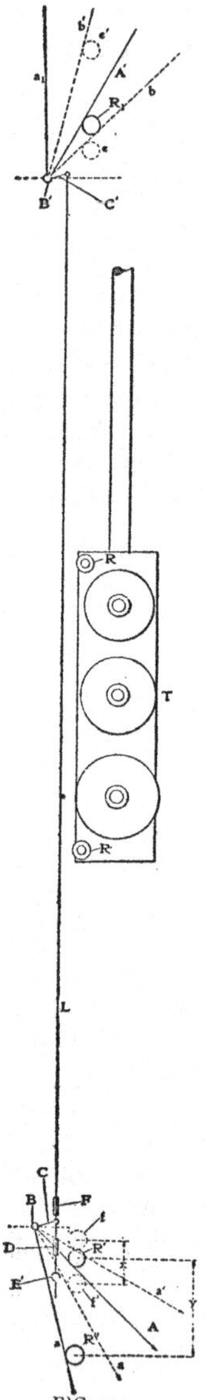

FIG. 193

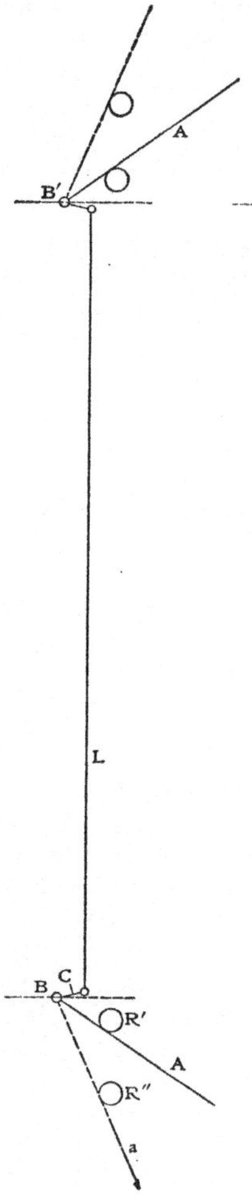

FIG. 194

FIG. 195

The turnbuckles by means of which the connections with the upper end of the valve and the lower end of the rod *L* are varied in length are represented by *D* and *F*. With *D* adjusted so that the lever *C* stands in the horizontal position when the valve is in the central position *E'*, if *L* is of the same length as the distance between *B* and *B'*, the upper lever *C'* will also rest in the horizontal position, the lower lever *A* will be in the position shown, and the upper lever *A* in the position of the line *b*. If now the traveling sheave comes down, the stop roller *R* will strike *A* and depress it to the position *a* to shift the stop valve to the stop position. If the traveling sheave runs upward, the upper stop roller *R* will strike the lever *A* in the position *b* and carry it to the position *b'* to move the valve to the upper stop position. Now suppose the turnbuckle *D* is lengthened; then since the valve *E* will be held in the position *E'* by the tension of the spring *S* (see Fig. 182), the lever *C* will be moved upward, and if rod *L* is not shortened, the upper lever *C'* will be forced up above the horizontal position, and so will the lever *A'*. If, with this change in the adjustment, the traveling sheave runs down the stop roller *R* will strike the lever *A* when it reaches the position *f*, because the lever *A* will now rest in the position of broken line *a'* and when the roller has reached the position *f* the valve will be moved to the stop position, thus stopping the movement of the traveling sheave within the distance *x*; while before the turnbuckle *D* was lengthened the roller *R* traveled from *R'* to *R''* to move the valve to the stop position, equal to the distance *y*.

Looking at the upper end of the diagram it is clear that, as the lever *A'* has been shifted to the position in which it is drawn, the upper stop roller *R* on the traveling sheave will have to reach the position *b'* before it begins to move the lever *A'*, and as the lever will have to move to the position A_1 to close the valve, the latter will not be closed, and as a result the traveling sheave will continue moving until stopped by some other means. From all this it can be seen that unless the turnbuckle *D* is lengthened very little, the upper lever *A* will be rotated so far around that the stop roller on the traveling sheave will not be able to move it far enough to close the stop valve *E*. If, however, the connection *F* is reduced in length so as to cause the lever *C'* to rest in the horizontal position normally, then the traveling-sheave stop roller *R* will be able to close the valve fully because the lever *A'* will have to be moved from the position *b* to the position *b'*.

By adjusting the lengths of the connections *D* and *F*, the distance the car will travel after the stop valve begins to move can be varied. Suppose *D* is made longer, then the lever *C* will be pushed upward, as shown in Fig. 194, and if *F* is shortened twice as much as *D* is lengthened, the upper lever *C'* wil rest below the horizontal line, so that the upper and the lower levers *A* will stand at the same angles above and below the horizontal; consequently the distance the traveling sheave will move at each end of its travel before the car stops will be the same, being equal to the distance between the circles *R′* and *R″*, which is less than the distance *y* in Fig. 193, because *D* has been made longer. Suppose now that *D* is made shorter; then the lever *C* will be drawn as shown in Fig. 195, and to cause the upper lever *A* to rest at the same angle above the horizontal line that the lower lever *A* does below this line, *F* will have to be lengthened out just twice as much as *D* was shortened. With this change in the adjustment, the traveling sheave will have to move the distance indicated by circles *R* and *R′* to shift the valve *E* to the stop position, and this distance is more than *y* in Fig. 193.

It will be noticed, however, that in Fig. 194 the valve will be closed sooner than in Fig. 195, this being clearly indicated by the position of the circle *R″* at the lower end of both diagrams. This simply means that if the change indicated in Fig. 194 is made the car will stop short of the floor at the top and bottom, and if the change indicated in Fig. 195 is made the car will run beyond the floor at both landings. From this it follows that if the car runs too far at one or both landings, it can be made to stop even with the floor by making *D* longer, being careful to shorten *F* the proper amount; and if it stops short of the proper point the defect can be remedied by making *D* shorter and *F* longer. If it is desired to effect the stop in a shorter distance, and still have the car stop even with the floor, the only way to do it will be by shifting the whole valve so that the distance between centers and *B* and *B′* may be increased in the proper amount. If it is desired to make the stop slower, the valve will have to be shifted so as to reduce the distance between *B* and *B′*.

DANGEROUS TO CHANGE ADJUSTMENTS.

It should be remembered, however, that unless one is thoroughly familiar with the operation of every part of an elevator system it is very dangerous to make changes in the adjustment, unless it is known that some part has changed from its original position and requires being restored to this position. To illustrate the importance of this point, take the matter of changing the distance in which the car is stopped, which can be accomplished by adjusting *D* and *F* in the manner just described. By this method the stopping distance can be changed, but if it is made too short, the passengers may get a bad shaking up, while if it is made too long there will be an unnecessary loss of time. The engineers of the elevator builders know as much as anyone about the proper adjustment of the stop-motion apparatus, if not a great deal more; and, generally, nothing can be gained but probably a good deal lost by trying to improve on their work. The knowledge of how to make these adjustments should only be utilized in restoring the parts to their original position if for any reason they become displaced.

While the stop-motion valve of Fig. 182 was shown only in connection with a vertical machine, it can be used just as well with the horizontal cylinder, and in like manner the stop valve shown with the latter type of machine can be used with the vertical cylinder. The adjustment of the latter type of valve is effected in the same way as that of the Whittier horizontal machine; that is, by changing the position of the stops that are moved by the operating arm attached to the traveling-sheave crosshead. If it is desired to have the car stop higher up at the top floor, the stop *E E* at the end of the machine is moved forward, the distance moved being equal to the additional travel required, divided by the gear of the machine. To cause the car to run farther down at the bottom floor, the stop *E E* nearest to the cylinder is shifted back toward the cylinder through a distance equal to the required increase in travel divided by the gear of the machine.

There is one difference between stop valves of the type shown in Fig. 165 and that of Fig. 182, and that is that with the former it is only possible to vary the point at which the elevator car will stop, while with the latter not only can the stopping point be varied, but also the distance in which the car will stop. With valves like that of Fig. 165 the stopping distance can be changed only by changing the form of the edges of the valve, or of the ports covered by the valve, so as to vary the angular distance through which the valve must be moved to stop the flow of water. While this might be regarded as a point in favor of the valve of Fig. 182, it is very doubtful if it is, because if the stopping distance is once made right it will always be right; hence, there is no need of providing means for adjusting it.

The main valve used with high-pressure machines is the same for either the vertical or horizontal type. In almost every case the pilot-valve design is used, but occasionally a simple valve for hand-rope or hand-wheel operation is installed. So far as the main valve is concerned, there is no difference between the two types, as can be clearly seen by reference to Figs. 174 and 172, the first being the simple hand-rope operated valve, and the second the more elaborate pilot-valve arrangement.

THE PACKINGS.

All the packings of the high-pressure valve are leather cups, and they are removed or inserted by removing the head and the cylinder linings.

To remove these cup packings it is necessary to draw out the valve, and this can be easily done by taking off the lower arm and freeing the valve from the upper arm. The lower head is then removed and the parts *M* and *N*, Fig. 174, are drawn down. The stuffing-boxes of the pilot valve and of the motor piston rod can be packed without removing any parts except the glands. To take out the piston it is easier to remove the lower cylinder-head and piston-rod guide than the upper ones, as in so doing it is not necessary to disturb the levers or the rods. The pilot valve can be easily removed by taking off the top cap, the lever *L* and the rod *M*. Whenever any work of this kind is done it is necessary, of course, first to close the hand valve in the supply pipe and then drain the water out of the cylinder and pipes. After the work is completed the supply-pipe valve is opened and the air is withdrawn from the cylinder and piping. To get all the air out it may be necessary to make a few trips with the car.

The lifting plunger is packed by means of a stuffing-box at the end of the cylinder, the construction of which for both the vertical and the horizontal machines was shown in the illustrations presented in previous chapters on high-pressure machines. Any kind of packing can be used in these boxes, but plain hemp with tallow works as well as can be desired, provided it is properly put in. As the pressure used in the cylinder is high, the packing must be put in the box evenly in order that there may be no soft spots, for if there are, the water will find them and force its way through, although the other parts may be pressed up hard. It is necessary to force the stuffing-box gland up tight to make a joint against the pressure used (750 pounds), but it is always desirable to compress the packing as little as possible in order to reduce the friction, and the extent to which the packing must be compressed to make it tight depends in a large measure upon the evenness with which it is placed in the stuffing-box. If it is crowded in tight on one side and is mushy on the other, no amount of tightening will keep it from leaking.

The cylinder may be packed with the car at the top of the building or at the lower floor. If it is at the upper floor, it must be secured to the overhead beams before the water is withdrawn. If it is at the bottom floor, and the traveling sheave comes so near to the stuffing-box as to interfere with free working, the car should be stopped a few feet above the floor and a temporary support put under it. The distance above the floor at which the car should be held can be determined by multiplying the gear of the machine by the additional room required to get at the stuffing-box easily.

An accumulator is packed at the upper end, by means of a stuffing-box, the same as the lifting cylinder. To do the packing the water is drawn out of the cylinder and the hand valve in the inlet pipe is closed. When the cylinder is empty the upper end projects above the accumulator weights, as shown in Fig. 177, and the stuffing-box can be easily reached. Just as much care is required in packing the accumulator as in packing the lifting cylinder, because they are both subjected to the same pressure. The stuffing-boxes of high-pressure machines are made deeper than those with low-pressure apparatus, but even with this extra depth they cannot be made tight and maintained tight unless the packing is done with care.

CHAPTER XXXIV

STOP VALVES USED WITH ACCUMULATORS, GENERAL ARRANGEMENT OF APPARATUS IN HIGH-PRESSURE SYSTEMS; THE PUMPS AND THEIR OPERATION

The automatic stop valves used with accumulators are practically the same in action as those used with the lifting cylinders, and require no special mention here. The lifting cylinder stop valves are in most cases made so that when they close the inlet, they also nearly close the outlet, this construction being used so that the elevator car may start up gradually on the return trip, and by the movement of the plunger through the first few inches of its travel the valve is opened to its full extent. The accumulator valves differ from this construction in being arranged so that when the inlet is closed, the outlet is left open, and in like manner when the outlet is closed, the inlet is left wide open. The stops on the actuating chain of the accumulator should be set so as to close the inlet valve before the plunger rises high enough to strike the shoulder at the upper end of the cylinder, and the outlet valve should be closed before the weights strike the lower buffer block, but in both cases the stop should be effected as near to the mark as possible, in order to utilize as much of the accumulator capacity as practicable.

If the position of the stop balls is changed, the pump stop must also be changed; otherwise, the pump will not act in time with the stop valve. The adjustment should be so made that the pump will stop just before the upper stop closes the inlet valve, and should start up again just after the inlet valve has returned to the central position. If the pump stops after the inlet valve is closed, there will be a momentary rise in pressure in the system, and if it starts before the inlet opens there will be another momentary rise, and this should be avoided because it will cause irregularity in the motion of the elevator cars. The adjustment of the pump stop is effected by changing the length of the rope that connects the pump valve with the weight lifted by the projecting shelf on the accumulator.

The auxiliary air chamber which was shown in Fig. 189 requires additional explanation in order that the object of its peculiar construction may be clearly understood. The lower end of this air chamber is closed by a valve attached to the lower end of a copper float. The object of this valve is to prevent any of the air in the chamber from escaping into the pipe and thus finding its way into the elevator-lifting cylinders. As the water pressure is so high, the air chamber has to be charged with compressed air, the pressure of which is about one-quarter of the water pressure, so that when compressed to the latter pressure it fills about one-quarter of the upper end of the air chamber. If the chamber were filled with air at atmospheric pressure this would be compressed into a space at the upper end of the chamber so small as to have little effect as a cushion.

As the chamber is filled normally to about one-quarter of its volume with compressed air, it would be possible for some of this to escape through the lower opening if there should be a sudden and great drop of pressure in the pipes, which is possible if there is an instantaneous increase in the draft on the pipe system or a corresponding decrease in the supply. If, however, the lower end of the chamber is provided with a check valve, the air cannot escape because the outflow of water will close the valve. It might be asked if this is the case why not use a simple check valve and not one attached to a float? The answer is that the simple check valve will not work. For example, suppose a simple check valve is used to close the outlet of the air chamber; then if there should be a sudden rise in pressure in the pipes, which could occur if several elevators were stopped at the same time, this pressure would lift the check valve and force water into the air chamber until the pressure became equal to that in the pipe, which might be, say, 800 pounds. From this time on the air chamber would be useless because the pressure in it would be greater than that in the pipe and the check valve, therefore, would remain closed; the cushioning effect of the air would then be lost. When the valve is attached to the lower end of a float the action is very different because when the level of the water in the air chamber rises above a certain point the float lifts the valve off the seat. The level at which the water should lift the valve is that which corresponds to a pressure considerably below the normal, so that except when the pressure drops very greatly the valve is open and the water in the air chamber can pass freely into the pipe. If, however, the pressure drops low enough, the level of the water in the air chamber will drop so low that the float will not be able to hold up the valve and then it will come down on the seat and prevent any further escape of water, and consequently there is no escape of air.

The valve seat is a brass ring and the valve itself is made of hard rubber. It might be supposed that hard rubber would not stand very well in a system using such high pressure, but it does, owing to the fact that the pressure to which the valve is really subjected is simply the difference between that in the pipe and that in the air chamber, which is comparatively small. Even if the valve should wear so as not to make a perfectly tight joint it would not matter, because the water could not escape fast enough through the leak to draw air with it.

GENERAL ARRANGEMENT.

The general arrangement of all the apparatus used in a first-class high-pressure elevator installation can be explained by the aid of Fig.

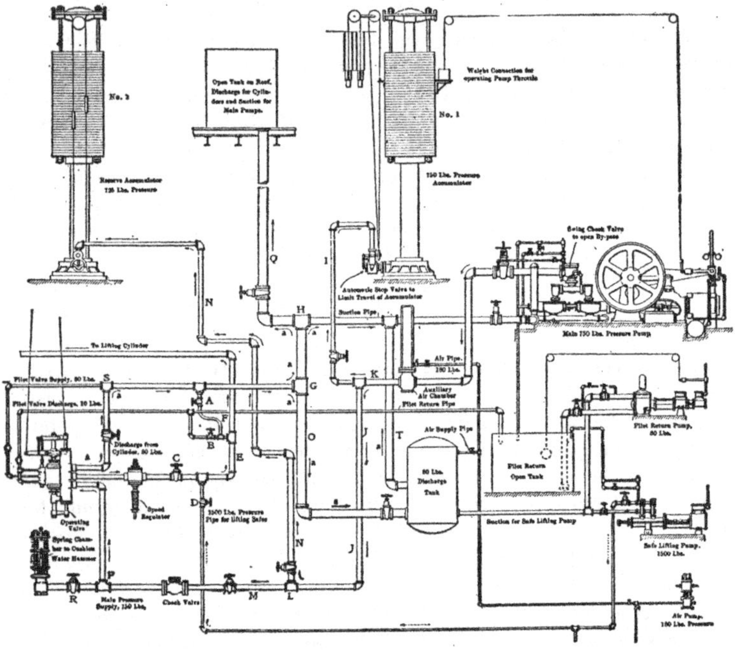

FIG. 196

196 and the simplified diagrams, Figs. 197, 198 and 199. Fig. 196 shows all the apparatus and the way in which the piping system is arranged, but is necessarily not drawn to scale, and in some cases the pipe lines are drawn in less direct paths than they would follow in actual practice, in order to avoid confusion by the frequent crossing of lines. Only one elevator is provided for in the pipe connections, but any number can be used up to the limit that the pump is able to supply. It will be noticed that there are two accumulators, No. 1 being arranged to stop the pump when it is full. No. 2 accumulator is provided with a stop valve only, and it is weighted so as to be lifted with a lower pressure

than No. 1 in order that it may act as a reservoir. Normally, this accumulator is full of water, but if at any time the drain on the system is more than the pump can supply, it delivers its contents to the system or as much of it as may be necessary to make up the deficiency. When the demand is reduced below the capacity of the pump, accumulator No. 2 is filled again, so as to be ready to help the pump when the next heavy drain comes. This reserve accumulator is connected as near to the elevator cylinders as the construction of the building will permit, in order to reduce the loss of energy due to forcing the water through the pipes. The pipe *N*, as shown in the drawing is not very short or straight, but in practice it is made as short and straight as practicable.

THE PUMPS.

The pumps are generally made so that when they stop pumping to the system they do not actually stop moving; they continue to run, but the water circulates through a by-pass and the steam valve is opened just enough to keep the piston moving very slowly. This arrangement is used in order that when the pump is called upon it may get under full headway quickly and thereby avoid the possibility of a drop in pressure.

The course of the water from the pump is indicated by arrows drawn in full lines; it passes first through the auxiliary air chamber (the one with the float-supported valve) and on reaching the junction *K* can run up the pipe *I* to the accumulator, or down the pipe *J* to the junction *L* whence it can flow either through *N* to the No. 2 accumulator or through *M* to the main valve of the lifting cylinder. If there are several elevator cylinders they can be connected with the supply pipe *J*. Examination of this piping will show that if accumulator No. 1 is not full, water may pass into it and also flow on through the pipe *J* to the pipes *N* and *M*; through the first of these it flows to accumulator No. 2 and through the other to the operating valve of the power cylinder. If the demand is greater than the pump can supply, water will flow out of accumulator No. 1 to make up the deficiency, but water will not flow out of accumulator No. 2 unless the drain is so great that accumulator No. 1, together with the pump, cannot keep the pressure up above that for which No. 2 is adjusted; in this case the latter accumulator will also discharge into the system. After the water passes through the operating valve it flows through the speed regulator and thence into the pipe *E* and to the lifting cylinder (not shown). Returning from the lifting cylinder it comes through the same pipe *E* and the speed governor to the operating valve, and through the latter to the discharge pipe, following the path of arrows *a*. At *G* it can flow up into the open tank on the roof through pipe *Q*, or into the pump-suction pipe, or it can flow down through the pipe *O* into the discharge pressure tank, and thence through the pipe *T* to the pump suction.

The object of providing the discharge pressure tank is to cause the discharge to flow out so freely as not to give the car a jerky motion. If this tank were not provided the discharge would have to pass to the junction *H* and then divide between the pump suction and the roof tank, going to the latter through the pipe *Q*. The latter path would offer considerable resistance, as the tank would be about 175 feet high, and a column of water of this hight cannot be forced upward suddenly. The discharge pressure tank is located in the basement, on the same level as the discharge pipe or possibly lower, and not far from the elevator cylinders; consequently, the water can flow into it with comparatively little resistance. The actual result of this arrangement is that the pressure discharge tank receives nearly all the water and the roof tank serves simply to keep the pressure constant. The pipe *T* is shown as tapping the tank rather high up in the side, but in practice it enters near the bottom so as not to draw air into the pumps.

From the tee *S* in the discharge pipe a connection runs to the pilot valve to actuate the motor piston that moves the main valve. The discharge from the pilot valve runs to the open tank shown to the left of the pressure tank, and from this tank it is pumped by the small pilot-return pump into the suction of a small extra-high-pressure pump that is used to lift safes and other heavy weights.

At the end of pipe *M* is placed a spring-cushion chamber to receive and subdue the impact of the stream of water when the operating valve is closed. This spring-cushioning device was described in connection with Fig. 192, and in the present diagram the way in which it acts is very clearly shown. When the flow of water through the operating valve is stopped, the current in the pipe *M* will send the water forward into the cushioning chamber, where its momentum will be absorbed in pushing the plunger out and compressing the springs. When the water comes to a state of rest, the springs force the plunger back into the cylinder, ready for the next blow.

The air pump shown is not kept in operation all the time; it is used only when it is necessary to replenish air in the pressure tank, or in the air chambers. When the air in the tank is not sufficient the pump is set in motion and then the valve in the air pipe is opened and air is allowed to flow into the tank until the water gage shows the level to be at the proper point.

ELEVATORS FOR HEAVY SERVICE.

In all large buildings, where there are several elevators, one of them is arranged so as to lift heavy weights, such as safes or large machines.

This elevator is generally capable of lifting about three times the load of any of the other elevators, and as it is only required to do this work occasionally, simple means are provided to accomplish the result. The means consist in providing a small pump that can develop a pressure sufficient to lift the required load and this is connected with the piping of the elevator in such a way that by manipulating hand valves the car may be caused to run either up or down. The arrangement shown in Fig. 196 is actuated as follows: If it is desired to lift the load, the valves *A, B* and *C* are closed and the valve *D* is opened, then the safe-lifting pump is started and the water follows the path of the dotted arrows to the lifting-cylinder pipe and to the cylinder, and the car moves upward. The velocity of the car will be very slow. If it is desired to run the car down, the valve *D* is closed and the valve *A* is opened; then the water

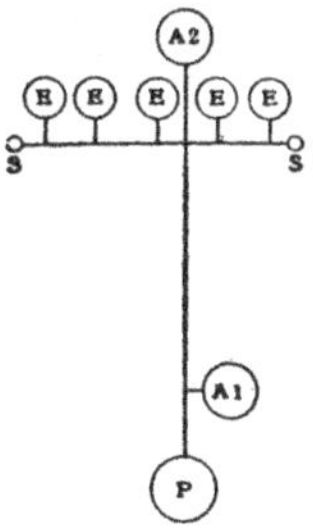

FIG. 197

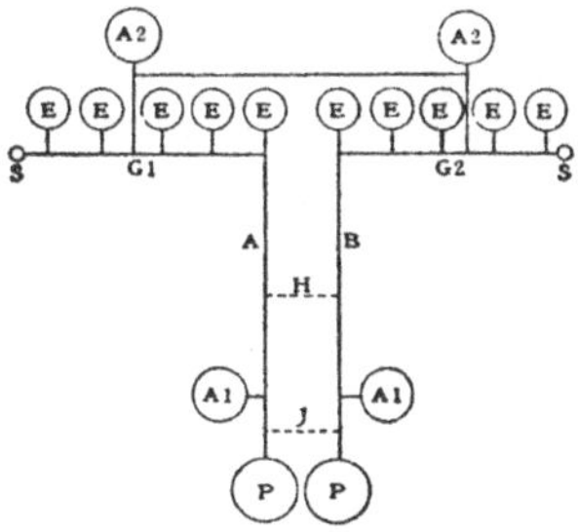

FIG. 198

from the lifting-cylinder pipe will escape into the discharge pipe and through it to the pressure tank. The car can also be run down by opening the valve *B* with *A* either open or closed; the speed will be greater if *B* is opened. It is not actually necessary to close the valve *D* in coming down. From the foregoing it will be noticed that when the safe pump is used to lift an extra heavy load, the main operating valve is not used; in fact it is cut out of service by the closing of the valve *C*, and the movement of the car is controlled entirely by the hand valves *A, B* and *D*.

While the diagram, Fig. 196, shows very well the general arrangement of the high-pressure system with only one elevator, it leaves considerable to the imagination when one considers an extensive plant containing a large number of elevators and a corresponding number of pumps and accumulators. To make the arrangement of such an installation clear, Figs. 197, 198 and 199 are given. Suppose that a building is equipped with twenty elevators, and assume that there are four main pumps and eight accumulators. Such a system can be arranged in many different ways; it can be connected all in one general piping system, or it may be divided into two, four or even more separate systems. Suppose it is divided into four separate systems, then each one would consist of five elevators, two accumulators and one pump, and these could be connected in the manner shown in Fig. 197, in which the circles *E* represent the elevator cylinders, *P* the pump, *A1* the accumulator that stops and starts the pump, and *A2* the reserve accumulator shown in Fig. 196. At each end of the pipe line that connects the elevator cylinders is shown a spring cushion *S* to subdue the water-hammer effects. Four independent systems like this could also be made so as to be connected with each other or disconnected at will. In Fig. 198 is shown an arrangement by means of which two of these units could be connected into one, by simply providing the pipe connections *H* and *J*. With the

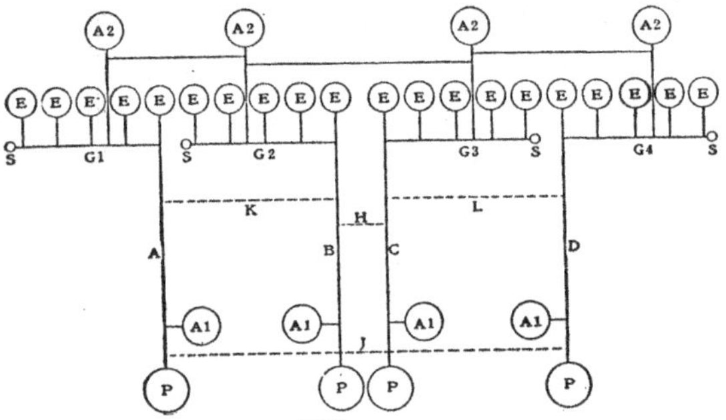

FIG. 199

first pipe connection either one of the pumps and its accumulator *A1* could be used to supply the entire ten elevators on an emergency, and by making the connection *J*. Both the *A1* accumulators could be put in service, which would be the best arrangement, as when the pump capacity is small the accumulator capacity should be large.

Fig. 199 shows how the four units can be arranged to be independent if desired, or to be connected two and two, or three in one system and one independent, or all four in one system. The pipe connection *H* will join the two center sections, and *K* and *L* will connect the side sections independent of each other if *H* is disconnected. Many combinations can be made by means of proper connections between the pipe lines *A, B, C* and *D*. This is also true of the connections between the *A2* accumulators. This way of arranging the piping is advantageous not because any economy can be gained by cutting up the plant into several sections (in fact this is not advisable as a rule), but because if there should be a breakdown in any part, the other parts can be made to take its place while repairs are being made. For example, if one of the pumps gives out, the pipes leading to it can be closed, and the elevators it supplies directly can be connected with the remaining pumps.

There are probably no two elevator installations of large size that are arranged exactly alike, and although the differences between them are due to a considerable extent to the notions of the architects of the building, the owners, the engineers and others interested, not a small portion of it is due to the difference in the conditions controlling the case, hence, an arrangement of an installation precisely as shown in Fig. 199 might be hard to find.

In these last three diagrams only the supply pipes have been shown, in order to avoid confusion and to present clearly the way in which the pumps, the elevator cylinders and the accumulators are connected with each other to obtain the best results.

CHAPTER XXXV

PLUNGER ELEVATORS

CONSTRUCTION OF PASSENGER ELEVATORS OF THIS TYPE; DETAILS OF THE CYLINDER AND VALVES; HOW WEIGHT OF PLUNGER IS COUNTERBALANCED

What is commonly known as a plunger elevator is a direct-acting machine in which the lifting plunger is located under the center of the elevator car and attached directly to it. The plunger works in a cylinder sunk into the earth to a depth somewhat greater than the rise of the elevator. When water is forced into the cylinder, the plunger rises and pushes the car upward. To make the downward trip the water is permitted to escape gradually from the cylinder, allowing the plunger to descend under the influence of gravity.

For many years this type of elevator has been used for short travels, but within the last ten or twelve years it has gradually come into use for higher rises, and at the present time is used in buildings where the rise is as great as 300 feet. As the elevator car can rise only as high as the plunger travels, it follows that when the rise is 300 feet, the cylinder must extend down into the earth several feet more than 300, because when the car is at the top of the elevator well the bottom end of the plunger must be some distance below the top end of the cylinder. Furthermore, it is necessary to provide sufficient length of plunger to carry the car a short distance above the upper floor, say, two feet, in order to avoid running the bottom of the plunger too high up in the cylinder if the elevator should overrun the upper limit of travel.

As a plunger elevator is a direct-acting machine, the diameter of the plunger need not be large, unless very low pressure is used. As a rule the plungers are made of 6-inch steel piping, which, when finished, has an external diameter of 6½ inches. The cylinder is also made of steel pipes, and of a diameter generally about two inches greater than that of the plunger. This diameter may be reduced for elevators of small rise and slow speed, but for high lifts and high speeds it is greater, as it is necessary to provide sufficient space between the plunger and the cylinder to permit the water to pass down when the plunger ascends without absorbing too much power by its friction against the walls of the cylinder and the plunger.

FIG. 200
OTIS PLUNGER PASSENGER ELEVATOR

The plunger passes through a stuffing box at the upper end of the cylinder, and is provided with guide shoes at the lower end to keep it in line and central. The well in the ground in which the cylinder is set is made about three or four inches larger than the outside of the cylinder. Any portion of it that passes through gravel or clay is lined with a retaining casing, but portions that run through rock are not lined. The space between the cylinder and the sides of the well is filled in with sand.

CONSTRUCTION OF THE PASSENGER ELEVATOR.

The construction and general arrangement of a first-class passenger elevator of the plunger type are indicated in Fig. 200, which is an elevation of the Otis machine. This illustration is broken at a point between the elevator car and the bottom of the elevator shaft in order to reduce its length, but the part broken away would only show the continuation of the guides, plunger, operating ropes, etc.; all the operating parts of the outfit are shown in the illustration. The car rests upon the upper end of the plunger *P*, and the latter runs down into the cylinder *C*, the upper end of which projects above the ground floor. From the top of the car a number of ropes *R* extend upward and over a sheave *S* and thence down to a counterbalance *W*. This counterbalance serves to reduce the pressure required to raise the elevator, and also to reduce the compression stress to which the plunger is subjected.

The pipe of which the plunger is made weighs about 22 pounds per foot, so that a plunger 200 feet long will weight about 4400 pounds; this is more than the car is likely to weigh, the latter ranging between 3000 and 4000 pounds. If the car weighs, say, 3600 pounds, and the plunger 4400 pounds, the two combined will weigh 8000 pounds, and with no counterbalance this weight would have to be raised in addition to the load. Consequently the plunger would be subjected to a compression stress of 3600 pounds plus the load at the upper end, and 8000 pounds plus the load at the bottom, the stress increasing from top downward at the rate of 22 pounds per foot. With a counterbalance weighing, say, 5000 pounds, the weight raised will be reduced to 3000 pounds plus the load, and as the counterbalance exceeds the weight of the car by 1400 pounds, it will actually hold up about one-third of the plunger, from the upper end downward, when the car is empty.

When the car is at the bottom of the shaft the plunger is immersed in the water in the cylinder, consequently a portion of its weight is balanced by the water it displaces. When the car is at the top of the shaft the plunger is out in the air and its weight is not counterbalanced to any extent by the water. This being the case, the weight lifted will be less when the car is at the bottom of its travel than when at the top,

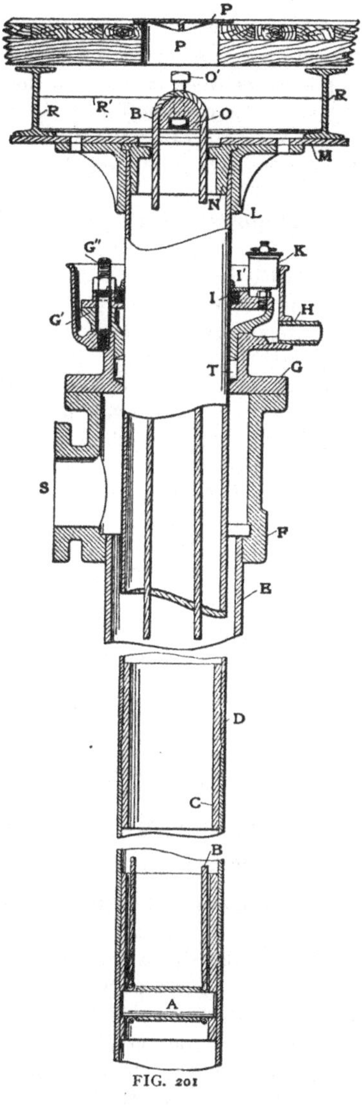

FIG. 201

the difference being equal to the weight of water displaced by the plunger. By properly proportioning the weight of ropes *R*, the load lifted can be made equal at all points, for when the car is at the bottom of the shaft these ropes will hang above the car, and thus will offset a portion of the counterbalance *W*, while when the car is at the top of the shaft the ropes will hang above the counterbalance *W* and balance a portion of the weight of the car.

The main valve for controlling the movement of the car is shown at *V*, and the pilot valve at *V'*. The two valves *A* and *B* are the automatic stop or limit valves, *A* being the top limit and *B* the bottom. The valve *A* is actuated by the rope *A"* which pulls up the lever *A'* and thereby closes the valve. This rope moves the lever *A'* through the motion of the elevator car. Looking at the illustration, it will be seen that rope *A"* runs over a sheave *D* mounted on top of the elevator car, and it can also be seen that when the car approaches the upper limit of travel, *D* begins to put a bend in *A"* and thereby draws up lever *A'*; by the time the car reaches the upper floor, *A'* will be raised enough to close valve *A*. By this arrangement the valve is closed gradually and the car is as gradually brought to a state of rest.

The valve *B* is actuated by rope *B"* in precisely the same manner that *A* is operated by the rope *A"*. The rope *B"* passes over the stationary sheave *D'* and under the sheave *D"* located under the car, and when the latter descends near enough to the lower floor, the bend put in rope *B"* by the sheave *D"* will raise the lever *B'* and gradually close the valve *B*.

The pressure water enters through valve *A*; hence, at the top landing the automatic stop arrests the movement of the car by shutting off the supply water. When the elevator car descends, the discharge water passes out through valve *B*; hence, the bottom limit valve stops the descent of the car by stopping the escape of water from the cylinder.

CONSTRUCTION OF CYLINDER.

The construction of the upper end of the cylinder is shown in Fig. 201. This drawing, which is a vertical sectional elevation of the top of the cylinder and plunger, also shows the way in which the plunger is fastened to the under side of the car, as well as the construction of the plunger. For the purpose of reinforcing the plunger, a steel cable *B* is strung inside, both of its ends fastened to a pin *A*, located some distance below the center of the plunger, and the loop or bight, at the top of the plunger, is passed around a tightening block *O*; this block is arranged so as to be drawn up by the bolts *O'* to put the desired tension on the rope *B*. The plunger D is made of as many lengths of piping of the proper size as may be necessary, these being connected by means of long internal sleeves *C*. The plunger sections are turned true and highly polished, and the screw threads at the ends are made with great accuracy, so as to hold the sections in perfect alinement when connected. The threads are also made extra long, so that the joints may be as strong as the other parts of the pipe. For the purpose of making the pipe sections come

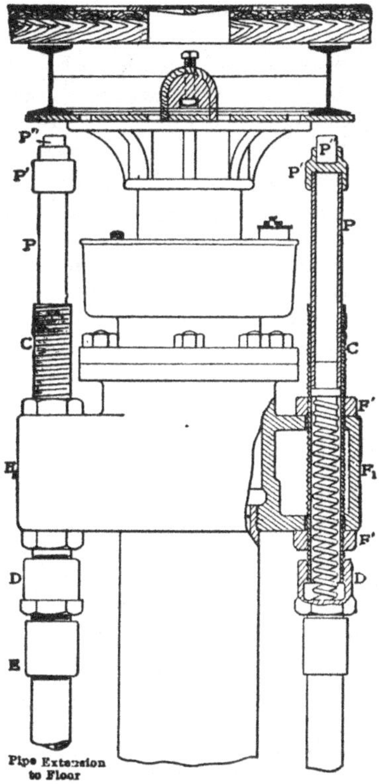

FIG. 202

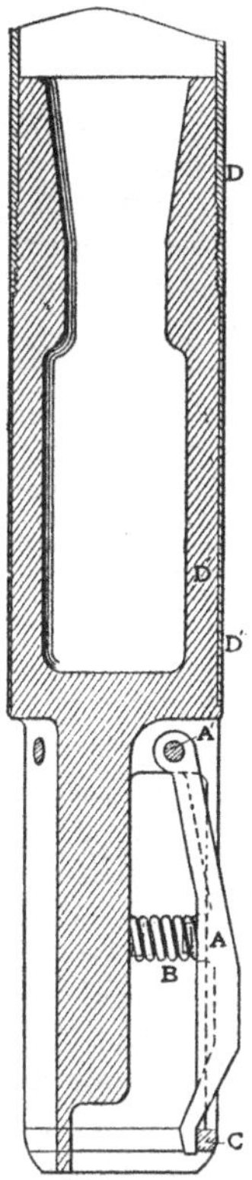

FIG. 203

together perfectly central when joined, the center portion of the sleeve is turned true, and the ends of the pipe are bored to fit this portion; when the parts are screwed up, the turned central portion of the sleeve slides into the bored-out ends of the pipes and brings them into line, so that there is no point around the joint where one part projects over the other.

The top of the cylinder is finished off with a casting *F* screwed to the top of the upper section of the cylinder barrel *E*. On top of the cylinder cap *F* is mounted a stuffing-box casting *G*, containing the usual packing space *T* and fitted with a compressing gland *G'*; the latter is constructed so as to form a space surrounding the plunger to hold oil, the latter being fed in from an oil cup *K*. Above this oil reservoir is a recess in which babbitt wiping rings *I* are placed for the purpose of scraping the oil off the plunger as it moves upward, and retaining it in the space in the gland *G'*.

In Fig. 200 it will be noticed that buffers *F* are provided for the car to rest upon when at the lower floor. Similar buffers are also provided for the counterbalance *W* to rest on, to prevent running the car up against the overhead beams. The construction of the car buffers is shown in Fig. 202, which is an external view of the upper end of the cylinder taken at right angles to Fig. 201. The buffer consists of a plunger *P* made of pipe, provided with a cast cap *P'* and a rubber cushion *P''*. The plunger *P* slides within a cylinder *C*, also made of pipe. Within this cylinder there is a spring that is compressed by the plunger, the lower end of the latter being provided with a flat head to press against the top of the spring. The cylinder *C* is held in position by a side extension *F*, forming on the top cylinder casting *F*. The nuts *F' F''* are screwed on the cylinder *C*, the latter being threaded, and by this means the hight of the buffer is adjusted. To furnish additional support, so that the buffer may not be pushed down and the thread of the nut *F'* stripped if the car should come down unusually hard, a pipe extension *E* is provided, extending down to the floor, or some other firm support. These buffers are set so as to be struck and compressed every time the car comes down to the lower floor, acting to stop the motion gradually. If the car descends at the normal speed, the buffer is compressed slightly, just a trifle more than is necessary to hold the unbalanced portion of the weight of the car, but if the car speed in approaching the floor is excessive, the buffers will be compressed farther, and the car will run a few inches below the floor.

PLUNGER CONSTRUCTION.

The construction of the lower end of the plunger is shown in Fig. 203. The main portion consists of a casting *D'* to the upper end of which is screwed the lower end of the plunger. This casting does not run up to the oil well in the stuffing box with normal operation of the elevator; therefore, its surface, if of iron, would soon become rusty. On this account it is covered with a brass casing *D''*. At the lower end of D' spaces are cored out to receive the guide shoes *A*, pivoted at the upper ends at *A'*, and held from moving too far outward by a ring *C* at the

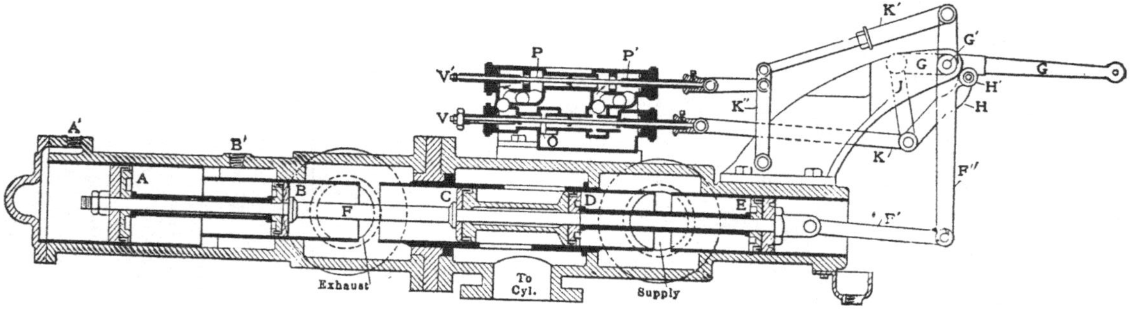

FIG. 204

extreme lower end of the casting D'. Provision is made for either three or four shoes like A. The springs B press the shoes out against the cylinder and hold the lower end of the plunger central.

The main operating valve used with the Otis plunger elevator for high-grade passenger service is made in several designs. The valve shown in Fig. 200 is constructed as shown in Fig. 204, the pilot valve part being more clearly illustrated in Figs. 205 and 206, drawn to larger scale. The pipe I shown in Fig. 200 connects the inlet P of the pilot-valve chamber and the inlet B^2 of main valve with the supply pipe, through the limit valve A. In like manner the pipe J connects the outlet

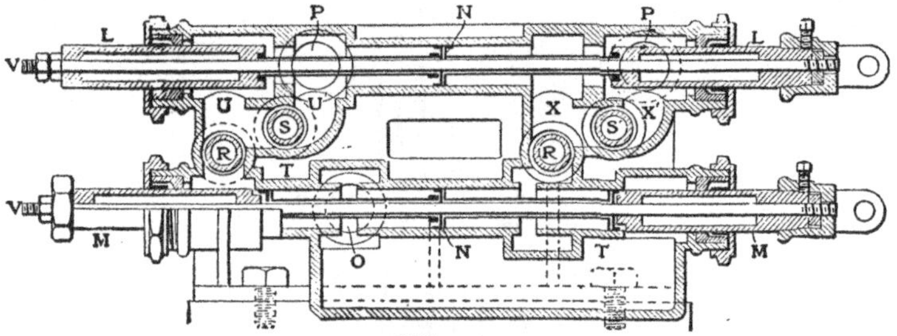

FIG. 205

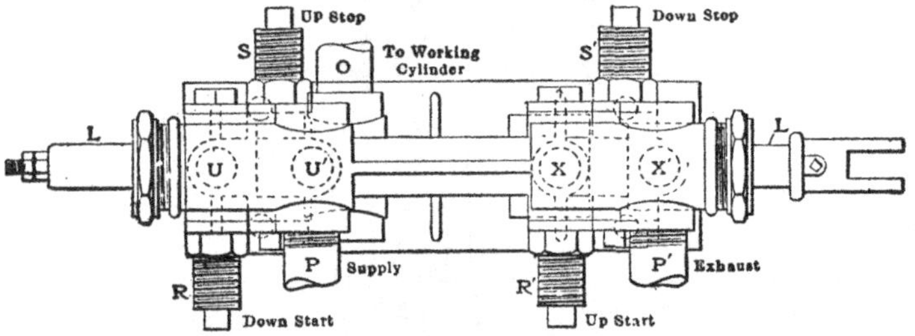

FIG. 206

P' of the pilot-valve chamber with the discharge, through the limit valve B. The inlet O on the pilot-valve chamber is connected with inlet A' at the back end of the main cylinder by means of the pipe marked K.

The operation of the valve is as follows: When the elevator is not in motion the valves are in the position in which they are shown in Figs. 204, 205, and 206. The space between the main valve pistons A and B is filled with pressure water, as it is permanently connected with the supply pipe. The space between the valve pistons D and E is permanently connected with the supply through the pipe M, shown in Fig.

200. The space between the valve pistons *B* and *C* is connected with the discharge through pipe *N*. The space back of piston *A* is filled with water through the pipe connection running from *A* to the inlet *O* on the pilot-valve chamber, and as the pilot valves are closed this water cannot escape, hence the valve *A* cannot move to the left. If it is desired to run the car upward, the operating lever *G* is raised and the through the connecting levers *H* and *J* the connecting rod *K* is drawn to the right, and the valve *T'*, Fig. 205, is thereby opened, and the water back of the valve piston *A* escapes into the pilot-valve chamber through the inlet *O* and out through valve *T'* and valve *R'* to the outlet port *P'*, following the path through the outlet *X*. As the pressure on the right-hand side of the piston *A* is greater than that on the left-hand side of *B*, the main valves will be moved to the left, and pressure water will pass by the valve piston *D* and into the pipe connecting with the lifting plunger cylinder, and the car will be moved upward.

The movement of the main valves to the left will cause the pilot valve to move in the same direction, as the valve rod *F*, through the connecting rod *F'*, will carry the lever *F''* to the left, swinging the fulcrum *H'* downward, and, through the link *H*, moving the rod *K*. The movement of the main valves to the left will continue until, through the motion of the lever *F''*, the pilot valve has been carried far enough to the left to close the valve *T*; then the outflow of water from behind the piston *A* will cease, and the main alve will move no farther. In this respect the operation is the same as that of every type of pilot valve that has been explained in preceding chapters. The distance through which the main valves will move is evidently proportional to the distance through which the pilot valve is moved by the lever *G*; for the movement of the main valve must be sufficient to return the pilot valve to the closed position, and no more.

THE PILOT VALVE.

Although the operation of this pilot valve is the same in general as that of all other pilot valves insofar as it has been explained, it possesses one feature not found generally on the pilot valves of the cable types of hydraulic elevator. It will be noticed that there are apparently two pilot valves, *V* and *V'*. In reality only *V* is a pilot valve, the valve *V'* being supplied for the purpose of controlling the distance in which the elevator can be stopped. Referring to Fig. 204 it can be seen that when lever *F''* is moved to the left, as the main valve opens to let pressure water flow into the lifting cylinder, the upper end of *F''* carries lever *K'* to the right, and this moves valve *V'* to the right while valve *V* is moved to the left. This is the action when the pilot is moved to start the car, but when it is desired to stop the car on the upward trip, lever *G* is moved downward, and then the pilot valve *V* is moved to the left and valve *T* is opened, allowing pressure water from the inlet *P* to pass through valves *R* and *T* into the pilot-valve chamber and through the outlet *O* to the inlet *A′* at the back end of the main valve, thereby equalizing the pressures on both sides of the piston *A*; the unbalanced pressure on the left-hand side of *B* will then force the main valves to the right and shut off the flow of pressure water from the supply pipe into the lifting cylinder.

Looking now at Fig. 205 it will be seen that when the pilot valve *V* is in the central position, valve *V′* will also be central, and the end valves *L* and *L′* will be opened, so that water can flow freely from inlet *P* to port *U* without passing through port *U′*, and also flow out from the passage *X* to *P′* without going through *X′*. The valve *S* in port *U′* checks the flow of water so that a much smaller quantity can pass through *U′* than through *U*; hence if valve *L* is closed, so that water has to flow in through *U′*, it will require much more time to fill the space back of the main valve piston *A* than if the water were flowing through *U*. Now when lever *G* is raised to send the car upward, the movement of the main pistons to the left will carry valve *V′* to the right, and thus carry valve *L* into the opening through which water flows into port *U*; therefore, when the pilot-valve lever *G* is depressed to stop the car, the main valve will not move back rapidly to the stop position, in which it is drawn, because the water flowing in from *P* to move the main valve has to pass through the restricted passage *U*; therefore the return movement of the main valve is much slower than the forward or starting movement.

It is evident that by adjusting the opening through the valve *S* the flow of water through *U′* can be made as slow as desired. As will be seen from Fig. 206 the valves *R R′* and *S S′* are simply plugs that can be adjusted for any amount of opening and held permanently in that position. The object of using these valves and the valve *V′* is to provide positive means to control the rapidity with which the elevator can be stopped. In a plunger elevator, if the car is stopped very quickly on the up trip, the momentum of the upward moving car and the downward moving counterbalance will tend to continue the car travel and as a result the plunger will move away from the water in the cylinder. As soon as the car stops, it will run back and stop suddenly when the plunger strikes the water. If the car is coming down and the operator stops it suddenly, the momentum of the car and its load will tend to buckle the plunger. As it is not possible for an operator to judge just how fast to move the operating lever to make as quick a stop as can be made without causing the plunger to leave the water on the up trip, or to buckle on the down trip, it is necessary to provide means for controlling the rapidity of stopping, and the valves described are the means provided

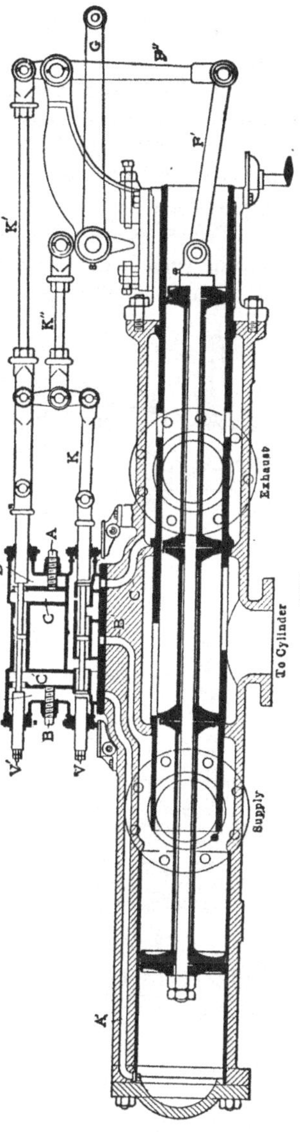

FIG. 207

for this purpose. With this arrangement of valves, the operator can make as slow a stop as he may desire, but if he moves the lever *G* instantly to the stop position, the elevator will be brought to a state of rest gradually in the shortest practical time, but no sooner. The time or distance in which the car will be stopped can be varied as desired by changing the adjustment of the valves *S S'*. The valves *R R'* are provided to prevent too rapid a start.

ANOTHER TYPE OF STARTING VALVE.

Another design of starting valve is shown by Fig. 207. This arrangement is provided with more simple pilot valves and there are no adjusting

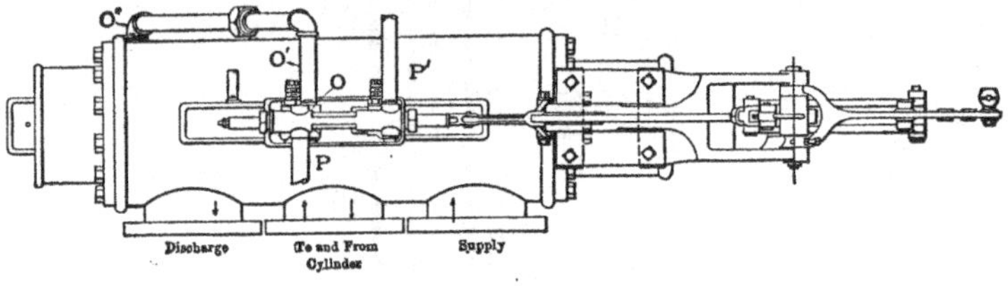

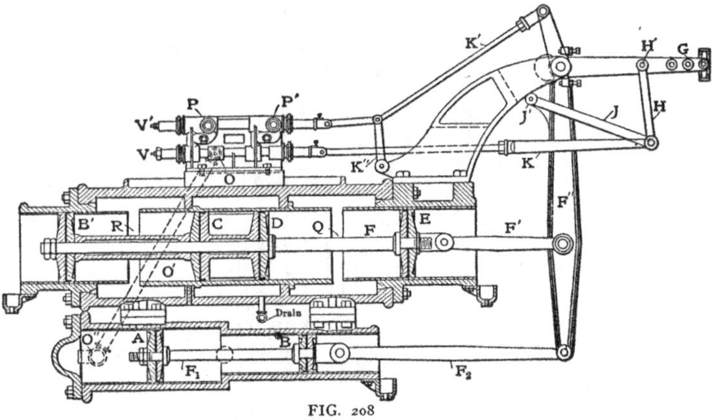

FIG. 208

valves to regulate the rapidity of acceleration in starting. The plug valves *A B* are for the purpose of adjusting the stopping distance. In starting, water can flow directly into the passage *D* or *D'* as the case may require, without passing through the holes connecting these passages with *C* and *C'* which are controlled by the plugs *A* and *B*. This is possible because the valve *V'* is in the central position, but in stopping, this valve is shifted in one direction or the other, just as in the valves of Fig. 205, and the water then has to flow from the passage *C* to *D* through the hole that is throttled by plug A, or from C' to D' through the hole throttled by B. In this design outside piping is not used to connect the pilot valve with the main valve; passages are cored in the castings, as shown clearly in the drawing. By comparing this valve with Fig. 204 the operation of every part will become perfectly clear. If the lever G, Fig. 207, is raised, K will be shifted to the left through the action of K''', because F'' holds K stationary. If G is moved downward,

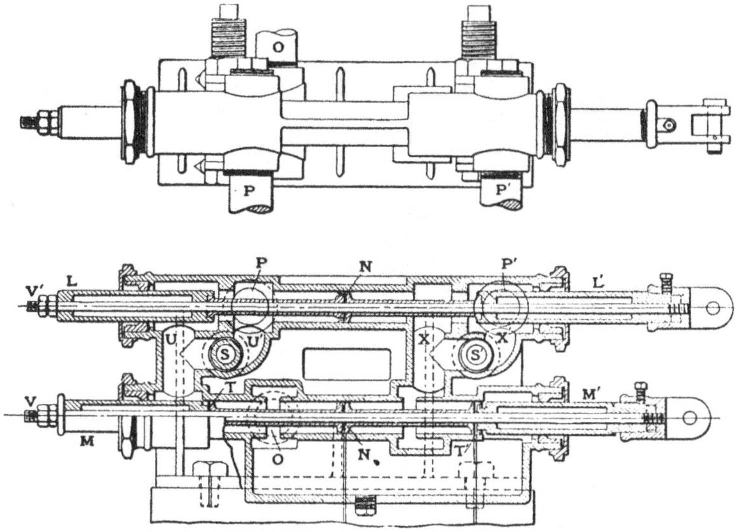

FIG. 209

K will be shifted to the right. Thus V' will be left in the central position when G is moved to start the elevator running in either direction. As soon, however, as the main valve starts to move, V' will be shifted from the central position because the upper end of F'' moves K'.

The main valve of Fig. 204 has five pistons, while that of Fig. 207 has only four. The former valve could also be made with four pistons by simply exchanging the supply and exhaust pipes.

The latest design of operating valves used with the Otis plunger is shown in Fig. 208. An enlarged view of the pilot valve and of the valve V' is given in Fig. 209. The principal difference in the main valve is that it is divided into two sections, these being located in separate cylinders, one below the other. On account of this construction, one more piston, B', is required, but on the other hand the pistons A and B are of much smaller diameter than in Fig. 204. The only function of these pistons is to move the valves and on this account *A* can be made considerably smaller than the valve pistons *C, D, E,* as all that is necessary is that the area of *A* be sufficiently in excess of that of *B* to provide all the force necessary to move the valves, and this difference in area is obtained by reducing the diameter of *B* to the proper point. Although two valve cylinders are provided, they are both of smaller diameter than the single one of Fig. 204 and in addition the length of the valve is considerably reduced.

The pipe *O'* connects the pilot-valve chamber at *O* with the lower valve cylinder at *O"*. The inlet *I'* between the pistons *A* and *B* is permanently connected with the supply pipe so that the full tank pressure is always in this space; hence, when the pilot valve is moved so as to

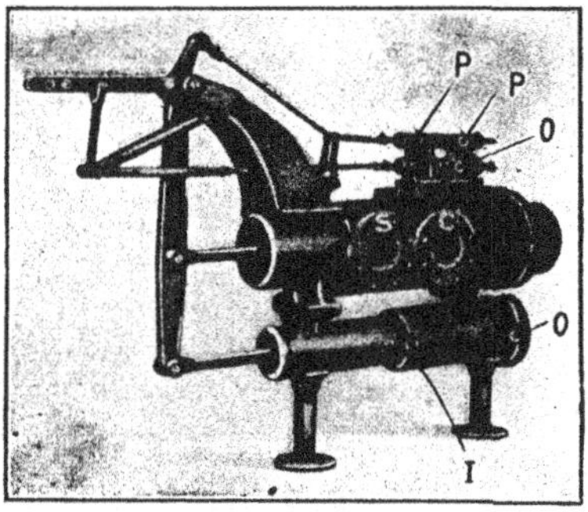

FIG. 210
THE PLUNGER VALVE

connect *O'* with the discharge, the piston will move to the left, as the pressure on the right-hand side of *A* is greater than that on the left-hand side of *B,* but if *O"* is connected with the supply pipe, through the opposite movement of the pilot valve, the pressure on both sides of *A* will be equal, while *B* will be subjected to the full tank pressure on the left-hand side, and only to the atmospheric pressure on the right-hand side; therefore, the valves will move to the right. The construction of the pilot valve and the valve *V'* is clearly shown in Fig. 209. Looking at the sectional view it can be seen that about the only difference between it and Fig. 205 is that the plug valves *R R'* are not used. Experience has shown that they can be omitted, as gradual acceleration can be effected in several other ways, as for example, by varying the distance between the valve *L* and *L'* so as to throttle the passage into *U* or *X* more or less, as may be required, or by varying the diameter of the spacing sleeves between these valves. The external appearance of this valve is shown in Fig. 210. This illustration also serves to make perfectly clear the construction of the levers F'', H, J, K, K', etc. The lever K' is actuated by an extension of F'' above the pivot on which this lever swings; hence, in starting, K'' remains stationary.

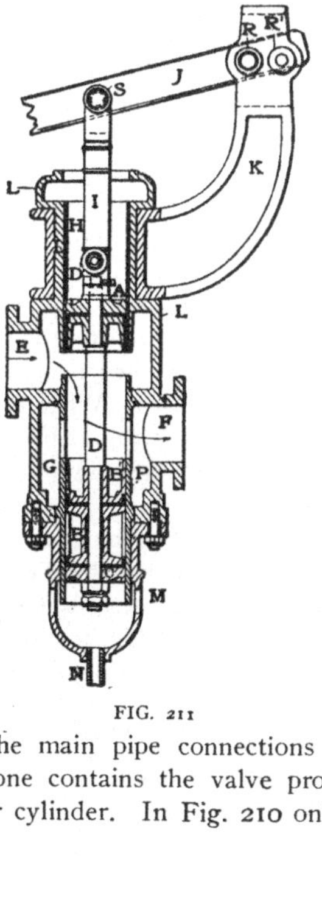

FIG. 211

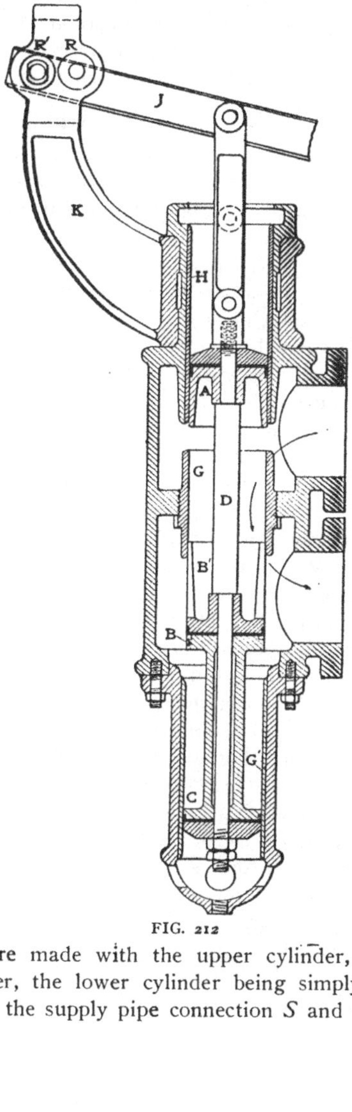

FIG. 212

The main pipe connections are made with the upper cylinder, as this one contains the valve proper, the lower cylinder being simply a motor cylinder. In Fig. 210 only the supply pipe connection S and the lifting cylinder connection *C* are shown, the discharge pipe connection being on the other side. The position of these pipe connections is varied to suit the location of the piping in each particular case, and this location is governed by the shape of the building and the location of the pumps, tanks, etc. Fig. 208 shows the three pipe connections all on the same side. In most cases the supply and discharge are more likely to be on one side and the cylinder connection on the other. The two inlets *O* and *O″* are connected with each other, and the inlet *I′* is connected with the supply pipe.

AUTOMATIC STOP VALVES.

The construction of the automatic stop valves *A* and *B*, of Fig. 200, is very simple. Two of these valves of different design are shown in Figs. 211 and 212. In Fig. 211, the water enters through inlet *E* and passes out through the opening *F*, whether the valve is connected in the supply or the discharge pipe; in Fig. 200 it will be noticed that in both valves *A* and *B* the water flows in through the top pipe. The piston *A* serves to balance the pressure on top of piston *B*, and also keeps the water from passing out through the upper end of the valve. The piston *C* acts to prevent the escape of water from the lower end. When the elevator is going up the valve in the supply pipe is in the position in which it is shown in Fig. 211, and the valve in the discharge pipe is raised so that piston *B* is some distance above the upper edge of port *P*. When the piston *B* is in this position, piston *C* is just below port *P*, so that if any part of the discharge pipe beyond *F* is higher than *P* the water in it cannot flow out through the bottom of the valve casing. To catch any drip that may occur, the drain pipe *N* is provided, and the opening *M* is for cleaning out this connection if it becomes clogged up. When the valve piston *B* is above the ports *P*, the water within the lifting cylinder cannot escape; hence, as water flows in through the open supply pipe the plunger is forced upward and the elevator car is raised. When the car reaches the upper floor, the limit valve *A*, Fig. 200, is closed by lifting piston *B* above the ports *P*, and the car stops. As it is necessary that the car be stopped gradually, the upper edges of the ports *P* are made to incline toward each other, so as to gradually stop the flow of water. When the car is stopped on the down trip at the lower floor by the action of the stop valve *B*, Fig. 200, the action is the same as described for the up stop, with the exception that instead of stopping off the flow of pressure water into the lifting cylinder, the discharge of water from the cylinder is stopped.

There is a slight difference between the arrangement of the actuating lever *J* in Fig. 211 and the corresponding lever *A′* in Fig. 200. In the latter the lever is pivoted between the ends and the valve connecting rod is connected with one end. When the valve is so arranged, the piston is depressed to stop the elevator, while in Fig. 211 it is raised. To arrange the valve so as to stop by depressing piston *B*, it is necessary that the latter be raised sufficiently to rest above the inlet *E* when the valve is opened; then if it is depressed below *E* the flow of water is stopped. To make the valve operate in this way the cylinder lining *G* is made longer so as to extend up above the inlet *E*, and the ports *P* are cut opposite this inlet. The piston *B* is raised so as to rest just below the bottom of ports *P* when the valve is closed and lever *J* is in the position shown; then when the valve is opened, by lifting it, the piston *B* is carried above the ports *P* and the piston *A* is raised to the top of the valve cylinder. The valve piston *C* remains in the same position as in Fig. 211. When this change is made, the cylinder linings *G* and *H* are virtually interchanged, although the same linings are not used; *H* is made long enough to contain the ports *P*, and *G* is shortened so as not to project up above the lower edge of the outlet *F*. The arm *K* that carries lever *J* is secured to the valve casing, so that it may be moved around into any radial position that may be desired. The arm *K* is provided with a clamping cap, and the two parts are bored to fit the upper neck *L* of the valve casing, so that in whatever position the arm may be set it will come into line with the valve rod *D* when the cap is tightened up by the clamping bolts.

The valve shown in Fig. 212 differs from that of Fig. 211 in the construction of the piston *B* and the cylinder lining *G*. The latter is a short tube without ports, these being formed in the upper portion *B'* of piston *B*. When the valve is fully opened, *B'* drops to the position in which it is here shown, and the ends pass nearly out of the lining *G*. When the valve is closed the piston *B* is raised until the cup packing enters the lower end of lining *G* far enough to make a joint and stop the flow of ater. The ports cut in *B'* are V-shaped, so that as the valve moves upward, the flow of water is gradually stopped.

THE PISTON VALVES.

The construction of the piston valves is clearly shown in Fig. 212. The arm *K* that supports the actuating lever *J* is made with two pivot holes *R R'*, in order to adjust the leverage of *J* for different movements of the actuating ropes. The travel of these ropes can also be adjusted by varying the point of attachment of one and the position of the lever *J* for the other. The effort required to move the limit valves can also be varied by means of the two pivot holes *R R'*, for if the one nearest the valve is used, the lift of the actuating rope will be increased and the pull will be reduced. With small valves, as the frictional resistance is less, the lever *J* can be pivoted in the hole *R*, and then less movement of the actuating rope will be required.

CHAPTER XXXVI

OPERATION OF MAIN AND PILOT CONTROL VALVES; COMPARISON OF DIFFERENT CONTROL AND AUTOMATIC STOP VALVES

The general arrangement of the main valve and pilot, the automatic stop valves, the actuating ropes, etc., of the latest type of Otis passenger plunger elevators is shown in the vertical elevation, Fig. 213. This drawing is made to scale, and shows all the parts in their true proportion. The reference letters are the same as in Fig. 200, so that the difference between the two arrangements, as well as the difference in the design of the valves, can be easily distinguished. The automatic stop valves *A* and *B* are of the design shown in Fig. 211 and the main valve is the type shown in Fig. 208. The flow of water through the entire system is indicated by the arrows. When the valve is set to ascend, water flows in from the supply pipe through the stop valve *A* to the inlet at the right-hand end of the upper cylinder of the main valve. Passing through the valve the water enters the vertical pipe leading from the central outlet of the valve chamber, enters the upper end of the lifting cylinder, and forces the plunger upward, thus lifting the car. When the valve is set to come down, the water in the lifting cylinder flows out and down through the vertical pipe to the center of the valve chamber, through the valve to the left side outlet, and thence through the pipe to and through stop valve *B*. In Fig. 200 the two automatic stop valves are made so that the water flows through them from top to bottom, but in Fig. 213 the water enters both valves at the bottom and passes out at the top.

The levers of the valves *A* and *B* are arranged as in Fig. 211. The rope that actuates valve *A* runs over and under the pair of sheaves shown just under the floor of the elevator pit. This arrangement is shown in order to be able to give a side view of lever *A'* and rope *A"*; in practice lever *A'* would be rotated around far enough to permit the rope to run up in a straight line back of the elevator car. The upper end of rope *A"* is secured to one of the overhead beams, as shown in Fig. 214, which shows the top of the elevator well, with the overhead sheave over which the counter-balance ropes run, and also the arrangement for holding the upper ends of the pilot valve operating ropes. The rope *B"*, that actuates lever *B'* of valve *B*, runs up and over a stationary sheave and

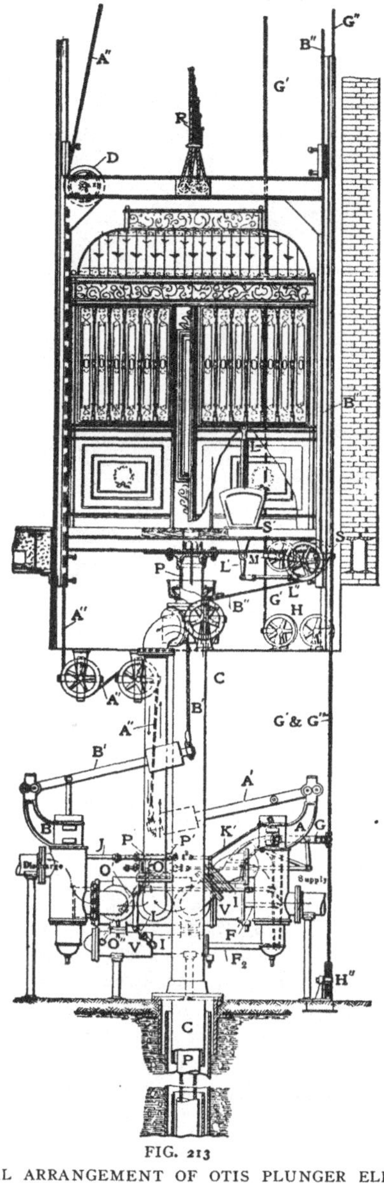

FIG. 213
GENERAL ARRANGEMENT OF OTIS PLUNGER ELEVATOR

thence under a sheave attached under the car, near the right-hand side. In this case the stationary sheave is necessary to prevent the rope from pulling sidewise on lever *B′* when the car approaches the bottom floor. The rope *B″* is secured to the overhead beams at one side.

The pilot valve is actuated by the ropes *G′ G″*, one side of which is

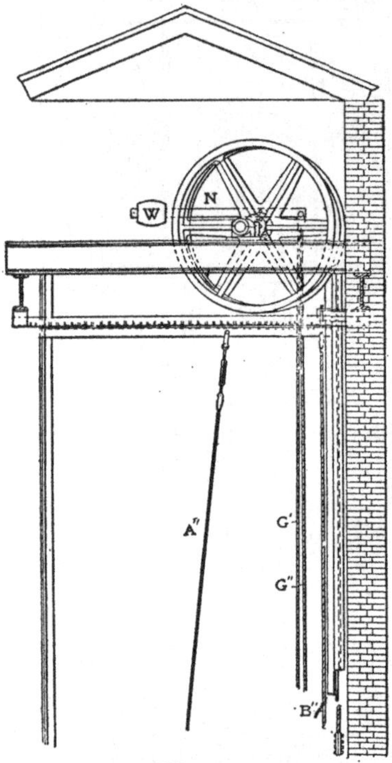

FIG. 214

secured to the end of the operating lever *G*. This arrangement is of the standing-rope class, but it is slightly modified, the modification being necessary by reason of the fact that the lower supporting sheaves *H* cannot be run down below lever *G*, on account of the space being taken up by the valves and piping. The stretches of rope *G′* and *G″* are a single continuous rope, the two ends of which are secured at the top of the elevator well to a tightening lever *N*, the weight *W* being set to put the proper tension on the ropes. Coming down to the car, these ropes pass around the two sheaves S, S_1 carried in a frame *M* mounted upon the car and connected with the car lever *L* through rod *L″*. The rope G' runs under the sheave S' and over the sheave S, thence down to lever G, and passes under the sheave H'' located below lever G. This part of the apparatus is more fully shown in Fig. 215, which is a side view of the lower part of Fig. 213. Passing up from the right side of sheave H'', the rope becomes G'', and upon reaching the sheaves H, passes over the one on the right-hand side and under the one on the left, and then runs up and over sheave S' and under sheave S, thence up the elevator well to the tension lever N.

Keeping in mind the arrangement of the ropes just described, it can be seen that if the car lever L is moved to the right, the connecting rod

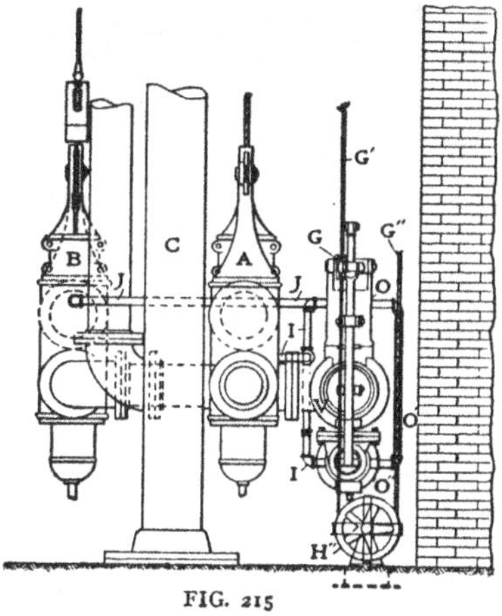

FIG. 215

L'' will move to the left, and swing the frame M clockwise, so that sheave S will move downward and S' upward. As rope G'' passes over sheave S' it will be drawn up and G', which passes under S', will be let out, the result being that the pilot-valve lever G will be raised. The sheave H'' and the extension of the ropes below H complicate the arrangement slightly, but do not change the principle of operation in the least.

SIMPLE TYPE OF OTIS FREIGHT ELEVATOR WITH HAND ROPE CONTROL.

The general arrangement of a simple type of Otis plunger elevator is shown in Fig. 216, which is a vertical elevation of a freight elevator provided with hand-rope control. The operating valve is shown with a long lever to the end of which the hand rope is attached. The valve, as will be seen, is very simple, the supply water reaching it through the pipe A, which enters the upper end of the valve cylinder. From the lower

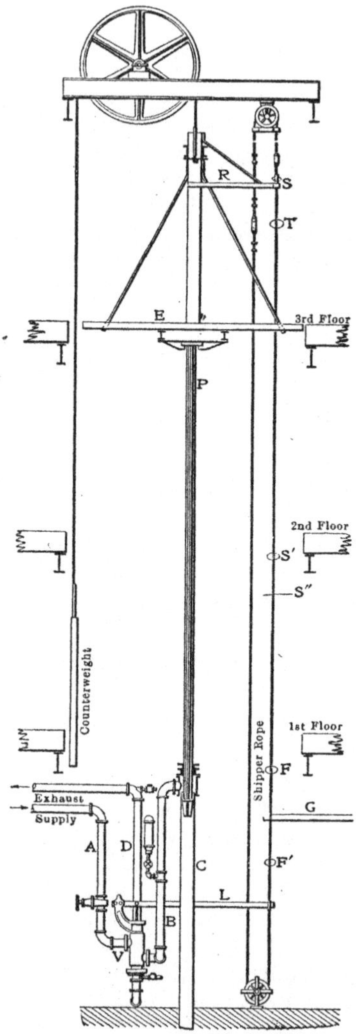

R
S
T
E
3rd Floor
P
2nd Floor
S'
S"
Counterweight
Shipper Rope
1st Floor
F
Exhaust
Supply
G
A
D
C
F'
L
B
V

FIG. 216

end of the cylinder a pipe *B* connects with the upper end of the plunger cylinder, and through this pipe the water enters when the elevator ascends, and escapes when it descends. The escaping water passes out of the main valve through the bottom, and is carried off to the discharge tank through the pipe *D*. The operating lever *L* is used when the car runs at very slow speed, as is generally the case in buildings of three or four floors. For higher velocities of car it is necessary to substitute for the lever *L* other means of moving the valve, whereby greater movement of the operating hand rope is required to open and close the valve. If such means were not provided, the starting and stopping would be entirely too rapid. If the car speed is moderate, say 100 feet per minute, the valve is operated by means of a rack and pinion, in precisely the same way as the hand-rope operated valves already described as being used with cable hydraulic elevators.

In some cases freight elevators are arranged so as to carry passengers also, and on that account are run at a higher speed, and occasionally passenger elevators in buildings where economy of construction is considered are arranged to be controlled by a simple hand-rope operated valve. For such installations, the simple rack-and-pinion arrangement is not satisfactory because it will close the valve with too small a movement of the hand rope. For such service the valve is made with a double set of gears, the pinion that meshes with the rack being mounted upon a shaft that carries a larger gear, and that in turn meshes with a second pinion that is mounted upon the hand-rope sheave shaft.

When a lever-operated valve is used, the travel of the hand rope is varied by varying the length of the lever *L*, which is made about 3 feet long for very low speeds and as much as 7 feet for the highest speeds. When the lever is 3 feet long, the movement of the hand rope required to make a stop on the up trips is about 15 inches, and on the down trips about 10 inches. With the longest lever, about 7 feet, the rope movement to make a stop on the up trips is about 3 feet, and on the down trips about 30 inches. With the single-geared rack-and-pinion the rope travel required to stop the car is about 5 feet, which can be varied to some extent by changing the diameter of the hand-rope sheave, say from 4 to 7 feet. When a double-reduction gear is used the rope travel varies from about 9 to 18 feet, according to the car speed. If only the manipulation of the hand rope by the operator were considered, it would not be so important to vary the movement of the hand rope with the speed of the car, because with a little practice an operator could learn just how to manipulate the rope to make satisfactory stops, but the car is also stopped automatically at the top and bottom landings by the movement of the hand rope and this movement is effected by the car itself; hence, the travel of the hand rope required to stop the car must be made to conform to the car speed.

In Fig. 216 the elevator is represented as being level with the third floor, and the stop ball *S* on the hand rope is shown resting against the upper side of the arm *R*. Before the car would reach the top-floor stop the stop ball would be in the position *T*, and would be carried from that position upward to where it is drawn in stopping the elevator. On the downward trip the arm *R* strikes the stop ball *S'* when it is in the position shown, and carries it down to line *S''*, to stop the car, and then the floor of the latter is even with the first floor of the building. If the length of the lever *L* is proportioned to the speed of the car so as to give the proper amount of hand-rope travel, the car speed will be cut down gradually until the car comes to a stop even with the floors of the building, without producing any noticeable shock. If, however, the hand-rope movement is too short for the car speed, the stop on the down trip will be very violent and on the up trip the car will not stop promptly, but will continue upward and lift the valve far enough, in an extreme case, to let water escape from the cylinder; as soon as the headway dies out, the car will drop back until the plunger strikes the water in the cylinder, when its backward movement will be stopped with a severe jolt. Even if the car does not run far enough above the floor to open the valve in the reverse direction, if it should pass above the floor any distance, the plunger will be drawn away from the water in the cylinder and on the return will be stopped suddenly when it strikes the water again. If the rope movement is made longer than necessary there will be no undesirable effect except that the time consumed in stopping will be unnecessarily increased.

ADJUSTMENT OF VALVE MOVEMENT.

When the elevator runs at a high velocity, it is generally necessary to provide means for adjusting the movement of the valve when it is opened, so that there may be no danger of imparting too high a speed to the car if the load is very light. With slow-running cars, it is not necessary to provide such adjustment, because even if the car should run 50 per cent. faster with a light load there would be no objection, but if the normal car speed is, say, 400 feet per minute, and it should run 600 feet with a light load, some damage might be done, either when the car is stopped at intermediate floors, or when it is stopped automatically at the top floor. The means employed to adjust the running speed consist of stops, placed on the hand rope, as shown at *F F'*, and a stationary stop *G*, Fig. 216. The stop *F* controls the opening of the valve for the up trip, and *F'* controls the opening of the valve for the down trips. If these stops are set close to the stationary stop *G*, the movement of the hand rope is limited, and as a result the opening of the valve is reduced. Contrariwise, setting these stops farther away from G increases the opening of the main valve, and, therefore, the car speed, with light loads. There is no great danger of the speed being too great with full load, as elevators are not often so generously proportioned as regards the driving power as to be able to run away when fully loaded, but when the car is empty the driving power is sufficient to develop a speed much

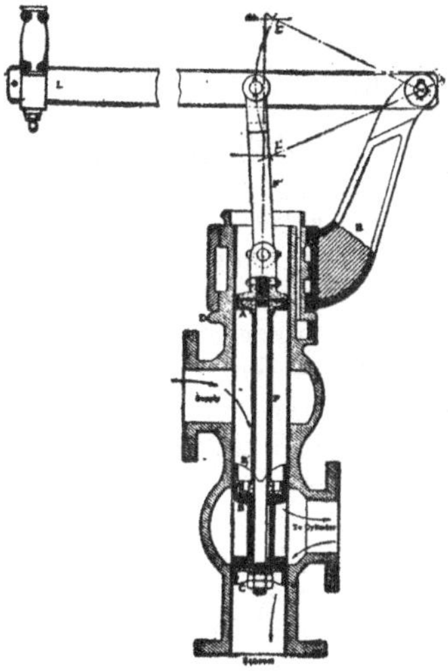

FIG. 217

higher than the normal. This is more likely to be the case with the cable hydraulic type than with plunger elevators, because with the latter the weight that has to be lifted when there is no load in the car is much greater than with the former.

CONSTRUCTION OF OPERATING VALVE.

The construction of the operating valve shown in Fig. 216 is illustrated in Fig. 217, which is a vertical section through the center of the valve and its cylinder. The supply water enters through the upper inlet and if lever L is depressed, it will pass by piston B and, following the path indicated by the arrow, will flow into the pipe that connects with the top of the plunger cylinder, marked B in Fig. 216. If lever L is raised, the water can flow out from the cylinder and past piston C to the discharge tank. The piston A serves to balance the valve and also to prevent the pressure water from escaping through the top of the valve cylinder.

The valve cylinder is made of cast iron and is provided with a brass lining within which the valve pistons slide, all of which is clearly shown in the drawing. The valve pistons are provided with leather-cup packings and heads to retain these in position. These heads are made with outer flanges cut away in a wide V-shape, as shown at *B'*, so that when the valve is moved to close the port, the flow of water is stopped gradually. For high-speed elevators these V-shaped flanges are made so as to stop off the flow of water quicker than is necessary, and when the elevator is installed in the building they are adjusted by actual trial, the edges being filed off until the proper degree of smoothness in stopping is obtained. For very slow-running elevators it is not necessary to go to all this trouble, as the shape of the flanges can be made near enough to accuracy at the shop.

The brass lining of the valve cylinder is provided with numerous narrow ports opposite the outlet so as to hold the cup packings of the pistons *B* and *C* in place. Similar ports are provided opposite the inlet, but these are necessary only to permit easy removal of the valve when desired, and also to make the whole lining in one piece. So far as the operation of the valve is concerned the lining could be made in two parts separated from each other opposite the inlet as much as might be necessary to permit the water to flow in. The arm *E* that carries lever *L* is made like the arm of the automatic stop valves described in connection with the passenger plunger elevator; that is, it is clamped to the upper end of the valve cylinder above the point *D*, this portion being turned to fit the bore of the arm *E* and its cap.

CHAPTER XXXVII

RACK-AND-PINION VALVES FOR OTIS FREIGHT ELEVATORS; TYPE OF VALVE USED FOR HIGH SPEED; NOTES ON THE CARE OF THE PLUNGER AND ROPES

The rack-and-pinion valve, which is used for elevators that run at a velocity too great for the lever valve, is shown in Fig. 218. The valve itself is the same as the lever valve, but the connecting rod *F'* is replaced by the rack R, and this meshes with a pinion *P* that is mounted on the shaft that carries the hand sheave *S*. The rack *R* is made of a round rod with the teeth cut on one side. A casing *D* is provided to cover the rack and pinion, and this is clamped against the upper end of the valve cylinder by means of the cap *D'*. Opposite the pinion *P* the cover *D* has a finished surface against which the rack bears so as to be held in mesh with the pinion. This sliding surface and the pinion shaft are provided with grease cups *E* and *F* to keep them properly lubricated. The rack is so formed at the ends that it cannot be run out of gear, even if there are no stops outside of the valve to limit the movement of the hand rope. When the pinion reaches the last tooth it strikes the curved end of the rack and can go no farther.

TYPE OF VALVE USED FOR HIGH SPEED.

The type of valve used when the car speed is so great that the plain rack-and-pinion valve will not give sufficient hand-rope movement, is shown in Fig. 219, which presents a vertical elevation in section and also a plan. The hand-rope sheave *S* is mounted on a shaft *A* that carries a pinion *B*, which meshes with a gear wheel *C* mounted on a second shaft *D*. This shaft carries a second pinion *E* that meshes with the rack *F* on the end of the valve rod. The relative positions of these parts is more fully shown in the plan view of the valve, at the top of the drawing. The pinion *B* and gear *C* are covered with a casing, and so are the rack *F* and pinion *E*. The rack *F* is provided with stops on the back at each end so that it cannot be run out of gear with the pinion.

All the valves shown in the preceding illustrations are balanced insofar as the pressure of the supply water is concerned, as this pushes up against the piston *A* and down against the piston *B*, and as both pistons are of the same diameter, the pushes are balanced, but if the discharge tank is higher than the valve, as is generally the case, the valve will not be in balance because there will be a pressure acting under the piston *C* that will be greater than the atmospheric pressure acting on the top of the piston *A*. For cases of this kind it is necessary to provide another valve piston under *C* so that the back pressure from the discharge tank may act upon it and thus balance the pressure acting under the piston *C*. The type of valve used for such installations is shown in Fig. 220. The only difference between this and the other valves is that a section is added

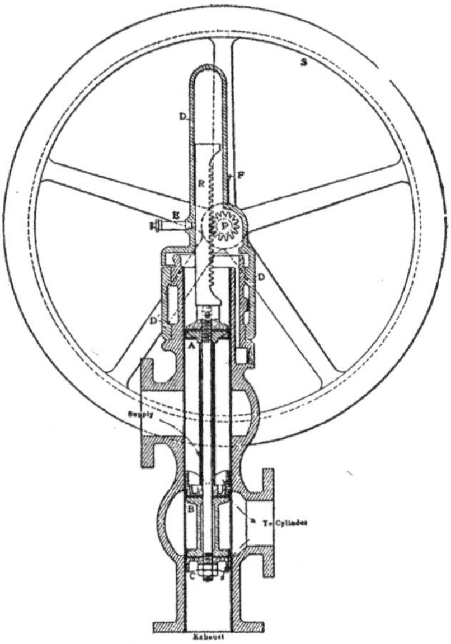

FIG. 218

to the lower end of the valve cylinder and the valve rod is lengthened out so as to carry an additional valve piston *D*. With this construction the two pistons *C* and *D* balance the back pressure from the discharge tank, and the pistons *A* and *B* balance the pressure from the supply pipe.

PROPER CARE OF THE ROPES.

Plunger elevators have fewer moving parts than cable hydraulic machines, hence they should be easier to keep in proper running order. The counterbalance ropes run over a sheave located at the top of the elevator well in the same position as the main overhead sheave of cablehoisted elevators. This sheave is of the same type as those used with cable elevators and is mounted in the same way, so that it requires the same attention. The ropes that run over it do not require as close inspection as the lifting ropes of cable machines because they are several times as large as is necessary to support the weight of the counterbalance, owing to the fact that they not only hold this weight, but also act as a

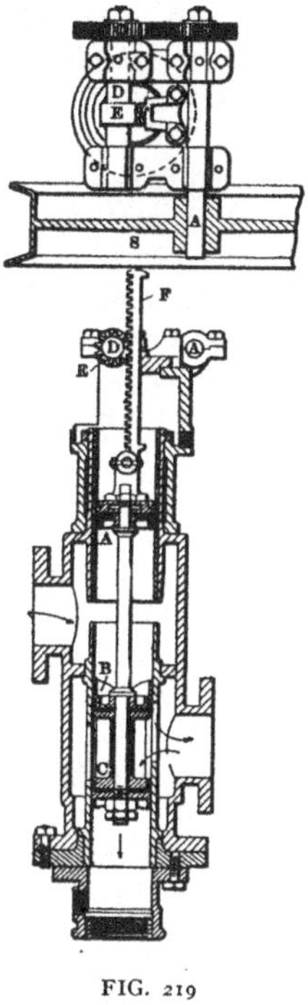

FIG. 219

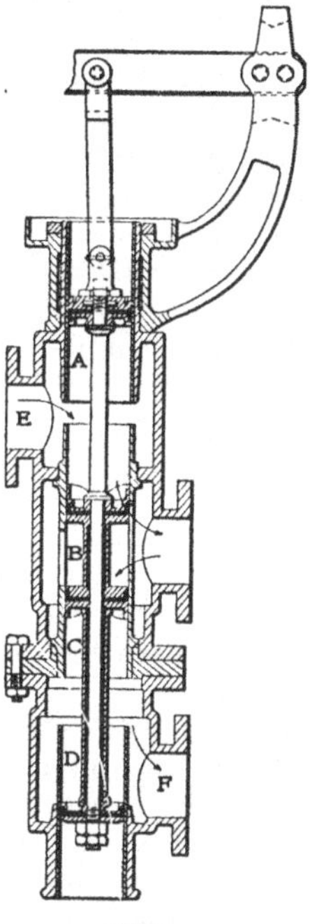

FIG. 220

compensating balance to offset the varying weight of the column of water displaced by the plunger. The diameter of the plunger is almost invariably 6½ inches, so that the water displaced for each foot of length is 0.2304 cubic foot, and the weight of this quantity of water is about 14.33 pounds. The weight of the ropes required to compensate for the varying weight of the water displaced by the plunger is just one-half this amount, as can be easily seen from the fact that each foot of rope hanging above the car adds its weight to the latter, while each foot of rope hanging above the counterbalance deducts its weight from that of the car; hence, to exactly balance the water displaced by the plunger the ropes must weigh one-half as much per running foot.

Although the ropes are far stronger than necessary, it is important to keep them all drawn up to the same tension, and this can be done with a sufficient degree of accuracy by pushing them sidewise at a point several feet from the end and noting whether they all offer about the same resistance to being moved. The end-fastening bolts must be held so that they may not work loose and slack the ropes. As a rule these bolts are provided with means for holding the nuts from turning, but even these locking devices can get out of order occasionally, and they should be inspected frequently.

The ropes that operate the valves should be kept in perfect condition, and the sheaves over which they run should have the bearings well lubricated; care should be taken that there is nothing nearby that can fall into them and cause the ropes to run off, or to be jammed. This applies especially to the automatic-stop valve ropes, for if these should become disabled the car would in all probability run far enough above the upper floor to do damage, or strike the lower bumpers violently the first time the operator permitted the car to run to the end of the trip without closing the main valve.

PLUNGER WILL GIVE LITTLE TROUBLE.

The plunger, if kept in proper condition, will give very little trouble, but if an elevator is out of service for some time and the plunger becomes rusty from want of proper protection, when it is started up the rough surface will act upon the packing like sandpaper, and soon cause it to leak. Whenever an elevator is put out of commission for any length of time, which is unusual, the plunger should be run slowly on the last downward trip that it makes, and the surface should be wiped dry and then covered with a hard lubricating grease, so that in passing down through the stuffing box into the cylinder, all the grease will not be rubbed off, but a thin layer will be left that will serve to protect the metal. If the elevator plunger is lubricated by a lubricating compound put in the water, it will not be necessary to grease the plunger, as it will receive a protective coating from the greasy water.

When an elevator that has been idle for some time is started up, the plunger should be carefully examined, and if it shows any rough spots they should be smoothed off, but emery should not be used for this purpose. As a rule use the softest polishing material that will do the work, such as whiting, pumice stone, and in extreme cases sandpaper. Draw-filing with a very fine file is the best means for smoothing the rough spots if they are very rough, but be sure the file is a dead smooth cut or next to it. After draw-filing a fine polish can be obtained by using the finest grade of sandpaper and finishing with rouge cloth. In fact, for ordinary roughness, the rouge cloth alone will answer every purpose.

TO RENEW PLUNGER PACKING.

Whenever it becomes necessary to renew the plunger packing, the first thing to do is to run the car downward until it rests solidly upon the bumpers. Then the hand valve in the supply tank should be closed, so that water cannot flow in from the pressure tank. If the discharge tank is higher than the bottom of the valve, which is generally the case, there will be a hand valve in the discharge, and this must also be closed so that water cannot flow back from the discharge tank. If any part of the pipe connections between the two hand valves just mentioned is higher than the top of the cylinder, there will be drain cocks through which the water

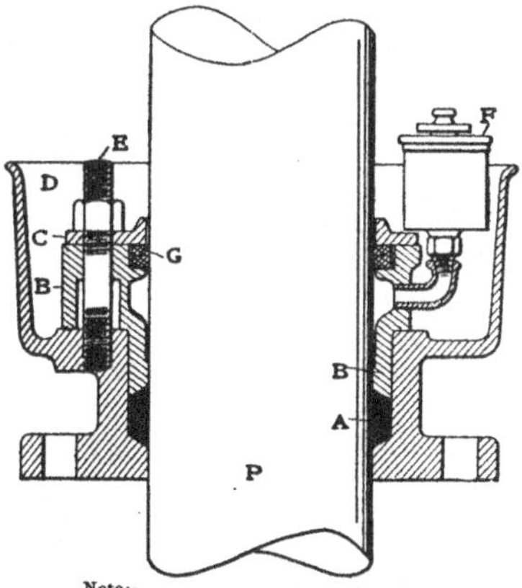

Note:- A = Actual Plunger Dia. + 1/32"

FIG. 221

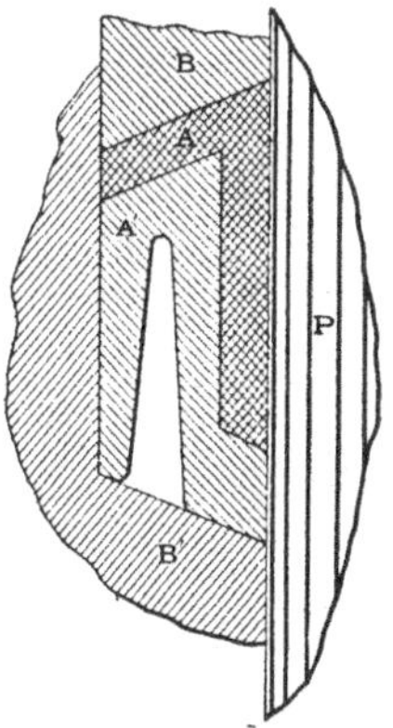

FIG. 222

may be drawn off, so that when the stuffing-box gland is removed water may not escape and flood the floor. The way in which the stuffing box of the Otis plunger is repacked can be understood by the aid of the detail drawings of the box and packing presented in Figs. 221 to 224. Fig. 221 is a vertical section through the several parts of the structure. The stuffing is shown at *A* and the compressing gland is marked *B*. Between the gland *B* and a ring *C* are two rings of babbitt metal which are provided to rub the oil off the plunger as it ascends and thus prevents it from overflowing when the plunger comes down. As an extra precaution, however, the basin *D* is provided to catch any drippings of oil.

The packing *A* can be any of the soft kinds suitable for steam-engine stuffing boxes, but the Otis company makes a special ring packing in two parts, one of leather and the other of rubber, and this lasts longer and remains tight until worn out. A sectional drawing of this packing is given in Fig. 222. In this illustration *P* represents a portion of the side of the plunger, *B* is the lower end of the stuffing-box gland, *B'* is a portion of the stuffing-box casting and *A A'* are the two parts of the packing rings; the ring *A* is made of leather and *A'* is made of rubber. These two parts are made so as to just fill the stuffing-box chamber when the gland *B* is screwed down to the lowest position, hence the packing is held securely in place but not compressed. Two rings are used because it is necessary to cut them on one side to put them in place without lifting the plunger out of the cylinder, and by using two rings and "staggering"

FG. 223

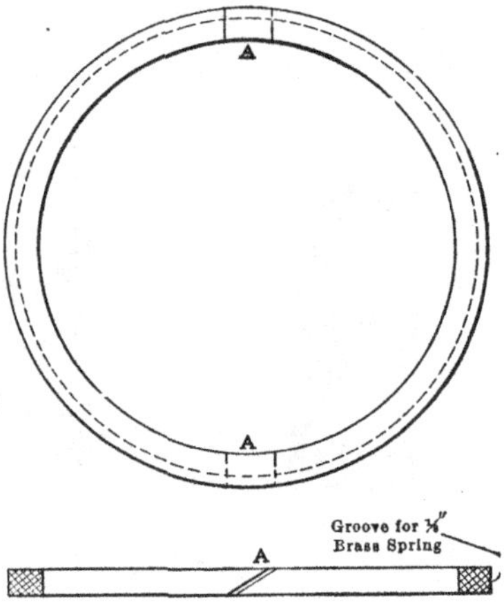

FIG. 224

the cuts, a tight joint can be made. The two rings are made to fit each other; the pressure of the water acts against the inner sides of *A'* and forces the outside part against the stuffing-box casting, making a tight joint at this point, and the inside part is forced against ring *A*, making a tight joint between the two rings, at the same time forcing *A* against the plunger to make a tight joint at this point.

THE STUFFING-BOX CASTING.

The construction of the stuffing-box casting is clearly shown in Fig. 221 and the construction of the gland *B* is more fully illustrated in Fig. 223, which shows a sectional elevation. This view shows that between the top and bottom edges of the casting there is an annular recess which forms an oil well that surrounds the plunger and keeps the surface of the latter well lubricated. Oil flows into this recess from the oil cup shown in Fig. 221. At the upper end of the gland there is another recess which contains the babbitt ring scrapers that wipe the oil off the plunger as it rises. To hold these in place the ring *C* shown in Fig. 221 is provided, and this is held in position by the studs *E*, that hold down the gland *B*. The babbitt rings are of square cross section, as shown in Fig. 224, and are divided into halves, the ends being cut on an angle as shown at *A*. On the outer surface of the ring is cut a groove that is large enough to accommodate a brass-wire spring, which presses the two halves of the ring against the plunger surface. Two of these rings are used, as shown in Fig. 221, and they are set so as to break joints.

To determine whether the plunger stuffing box needs repacking is a very simple matter. If it leaks it must be repacked, unless it is filled with an ordinary packing in which case it may possibly be made tight by screwing down the gland *B*. As the top of the cylinder is in sight, water running out of the stuffing box can easily be seen. If it is not known whether the packing is of the type shown in Fig. 222 or not, this can easily be determined by feeling around the outside of the upper flange of the gland *B* to ascertain whether it is resting against the lower casting or not. If it is not down tight, the packing is probably hemp, and a few turns of the studs may be sufficient to prevent the escape of water. If gland is down against the lower casting, the packing may be rings that are worn out, or hemp that is compressed as far as the gland will force it. If it becomes necessary to renew the packing and there are no packing rings at hand, use hemp well greased, and if this works well keep on using it. It is by no means certain that the rings will work better than hemp, although they are considered to be able to stand up better against the wear of high-speed plungers.

If a plunger elevator settles when standing at a landing above the first floor, and no water escapes from the plunger stuffing box, the valves leak, that is, providing the valve is properly closed. To renew the packing it is necessary to remove the valve from the casing, and before this is done the hand valves in the supply and discharge pipes must be closed and the water drawn from the valve casings. The car must be run down so as to rest on the bumpers at the lower floor before this work is begun. The hand-rope valves used with freight elevators are best removed through the upper end, although it is also possible to remove them from the lower end, providing there is space between the bottom of the valve cylinder and the floor for their removal. This way of taking out the valves is not advisable, however, because it involves disconnecting the discharge pipe, which is considerable work; nevertheless, there may be cases where the location of the valves renders this method necessary.

When the valve is removed through the upper end, if it is of the lever type, the pivot pin around which the lever swings is taken out so that the lever may be removed, and then the valve can be lifted out of the cylinder. If the valve is of the rack-and-pinion type, the casing that covers the rack is removed together with the pinion and shaft, and then the valve can be drawn out. If the valve is of the double-gear rack-and-pinion type it is only necessary to remove the shaft that holds the pinion that meshes with the rack, and also the covering of the latter; with these parts removed, the valve can be drawn out. If the brass lining of the valve cylinder happens to be in two parts there may be some difficulty experienced in passing the cups of the lower valve pistons up into the upper lining, but if this occurs the valve should be lowered so as to smooth up the cups, and then tried again with more care. There is very little liability of having trouble in this direction, because the lower end of the top cylinder lining is made with the inner corner well rounded off and the top end of the lower lining is made in the same way. There is no difficulty in getting the valve back into the cylinder, because the cup packings are not forced in edge first, or against the grain, as it might be expressed.

REMOVAL OF THE VALVES.

The main valve of passenger elevators is removed from the back end, as it cannot be removed from the front end, owing to the fact that the rear piston is larger than the others. In the valve Fig. 208, the upper valve pistons can be removed from either end, as all the pistons are of the same size; but it is less trouble to remove them from the back end, as then it is not necessary to disturb any of the connecting levers.

When new leather cups are to be put in the valves they should be obtained from the makers, so that they may be of the proper depth. In some cases it matters little what the depth of the cup is, so long as it makes a tight joint; this is the case with the pistons in the lower cylinder of the valve, Fig. 208, because the packings have only to maintain a tight joint. There are cases, however, in which the edge of the cup forms the edge of the valve, and when it passes over a port, closes it and stops the flow of water. To make the elevator work well it is necessary for such valves to have a certain amount of lap, and therefore the cup packing should be of the proper proportion. In the several valves of the Otis plunger elevators, for example, it will be noticed that in the valve piston which shuts off the flow of water from the supply pipe into the cylinder the cup is turned toward the supply pipe, hence its edge forms the edge of the valve and its length determines the amount of lap, so that if the cup is too deep there will be too much lap and if it is too shallow there will not be enough lap.

Of the two evils, too much lap is the lesser, so that if for any reason a cup is used that is not of the proper shape, take one that is deeper than it should be. If the lap is reduced the liability of reversing the elevator whenever it is stopped is increased, because a slight movement beyond the point where the valves are closed will open them for the reverse direction of movement. These observations relative to the proportions of cup packings apply to simple hand-rope valves as well as to the pilot-valve type.

The pilot valve and the *V'* valve can be removed from either end when it becomes necessary to renew the packings, but generally it will be found easier to draw the pilot out through the back end and the other valve through the front. In replacing the valves they can best be put in through the back. If any trouble is experienced in getting the front cup of the pilot valve past the two port openings, the trouble can be avoided by putting the valve in from the back with the two rear cups in place, and then the front cup can be put on from the front end. In doing this, care must be taken that the end fork which holds the connecting-rod end is screwed upon the valve rod tight, so as to force the packings firmly into position. The safest course is to note the position of the end of the valve rod in the fork before taking it apart, and to screw it up to that same position when putting it back. This plan should also be followed with the nuts at the back end.

Referring to the drawings of the Otis plunger-elevator pilot valves, as well as the main valves, it will be found that no means are provided for adjusting the position of the cup packings lengthwise of the piston rod; such adjustment is not provided because it is not required. When the valves are made the various parts are made of the proper length to bring all the cups in the proper positions, and the only way in which the adjustment can be made wrong is by using cups that are not of the proper length when replacing those that are worn out.

AUTOMATIC-STOP VALVES.

The automatic-stop valves of the Otis plunger elevators are very similar to the main valves used with hand-rope control, and are taken out of the valve cylinder through the upper end in the same way as the latter. These valves have the cup packings set in the right position when first installed and have no means for adjusting their position because such adjustment is not necessary. In repacking them all that is necessary is to use cups of the same shape as the original cups. All that has been said in relation to the main valves applies equally well to the automatic-stop valves.

CHAPTER XXXVIII

REPLACING WORN SHOES; HOW TO DISCONNECT THE PLUNGER; CARE OBSERVED IN HANDLING PARTS

The only point that remains to be considered in connection with the Otis plunger elevator is the lower end of the plunger. As the cylinder is from 2 to 3 inches larger in diameter than the plunger, the lower end of the latter has to be guided so that it may not wobble around when the elevator is in motion. The guides used for this purpose are shoes made of hard brass, their shape being shown in Fig. 225. In some cases three shoes are used, as shown in this drawing, the shoes being spaced equally around the circle. The lower part of Fig. 225 is a horizontal section through the lower end of the plunger, showing the position of the three shoes *S*, as well as the sides *T*, of the slots within which they move. In some plungers four shoes are provided, as shown in Fig. 226. In both arrangements the shoes are pivoted at the upper ends and are held from swinging out too far by a stop ring at the lower end, as shown in both drawings. The cylinder itself prevents the shoes from spreading out beyond a certain point, but if the retaining ring were not provided at the lower end the plunger would not be held central unless all the springs *A* were of equal tension.

As the shoes slide against the wall of the cylinder they wear out and from time to time have to be replaced. The rapidity with which they wear out depends on several things, such as the character of the cylinder surface, the straightness of the cylinder, the hardness of the metal, etc. In the early days of plunger elevators the lengths of pipe that form the cylinder were put together in the same way as a line of piping so that the joints presented the appearance illustrated in Fig. 227, which represents a section through one side of the cylinder at a joint. With a joint of this type the shoes will wear out rapidly, as the edges *a a* cut them away. This, however, is not the only objection to such cheap construction, for the constant passing of the shoes across the joints results in wearing away the edges *a a*, so that in time the cylinder is enlarged at this point, with the result that when the plunger end passes it it swings from side to side. All plunger cylinders are put together now with great care, the ends of the pipes being turned true in a lathe and screwed into the coupling sleeves until the ends come together as is shown in Fig. 228. Even with this construction, however, there is a possibility of having uneven surfaces at the joints, because the thread may not be central with the pipe, in which case the metal will be thicker on one side than the other, and

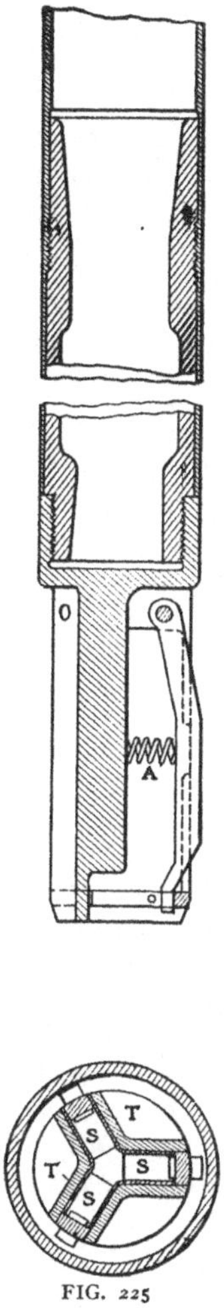

FIG. 225

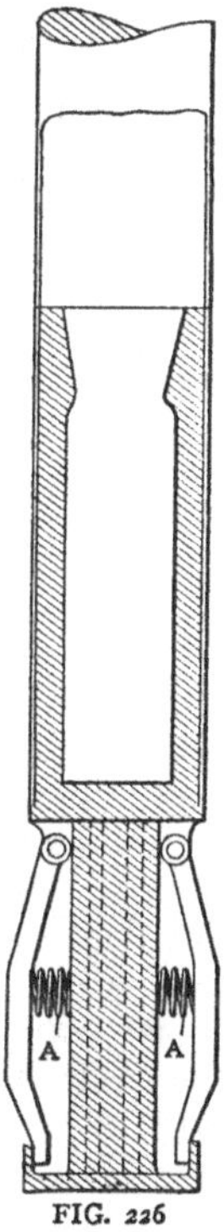

FIG. 226

if the thin part of one pipe comes opposite the thick part of the other, the result will be as shown in Fig. 229, and thus a cutting edge will be formed that will cause the shoe to wear away faster than if it rubbed against a smooth surface. The life of these shoes is therefore very variable, and according to the opinion of those who have had the greatest experience with plunger elevators may range all the way from five or six years down to as many months.

REPLACING WORN SHOES.

When the shoes wear out to such an extent that they do not all touch the sides of the cylinder at the same time they must be replaced, and it is needless to say that to replace them it is necessary to draw the end of the plunger out of the cylinder. There are two ways in which the end of

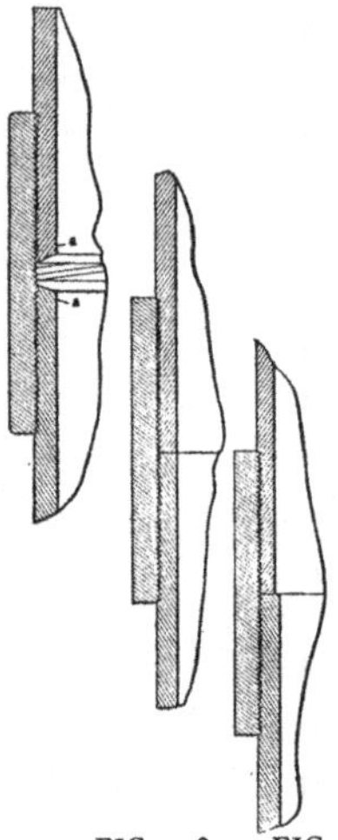

FIG. 227 FIG. 228 FIG. 229

the plunger may be drawn out of the cylinder; one is by lifting the car and plunger high enough to raise the lower end of the plunger above the top of the cylinder. This can be done in some cases where the overhead beams are high enough, but in most buildings if the car is raised as far as it can be the end of the plunger will still be within the cylinder. In such case the only way to get the end out is by disconnecting the plunger at one of the joints and then lifting the lower portion.

The first thing to do in either case is to run the car up as far as the pressure of the water will carry it. If the plunger can be lifted out whole, the car is made fast to the overhead beams by means of strong ropes that are abundantly able to hold the load. Next a powerful differential chain block is put in position to lift the car, being secured to the overhead framing and to the top car frame. The water is then shut off from the cylinder by closing the hand valves in the supply and discharge pipes, and the water in the valves and piping is drained if these are above the top of the cylinder, but not otherwise. Having done this, the stuffing-box casting is unscrewed from the top of the cylinder casting, and is made fast to the plunger so that it will not slide down when the plunger is raised. All this being done, the car and plunger are raised until the end of the latter comes out of the cylinder. As the car is raised, the slack in the ropes that hold it to the overhead beams is taken up so that if the hoisting tackle should break, the car would drop only a few inches.

When the end of the plunger is raised above the cylinder, if it is found that all the shoes are not badly worn they need not all be replaced, but as their cost is not great, and the trouble of replacing them is considerable, it is wise to not be too economical; hence, unless the wear is only trivial, they should be discarded. After the new shoes are put in place and the pins and springs set securely in position so that they cannot work out, the plunger is lowered into the cylinder until the stuffing-box casting rests on the top of the cylinder casting. Then the bolts are put in and screwed up so as to make a tight joint. The next thing to do is to remove entirely the ropes that hold the car to the overhead beams and lower the chain tackle until it slacks up, which shows that the plunger is resting on the water in the cylinder. The tackle is then removed and the valves in the supply and discharge pipes are opened and the elevator is ready to run. After it starts up, examine the joint between the stuffing box and top of cylinder to see if it is tight; if it is not, tighten the bolts, and if this does not make it tight, the car will have to be fastened up to the overhead beams again and the water turned off so as to loosen the bolts, separate the joint, clean it out and make it over again, using more care than before so as to be surer of making a good job of it.

TO DISCONNECT THE PLUNGER.

When the distance between the top of the car and the overhead beams is not sufficient to lift the car high enough to draw the lower end of the plunger out of the cylinder, and it becomes necessary to disconnect the plunger at one of the joints, the work is done as follows: After the car is run up to the top floor and made fast to the overhead beams, the water is shut off. The next step is to place wooden clamps on the plunger above and below the bottom joint, which is at the top of the first length of pipe. The illustration, Fig. 230, shows the position of these clamps, which must be made to fit the plunger accurately in order not to spring it out of shape. The clamps *A* are to be made long enough to reach from one side of the elevator well to the other, so that they may be firmly secured to the car guides, or some other suitable support. The object of these clamps is to hold the upper part of the plunger from turning so that when the lower joint is unscrewed the upper ones may not be disturbed. It is necessary that the clamp *A* be so tight that it cannot turn when the pipe is unscrewed. To make sure that it does not turn, make a chalk mark on the plunger and one on the clamp opposite and watch them when the clamp *B* is turned. Before starting to turn the latter clamp, look at the

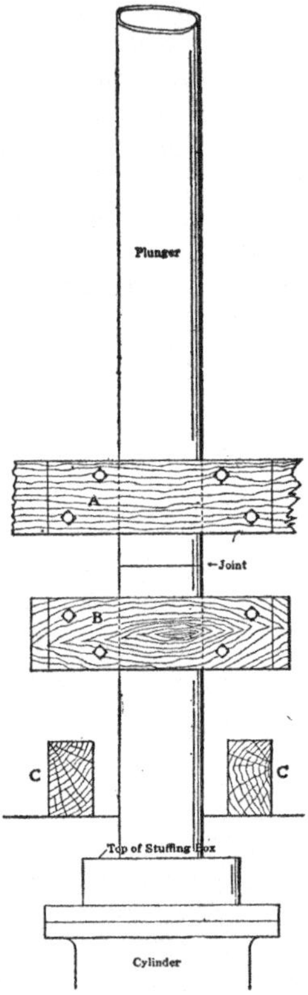

FIG. 230

joint and see if it is marked to show the position of the two parts when tightly screwed up. If not, make one with a cold chisel that has a sharp, straight edge. Then turn the clamp *B* to unscrew the pipe. If it sticks at first, it can be easily started by rapping the clamp rather hard with a large hammer or a sledge while pulling on the clamp.

After the pipe has been worked off, say 3/4 of an inch, place under the clamp *B* supports as shown at *C C*, so that it may not be able to drop when nearly loose and thereby strip the last few threads of the pipe or the coupling. The supports *C C* are here shown a considerable distance below the clamp *B*, but they should be held against it to carry the weight and relieve the screw thread from strain. It might be supposed that this is not necessary as the buoyancy of the plunger, together with the fric-

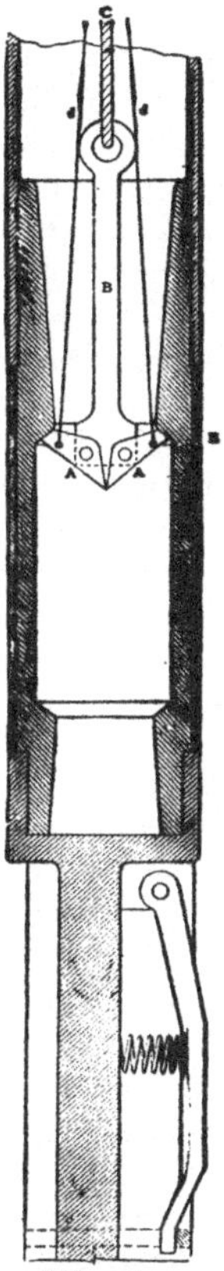

FIG. 231

tion of the stuffing, might be sufficient to hold the weight; but it is not safe to depend on this, because the water displaced only weighs about 14 pounds per running foot of the plunger while the lower end of the latter, on account of the weight of the lower casting, will probably weigh 30 pounds per foot. After the joint has been entirely unscrewed, the stuffing-box casting is unscrewed and then the lower section of the plunger is lifted out of the cylinder, the upper portion of the plunger being sprung to one side to make room. Unless the elevator is of very low lift, say less than 100 feet, no damage will be done to the plunger if it is sprung over to the side of the well, as this much bend is far within the elastic limit of the metal.

After the old shoes are removed the lower section of the plunger is put back in the cylinder and lowered until the end of the upper part can be brought into alinement with it; the thread is then covered with red lead for about half its length to make a water-tight joint, and the sections are screwed together a trifle farther than originally. On the last turn, when the chisel marks come in line, the two parts will be in their original relation, but the joint will not be as tight as before, because every time a screwed joint is taken apart and put together, the metal wears a trifle, leaving room for slightly more "take-up" when tightening it "home."

In the foregoing methods of taking out the lower end of the plunger, the stuffing box is raised with the plunger. This is done because it would be likely to damage the packing if the guide shoes at the lower end of the plunger were pulled and pushed across it. The type of packing shown in Fig. 222 could stand this treatment, but hemp packing could not.

When the lower end of the plunger is disconnected, care must be taken not to allow it to get away and drop into the cylinder. There is not a great deal of danger of such a thing happening, because if the clamp *B* is not removed the plunger end cannot drop. There is no reason for removing the clamp and every reason for keeping it in place, because it is an excellent point of attachment for the tackle used to pull the plunger end out of the cylinder. Notwithstanding all this, it is possible for something to occur that would cause the plunger end to get loose and slip down into the cylinder, and the possibility must be kept in mind. If one should ever get into such a scrape, he could get out of it by using a grapple of the type shown in Fig. 231, which consists of dogs *A A* mounted on the end of a rod *B* and arranged to swing out so as to catch against the under side of the shoulder *E* in the lower end casting of the plunger. After the plunger is raised out of the cylinder and made fast, the grapple can be taken out by pulling up the cords *d d*, thereby drawing the ends of the dogs into a position where they cannot catch against the shoulder *E*.

CHAPTER XXXIX

THE "STANDARD" PLUNGER ELEVATOR; PRINCIPAL DIFFERENCE BETWEEN THIS AND THE OTIS TYPES; VALVE CONSTRUCTION AND OPERATION

The plunger elevator made by the Standard Plunger Elevator Company differs in many details from the machine made by the Otis company. The principal differences are in the construction of the valves, and their location in the piping system connecting the lifting cylinder with the pressure tank. The general arrangement of the valves and piping is shown in Fig. 232, which is an elevation of the entire elevator apparatus. The valves are contained in two horizontal cylindrical chambers, one placed above the other, as shown. The upper cylinder contains the main operating valve and the pilot valve, the latter located at the right-hand end. The lower cylinder contains the two automatic stop valves, the top stop valve being in the left-hand end and the bottom one in the right-hand end. From the center of this chamber a pipe runs to the upper end of the lifting cylinder. The two valve chambers are connected with each other by two vertical pipes on opposite sides of the center, one for conveying the pressure water from the main valve to the lifting cylinder and the other for conveying the discharge water from the lifting cylinder to the discharge tank.

The pressure water enters the upper valve chamber through the supply pipe on the left, and passing through the main valve flows down through the left-side pipe connection to and through the top automatic stop valve and into the pipe that leads to the lifting cylinder, and thus forces the plunger and the car upward. On the down trip the water returns from the lifting cylinder to the lower valve chamber, and then passing through the down automatic stop valve on the right enters the right-side pipe connection that leads to the main valve chamber, and passing through the main valve reaches the discharge pipe. In the Otis machine, the automatic stop valves are placed in the pipe lines outside of the main valve, so that the pressure water passes through the top automatic valve before it reaches the main valve; and when the water is discharged as the elevator runs down, it first passes through the main valve and then through the down automatic stop valve. In the Standard elevator the arrangement is just the opposite of this, the pressure water passing through the main valve first and then through the top automatic stop valve, thence flowing into the cylinder; from the cylinder it flows through the down automatic stop valve and then through the main valve. One arrangement is as good as the other, and neither one would be improved or injured by changing it to the other arrangement.

In the Otis plunger the automatic stop valves are actuated by stationary or standing ropes, the upper ends being fastened to the overhead beams of the elevator well, and the lower ends to the valve levers. In Fig. 232 the stop valves are actuated by running ropes, the ends of which are fastened to the top and bottom of the elevator car. The rope *A*, which is attached to the under side of the car on the left-hand side, runs down to and around a sheave mounted on the end of the lever *L'* that actuates the down stop valve. From this sheave the rope runs up to the top of the elevator well to and around a sheave, and then runs down to the top of the car to which it is fastened on the right-hand side. The rope *B* is secured to the under side of the car on the right, at *B'*, and then runs straight down to and around the sheave on the end of the lever *L* of the top stop valve, and thence up to the top of the elevator well, over a sheave there and down to the left side of the top of the elevator car, where it is made fast. From this arrangement of the ropes it can be seen that when the car runs down the down-stop rope *A* will draw up lever *L'*, and when the car runs upward, rope *B* will draw up lever *L* of the top stop valve. The distance through which the levers *L* and *L'* are raised by the ropes can be varied by shifting the left-hand points of attachment of the ropes on the car in one direction or the other; shifting them farther to the left, as the drawing is made, increases the movement of the levers.

The pilot valve is operated by a standing-rope system, the ropes passing around sheaves *C* at the bottom of the well and other sheaves *C'* at the top. These latter sheaves are mounted on a frame that is forced upward by a spring contained in the cylinder shown below the sheaves. This maintains the proper tension on the ropes. The ropes pass around the sheaves *D* carried in a frame mounted on shaft *E* under the car, and to this shaft the car lever *F* is secured; by moving this lever in one direction or the other the pilot-valve lever *D'* is raised or depressed, and the car travel is thereby controlled.

CONSTRUCTION AND OPERATION OF VALVES.

The construction and operation of the valves of the system shown in Fig. 232 are indicated by Fig. 233, which is a sectional elevation parallel with the axis of the valve rods, and shows the general construction and positions of all the valves. The main valve consists of five pistons *Q*,

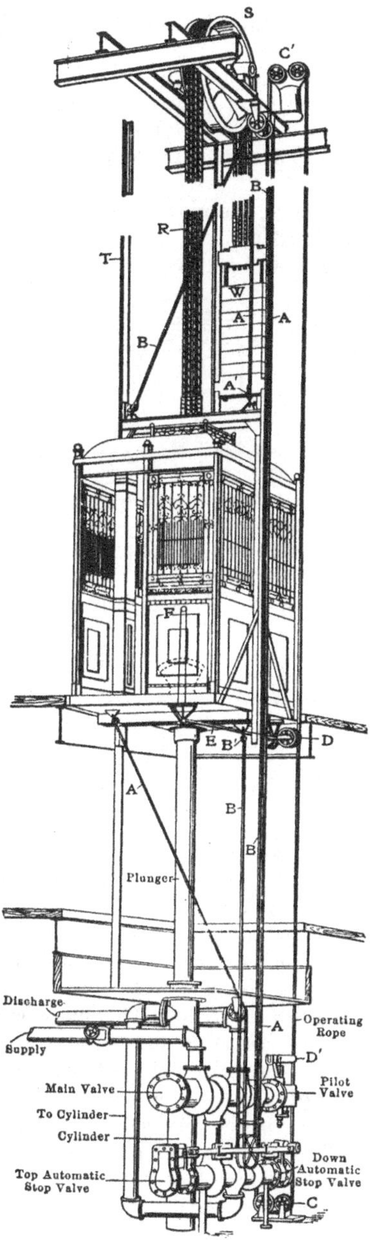

S
C′
B
R
T
W
A
A
B
A′
F
E
B′
D
A′
B
B
Plunger
Discharge
Supply
A
Operating Rope
D′
Main Valve
Pilot Valve
To Cylinder
Cylinder
Down Automatic Stop Valve
Top Automatic Stop Valve
C

R, *R'*, *T* and *T'*, the first of which acts as a motor to move the valve, in the same way as in all other pilot-valve constructions. The main valve is shown in the closed or stop position. The automatic stop valves in the lower cylinder are indicated by the letters *Y* (the down stop) and *Z* (the up stop).

In all types of pilot valve, the movement of the main valve acts to return the pilot to the closed position, and in most of the arrangements thus far shown the way in which this result is produced is easily seen,

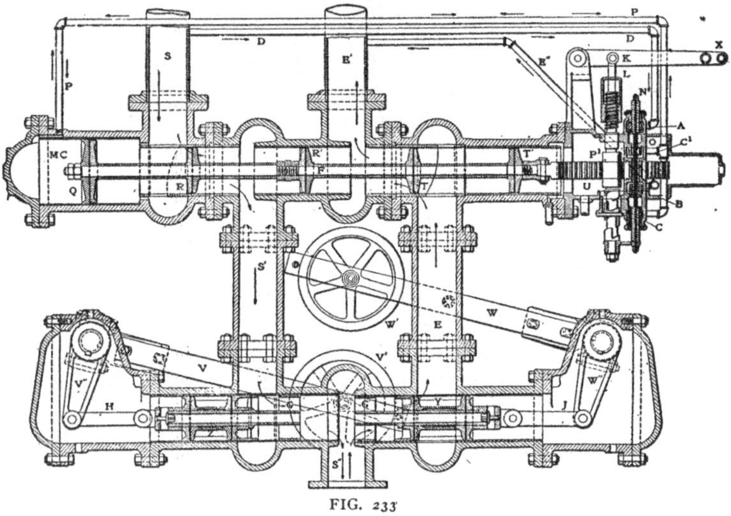

FIG. 233

but in this valve it is not so apparent. The part *U* at the right-hand end of the main valve stem is a rack, and *P'* is a pinion meshing with this rack. If the pilot-valve lever *X* is raised, the connecting rod *K* will lift the nut frame *L* and thereby raise the pilot valve, allowing water to flow in from the main supply pipe *S* through pipe *D*; passing through the pilot valve, it flows through pipe *P* to the end *M C* of the main valve chamber, where it forces piston *Q* to the right. This movement will rotate the pinion *P'* so as to run the screw at the upper end out of nut *L* and force *P'* down and with it the pilot valve, until the latter reaches the closed position when the movement of *Q* to the right will stop as no more water will flow into the space *M C*. If the pilot-valve lever *X* is depressed, the water in space *M C* will flow out through pipe *P* in the direction indicated by the dotted arrows and upon reaching the pilot valve will pass through it and into pipe *E"*, which leads to the main discharge pipe *E'*. The movement of the rack U will then be in the opposite direction and pinion P′ will be rotated so that the pilot valve will be raised to the closed position.

When the main valve pistons are moved to the left by the downward movement of the lever X, the ports of the supply pipe S will be opened by piston R and water will flow through as indicated by the arrows, to

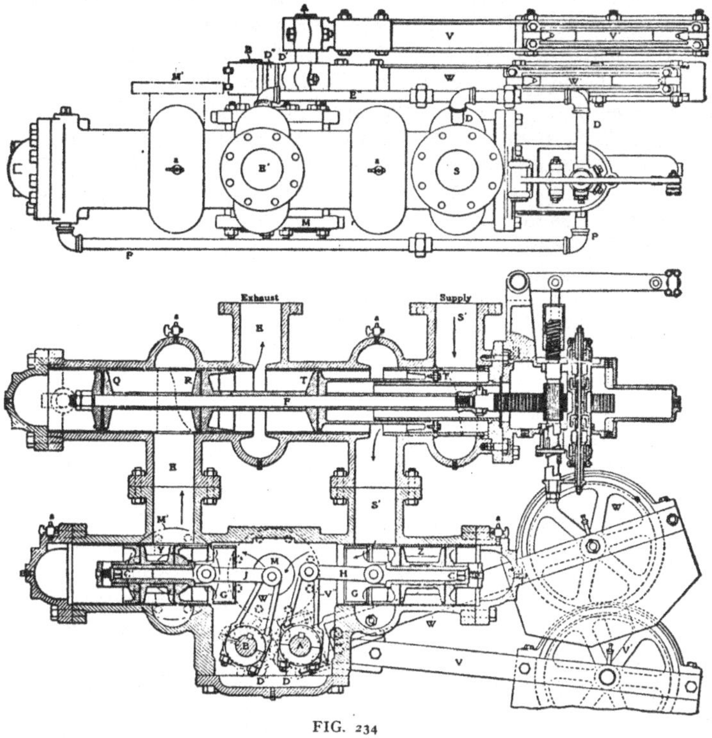

FIG. 234

the pipe S' and thence through the uncovered ports of the lower chamber to pipe S'' and to the cylinder, causing the car to run upward. The top stop valve Z will be in the open position as shown, and if the car is at the lower floor the down stop valve Y will be closed. As soon as the car begins to move upward, lever W of the down stop valve will descend and gradually open valve Y while the car is rising through the first ten or fifteen feet. When the car reaches the top of the well, if the operator does not move the car lever to stop the car, lever V will be gradually raised and the stop valve *Z* will slowly close and stop the car at the top floor. On the down trip the down stop valve *Y* will be opened, and as soon as the main valve is shifted to the right far enough to cause piston *T* to uncover the ports opposite pipe *E*, the water will flow out of the cylinder through pipe *S''* into pipe *E*, through the main valve to pipe *E'*, and the car will run down. When it approaches the lower floor the stop valve *Y* will be slowly closed by the lifting of lever *W*, being closed entirely when the car reaches the level of the lower floor.

Fig. 233 is one of the early designs used by the Standard company, and is considerably different from the valves used with elevators installed at the present time. The latest design of valve is shown in the drawings Fig. 234, which gives a plan view and a sectional elevation. The pilot valve is substantially the same as in Fig. 233, but the main valve is materially different and the automatic stop valves are turned around so that the cranks that move them are in the center, instead of at the ends of the valve chamber. The main valve consists of four pistons, and the force for moving them is obtained by making the piston *T'* of smaller area than the other three, so that this piston virtually takes the place of piston *Q* in Fig. 233. The flow of water through the valves is as indicated by the arrows; the port *M* at the center of the automatic stop valve chamber leads to the lifting cylinder. The two levers *W* and *V* which move the automatic stop valves swing around shaft *A* but only the lever *V* is keyed to it. The other lever swings freely around shaft *A* and through the gear segments *D'* and *D''* moves shaft *B* and thereby actuates the stop valve *Y*. This arrangement of the levers is made somewhat clearer by looking at the plan view in Fig. 234, which also shows the actual position of the pipe connections between the pilot valve and the main valve chamber, and as these are lettered the same as in Fig. 233, the direction of flow of water through them can be readily understood.

Each stop valve is provided with a cylindrical extension marked *G G'*, the same as in Fig. 233, and this extension acts as a throttle when the valve moves far enough backward. The valve *Y* is shown closed and valve *Z* open; through the latter water passes through a port that separates it from the extension *G*. If the valves are in perfect working order, the valve never moves farther away from the center than the position in which it is drawn, but if the rope that actuates the lever *V* should break, the lever would drop and carry the valve *Z* away from the center until its end touched the head of the valve chamber, and then the throttle *G* would cover the ports leading in from pipe *S'*. From this it is evident that the throttles *G G'* are simply safety devices that never come into action except in the case of some disarrangement of the ropes that lift the levers *V* and *W*; they are not made a tight fit, but are loose enough to permit water to leak through them fast enough to enable the car to move very slowly, so that if the actuating rope should break when the car is half way up the elevator well, it would not be held there but would slowly run to the end of the trip.

The valves *Y* and *Z* are made with an opening through the center so that water may circulate through them between the end spaces and the crank chamber at the center of the valve chamber. This construction is provided so that the valves may move freely. In each one of these openings a check valve *C* or *C'* is provided to prevent the valve from moving away from the center so rapidly, if the rope that actuates the lever should break, as to cause the throttle to stop the flow of water too suddenly. The check valves are so proportioned that when the water flows through them to the space back of the stop valve, the opening is large enough to afford it free passage, but when the water is forced out of this space, it has to pass through small holes made in the check valves and these are of such size that the weight of the lever *V* or *W* cannot move the throttle *G* or *G'* back over the ports so rapidly as to produce a too sudden reduction of the car speed.

It will be noticed that an outlet *M'* is shown in broken lines. This outlet is not always used, but when used its object is to economize pressure water, and also to prevent the plunger from rising above the water in the cylinder if a too rapid stop is made on an up trip.

CHAPTER XL

OPERATION OF THE PILOT AND MAIN VALVES; HOW THE ADJUSTMENTS ARE MADE

In order to explain the operation of the pilot valve shown in Figs. 233 and 234, larger drawings, Figs. 235, 236 and 237, are shown, the

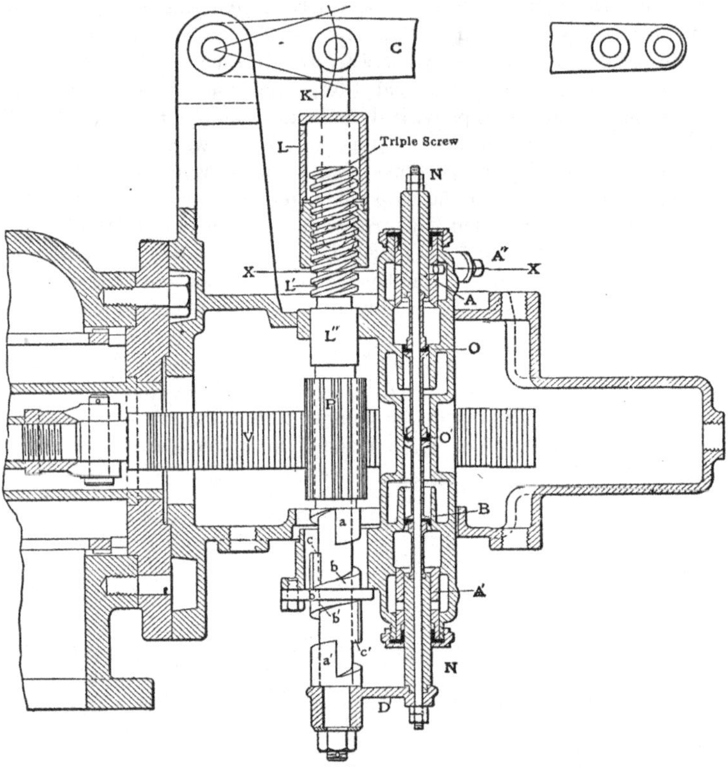

FIG. 235

first being a vertical longitudinal section, the same as that shown in Figs. 233 and 234, but on a larger scale, the second a cross-section through the center of the pilot valve, and as seen looking at Fig. 235 from the right side Fig. 237 is a top view showing a section through the upper end of the pilot valve on lines *X X* in Figs. 235 and 236. To the operating lever *C* is attached a connecting rod *K*. This rod guides a nut *L* that runs on a screw *L'* cut on the upper end of a shaft *L''*. This shaft also carries a pinion *P'* that meshes in a rack *V* mounted upon the end of the main valve stem. Below this pinion are two spiral cams *a* and *a'*, also mounted on the shaft *L''*. Between these cams there are two similar cams *b* and *b'* fastened to the plate *b''*. The shaft *L''* is connected with

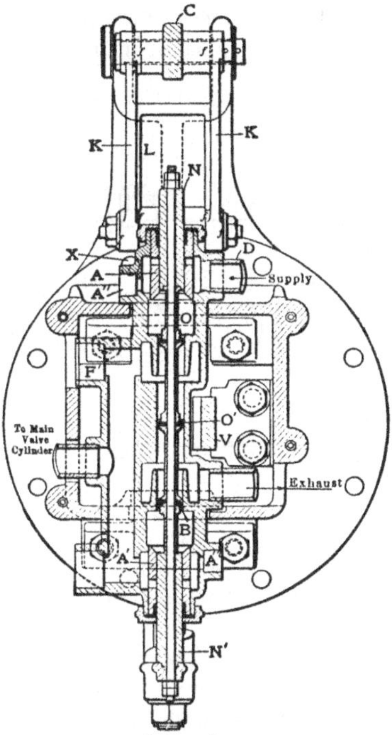

FIG. 236

the lower end of the pilot valve by an arm *D;* hence, if the lever *C* is raised the pilot valve is raised, and if the lever is depressed, the valve is likewise depressed. When lever *C* is raised, the cup packing *O* at the upper end of the pilot valve will be raised, and water from the supply pipe will pass through the opening made by lifting the cup *O* and pass to the central pipe that connects with the back end of the main valve cylinder; that is, to the pipe *P* of Figs. 233 and 234. This water will move the main valve to the right, and the rack *V* will rotate the pinion *P'* so as to cause the screw *L′* to run down in nut *L,* carrying the pilot valve downward and returning the cup packing *O* to the position in which it is shown in Fig. 236. This will stop the flow of water into the main valve cylinder and thereby stop the movement of the valve to the right.

The movement of the rack *V* must be sufficient to return the packing *O* to its seat, so that if lever *C* is raised a short distance the movement of rack *V* to the right will be small, and if the lever is raised as high as it can be moved, the rack will travel to the right its maximum distance. In other words, the amount of opening of the main valve must be directly

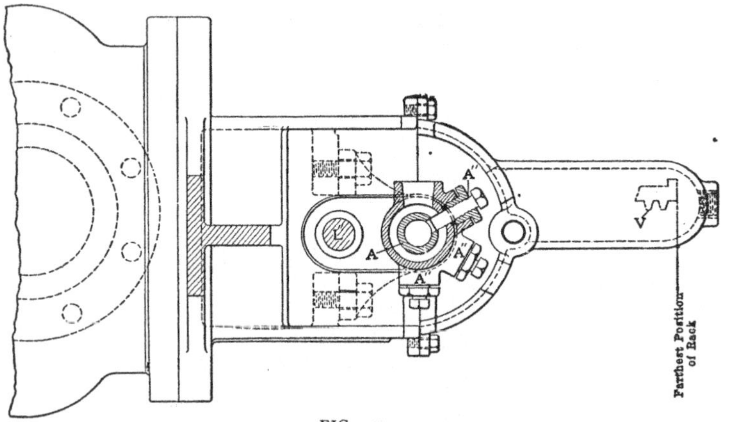

FIG. 237

proportional to the distance through which lever *C* is moved. Movement of lever *C* downward will have the same effect, but in the reverse direction, for then the pilot valve will be depressed and the lower cup packing *B* will be drawn down so that the water in the back end of the main valve cylinder, returning through pipe *P,* will be able to flow into the pilot valve through the central inlet, and passing down to the lower end will enter the valve cylinder and run up past cup *B* and out into the discharge pipe. As soon as water begins to flow out of the back end of the main valve cylinder, the main valve will move to the left, and rack *V* will rotate pinion *P′* in the direction to work the screw *L′* upward in nut *L,* thereby lifting the pilot valve and returning the cup packing *B* to its seat to close the outlet and prevent further movement of the main valve.

The stationary cams *b* and *b′*, in conjunction with the cams *a* and *a′* on the shaft *L″*, are provided to prevent sudden reversals of the elevator.

If these were not used, the operator could swing the car lever over from the full-speed position upward to the downward full-speed position almost instantly, and as the car motion could not be reversed in so short an interval of time, the plunger would be drawn away from the water, with the effects already explained. With the cams arranged as shown, the operator can only move the car lever back to the stop position; it will not move farther until the pilot and the main valves have been brought back to the stop position. The cams a, a' and b, b' are made of the same pitch as the screw L', and the distance between a and b or a' and b', when in the stop position, measured parallel with the axis of the shaft L'', is equal to the full stroke of the pilot valve; therefore, if lever C is moved its full swing in either direction, the corresponding cam on the shaft will bring up against the stationary cam opposite it, and the other cam on the shaft will be drawn to double the initial distance from the other stationary cam. As soon as the main valve begins to move it will rotate the last-named shaft cam around over the high point of the stationary cam, and as the thread of screw L' and the cam pitch are the same, the distance separating the cams will be just equal to the distance through which the shaft L'' has been raised; hence, if the operator moves the car lever in the reverse direction, he can only move it as far as he did when starting, that is, back to the stop position, but no farther. If, after the operator has moved the car lever to the central position and finds that it will go no farther, he persists in keeping up the pressure on it, when the shaft cam passes beyond the high point c or c' of the stationary cam, the lever will be freed and the operator will be able to move it until the shaft cam strikes the stationary cam, but this will be after the pilot and main valve have been brought to the stop position, and the elevator has likewise stopped.

In addition to the cams a, a' and b, b', several adjusting plugs are provided to regulate the flow of water through the pilot valve. Looking at Figs. 235 and 236, it will be seen that at the upper end of the pilot valve there is a plunger N that slides through a sleeve A, and at the lower end there is another plunger N' that slides through a sleeve A'. These sleeves have three port holes set on a diagonal line and opposite each port there is an adjusting plug A''; the positions of these are shown in Fig. 237. By means of these screws the opening through the port holes can be varied to any extent that may be desired. In addition to these adjusting plugs there are two others that screw into the holes F', shown in Fig. 236. The upper plug F' controls the flow of pressure water into the space at the back end of the main valve cylinder, and the lower plug controls the flow of the discharged water. From Fig. 234 it can be seen that in order to start the elevator on the up trip with this type of valve it is necessary to move the main valves to the right to allow water from the supply pipe to pass into the cylinder. To move the main valve to the right, pressure water has to be admitted to the space back of the piston *Q*. To start the car on the down trip the main valves have to be moved to the left, in order that the piston *R* may uncover the ports opposite the pipe *E* and permit water to escape from the cylinder. To move the main valves to the left, water must escape from the space back of the piston *Q*. To stop the car on the up trip the main valve must move to the left, hence water must be drawn from the space back of the piston *Q*, and to stop on the down trip the main valve must be moved to the right; hence water must be admitted to the space back of the piston. The rate of flow through the pilot valve from the supply pipe to the central outlet that connects with the pipe leading to the space back of the piston *Q* is controlled by the adjusting plugs *A″* opposite the port holes in the upper sleeve *A*, Fig. 236, and also by the plug in the hole *F′*. In starting the car on the upward trip it is necessary to open the main valve quickly, to get the full pressure in the lifting cylinder, because the load has to be started from rest. In stopping on the down trip it is necessary to shut off the flow quickly because the momentum of the downward moving car and plunger will force a large quantity of water through a small opening; therefore, the adjustment that is proper for giving the car rapid acceleration in starting on the up trip is also proper to give rapid retardation in stopping on the down trip, and this adjustment is made by the same set of adjusting plugs, namely, those at the top of the pilot valve.

In starting on the down trip it is necessary to open the valve slowly, so that the weight of the car and plunger may not force the water out of the cylinder so fast as to cause the car to run down too rapidly; in stopping on the up trip, also, it is necessary to close the main valve slowly, in order not to stop the flow of water into the cylinder faster than the momentum of the counterbalance will permit the car to stop, otherwise the plunger will be lifted from the water. As already explained, the car is started on the down trip and stopped on the up trip by letting out the water in the space back of the piston *Q*, and in escaping it passes by the adjusting plugs *F* and *A″* at the lower end of the pilot valve; hence, the adjustment of these that is proper for making stops on up trips is also proper for starting on down trips. When the lever *C* is moved the full distance the pilot valve is given its full stroke, and the three port holes in the sleeve *A* opposite the plugs *A″* are uncovered by the plunger *N*, so that the maximum quantity of water flows through the pilot valve. When the lever is moved less than the full distance, the pilot valve is given less than the full stroke, and all the port holes in *A* are not uncovered by the plunger *N*. Therefore, if the operator desires to get under full headway in the shortest time possible all he has to do is to swing the operating lever all the way over, and the speed of the car will be accelerated at the highest rate for which the pilot valve is adjusted. If the operator does not desire to run at full speed he moves the car lever part of the distance, and the plunger *N* is not raised high enough to uncover all the port holes in the sleeve *A*, so that the car will not only run at a reduced speed, but will also get under way more slowly.

THE VALVES OF THE "STANDARD" PLUNGER FREIGHT ELEVATOR.

For the operation of freight elevators the Standard Plunger Elevator Company provides simple hand-rope-operated valves. These valves are made to be moved by a lever if the car speed is very low, by single-geared rack and pinion for moderate speed and by a double-geared rack and pinion for high velocity; they are also of the balanced and unbalanced types. An unbalanced-type valve with double-geared rack and pinion is shown in Fig. 238, and a balanced valve of similar design in Fig. 239. The unbalanced valve is not, strictly speaking, unbalanced; it is only so when used in an installation where the discharge tank is located higher up than the valve. If the pressure acting upward against the under side of piston *B* is the same as the pressure acting downward on piston *D* the valve will be perfectly balanced, because the pressure from the supply tank acts equally against the under side of *D* and the upper side of *C*. The pressure of the atmosphere acts on top of *D*, and if the discharge tank is on a level with the valve, the same pressure, or nearly so, will act under *B*; therefore, the valve will be fully balanced. If, however, the discharge tank is several feet above the valve, the pressure acting under *B* will be greater than that acting down on D, and the valve will not be fully balanced. The valve in Fig. 239 is fully balanced, no matter whether there is a back pressure from the discharge tank or not, because this pressure acts equally against the under side of piston *B* and the upper side of piston *A*; and the pressure of the atmosphere acts equally against the under side of *A* and the upper side of *D*. For slow-speed cars this type of valve is better than the complicated pilot valve, with its accompanying automatic stop valves, because it accomplishes all that the more complicated and expensive construction can accomplish, and, being far more simple, is not as liable to get out of order. It is not desirable for fast-running elevators, however, because the movement of the car cannot be controlled with as great precision by means of the hand rope, owing to the rapid motion of the car and the long distance through which the rope has to be pulled to effect a stop. This is the only advantage of the pilot valve with car-lever control. With it a fast-running car can be stopped even with the floors of the building by anybody after a few days' practice but with the hand-rope control only

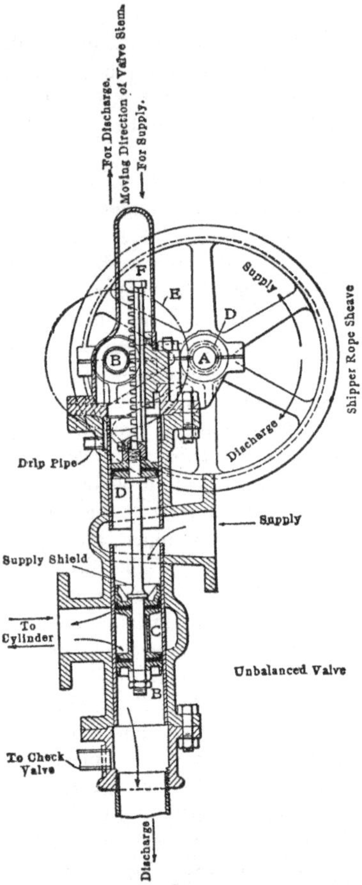

FG. 238

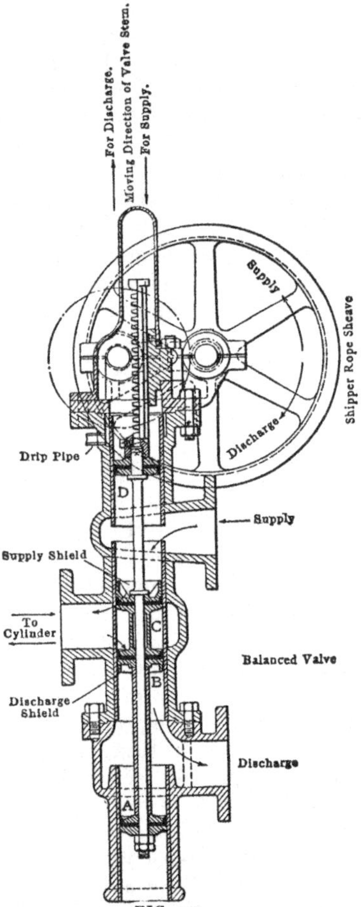

FIG. 239

the most experienced car operators can obtain results that are at all satisfactory in large office buildings.

LIFTING-CYLINDER DESIGN.

The casting that forms the upper end of the lifting cylinder is made in several designs by the Standard Plunger Elevator Company, one design being shown in Fig. 240, which is a vertical sectional view. The main casting is marked *A;* at *B* is the stuffing-box, and *C* is the upper end of the top-pipe section of the cylinder. The casting *A* is provided with a brass sleeve D that fits the lifting plunger and serves as a guide for it. This sleeve fits tightly at the upper end all the way around the

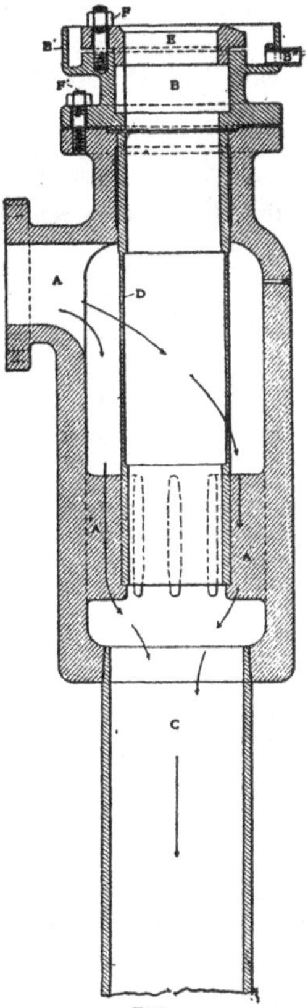

FIG. 240

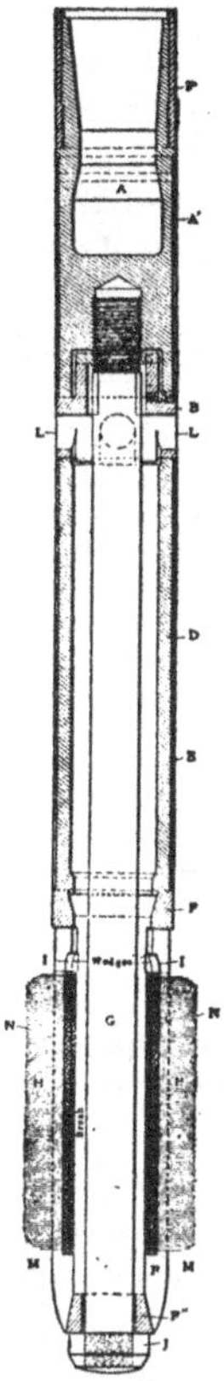

FIG. 241

circle, but at the lower end it is held in the central position by means of radial webs *A′ A′*, which are narrow enough to afford free passage for the water, but at the same time firm enough to give the sleeve proper support. The stuffing-box *B* is provided with a gland *E* pressed down by studs *F*. The box itself is secured to *A* by studs *F′*. The packing may be of hemp, or any good, soft packing material, but usually a special design of double cup packing is used. The stuffing-box is made with a rim B' which forms a basin to catch any water that may leak out of the cylinder. A drain pipe B'' is tapped in on one side to remove the water as fast as it accumulates.

Fig. 241 is a vertical section of the plunger end used in connection with the cylinder top shown in Fig. 240. This end is made up of the parts *A, B, D* and *F*, which are held together by a long central bolt *G*. The upper part *A* is screwed into the lower section of the plunger *P*. The parts *B, D* and *F* are pressed tightly against each other by the bolt *G* and nut *C*, and all these parts are held firmly against *A* by screwing the end of *G* into *A*, as shown. The parts *A* and *D* are made of cast iron, which would rust in time, as this part of the plunger does not ordinarily

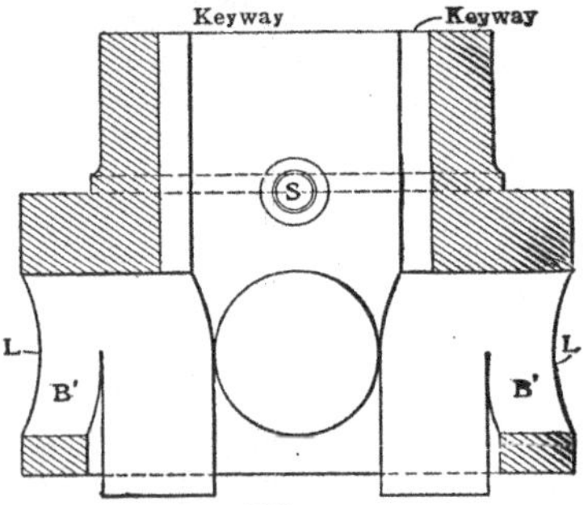

FIG. 242

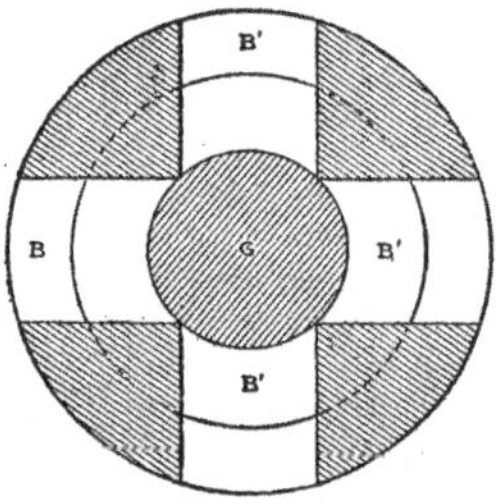

FIG. 243

run up into the sleeve *D* of the cylinder-top casting. On this account these parts are incased in brass, as shown at A' and *E*. The construction of the upper part *A* is simple, but the part *B* is better illustrated in Figs. 242 and 243, the first being a view similar to that in Fig. 241, the second a horizontal section through *L L*. This piece, it will be noticed, has four holes marked B' that radiate from a central opening larger in diameter than the bolt *G* opposite and below these holes. Above the holes the center hole of *B* fits the bolt *G*, and the latter is kept from turning in it by two keys.

The part *D* is simply a cylindrical piece shaped at its ends to fit over a projection depending from the under side of *B* and into a recess bored in the upper end of *F*, this construction being designed to bring the parts central when the bolt *G* is screwed up into the part *A*. It will be noticed that a screw *S* is run into the joint between *B* and *A* so these two parts cannot turn around with reference to each other and work the bolt *G* loose. The keys prevent *G* from turning in *B*, so all these parts are securely locked; therefore, the nut *C* cannot turn, but even if it did it could do no harm because after bolt *G* is screwed up tightly in *A* the nut is not depended upon; in fact, its principal object is to hold the lower parts together when they are disconnected from part *A*. The lower casting *F* has a longitudinal opening through it considerably larger than the bolt *G*, and this opening has lateral connections with the exterior of the casting. As the part *D* is also hollow, there is a free passage through the end of the plunger from the bottom of the casting *F* to the holes *B' B'* in the part *B*. The object of this construction is to provide positive means for stopping the upward movement of the elevator car before it

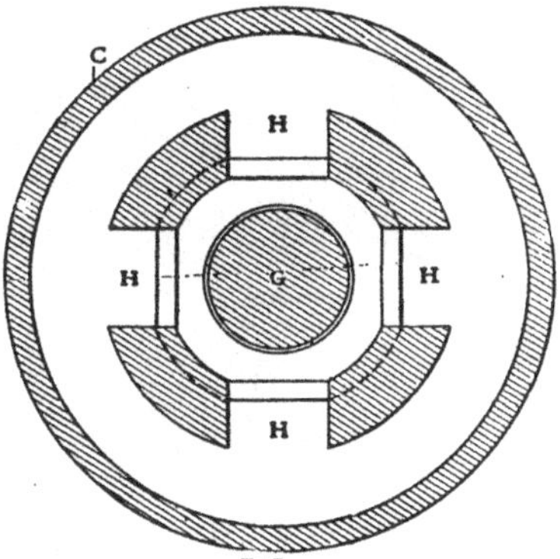

FIG. 244

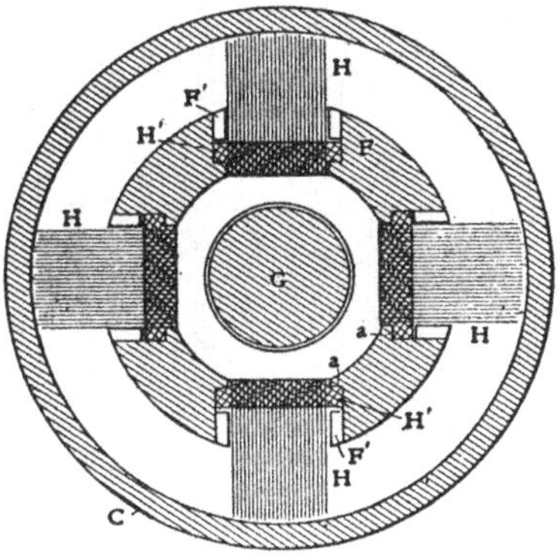

FIG. 245

reaches the overhead beams, if for any reason it should fail to stop at the upper floor. When the elevator is in perfect running order, the top automatic valve will stop the car even with the upper floor, and then the holes *B' B'* will be some distance below the stuffing-box, but if the stop valve fails to operate and the car continues upward, it will not rise far enough to strike the overhead beams before the holes *B'* will pass above the stuffing-box, and the water in the cylinder will find an outlet, the plunger will rise no farther.

CONSTRUCTION OF PLUNGER LOWER CASTING.

The lower casting *F* of the plunger is arranged to carry the guide brushes *H* that hold the plunger in the center of the cylinder. The construction of this casting and the way in which the brushes are held in place may be fully understood by the aid of the two horizontal sections, Figs. 244 and 245, taken on lines *N N* and *M M*, Fig. 241. These two sectional views also show a section of the cylinder *C*, to present more clearly the relative positions of the several parts. In Fig. 244 it will be seen that the brushes are held in grooves cast lengthwise of the casting *F*, and that these grooves are provided with flanges *a* along their inner edges, to prevent forcing the brushes too far in toward the center, and other short flanges *F'* to lock them in position. The brush back is made with short flanges *H'* that slide in back of the flanges *F'*. In putting the

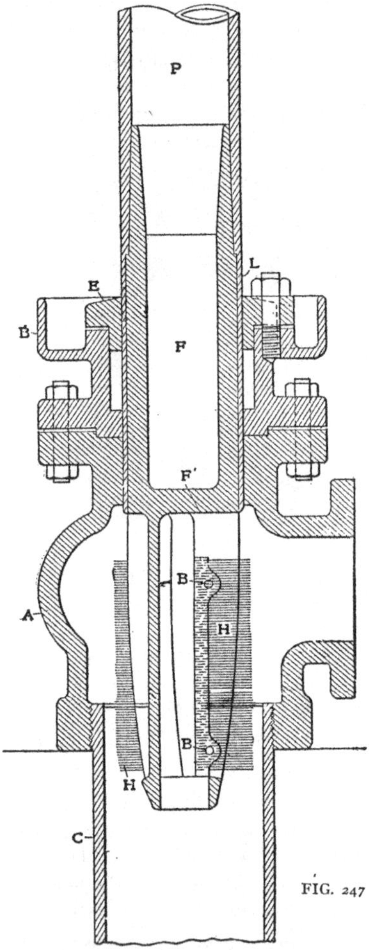

FIG. 247

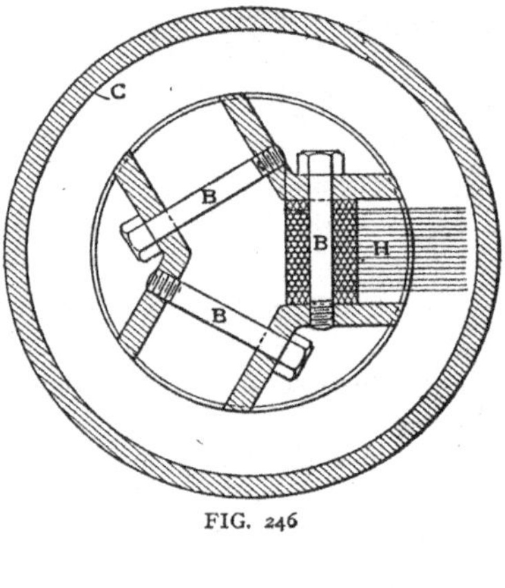

FIG. 246

brush in position it is raised to the top of the groove and then pressed in until the flanges *H'* can be forced down back of the flanges *F'*, then the brush is driven down and a key *I*, Fig. 241, is put in above the brush to prevent it from jumping up. The brush is forced down until the back rests hard against the bottom F'' of the side grooves in casting *F*. The keys *I* are not driven in endwise, but sidewise, that is, toward the center of the casting, and, when in position, are clinched so they cannot work out.

The brushes are made of hard spring-brass wire, about No. 22 gage. The back is of babbitt metal and is cast around the wires to hold them firmly in position. The grooves in the casting *F*, into which the brush backs fit, are not machined, but are simply carefully cast, and the burs well cleaned off. As the brush back is soft, there is no difficulty in forcing it into place. If it should fit too tightly, it can be easily shaved off where it binds. When the brushes are in place in the casting *F* the water in the cylinder can reach the central space through the openings above and below them, and also through the joints between the brush back and the casting, as these are not tight fits.

ANOTHER DESIGN OF PLUNGER END.

Another design of plunger end made by the Standard company is shown in Fig. 246, which is a vertical elevation in section, showing the plunger at its highest position, that is, in the position it reaches when the car is even with the upper floor of the building. The brushes in this case are held by the bolts *B*. A horizontal section through the lower end of the casting *F* is shown in Fig. 247, from which it will be seen that there are only three brushes. This design is simpler than that of Fig. 241, but it is not as perfect. In the latter if the car overruns the upper limit of travel the holes B' in the piece *B* will pass above the stuffing-box and let the water in the cylinder flow out before the brushes reach the packing, but in Fig. 246 it can be seen that for the water to escape the plunger must run up until the part F' of the casting passes above the gland *E*, and this will carry the upper end of the brushes up into the stuffing. If the latter is of the cup type it may not be damaged to any extent, but if it is hemp it is liable to be pulled out of place. This plunger end cannot be used with the cylinder top shown in Fig. 240, unless there is so much head room above the elevator car, when even with the top floor, as to permit running it several feet higher before the casting *F* is high enough to permit the water to escape. If with this cylinder top the plunger should run normally as high as it is drawn in Fig. 246, the brushes would be carried up into the brass lining *D* and, by being bent back and forth at every trip, would soon become useless. The cylinder top in Fig. 246 is very much shorter, so the plunger can rise just as high as the plunger in Fig. 241 can rise in the top in Fig. 240, without running the brushes up into the bore of the casting.

PIPING CONNECTIONS.

The pipe connections between the pump, tanks and lifting cylinder of a plunger-elevator system are generally very simple, but in some of the

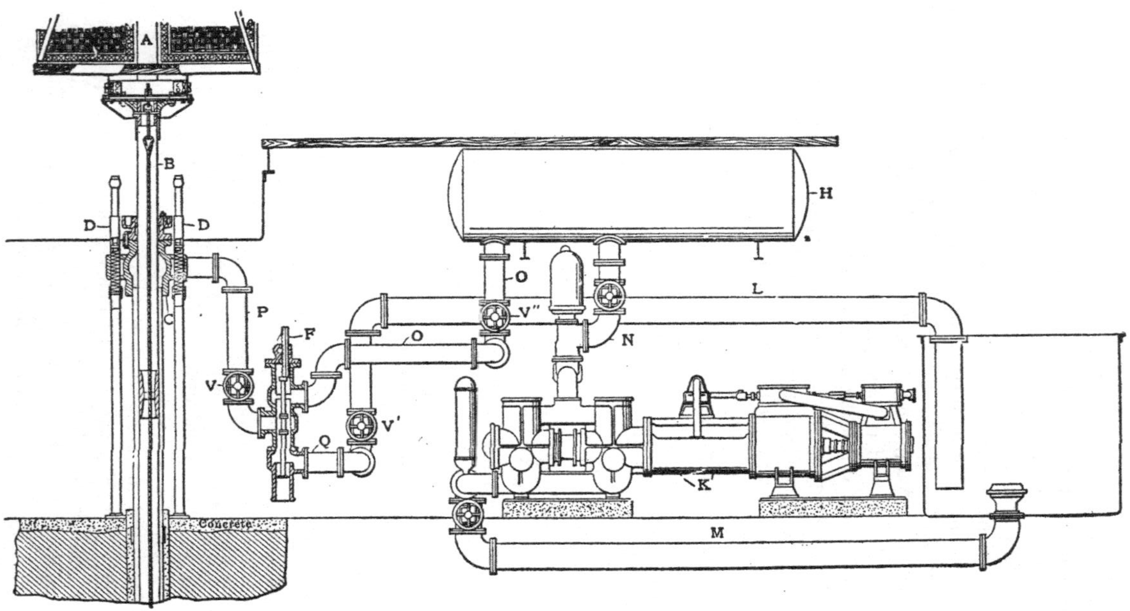

FIG. 248

higher-grade passenger-elevator installations they are very elaborate. The arrangement most commonly used is shown in Fig. 248. In this diagram *A* represents the lower portion of the elevator car, *B* the plunger, *C* the cylinder and *D D* spring buffers provided for the car to rest on when at the lower floor. The main valve is shown at *F*, and is represented as of the simple rack-and-single-gear type. The discharge tank is at *G*, and *H* is the pressure tank. The water in the lifting cylinder *C* is discharged into tank *G* through pipe *L*, and from this tank the pump draws its supply through the suction pipe *M*. The discharge pipe *N* of the pump leads to the pressure tank *H*, and from the latter the water is carried to the lifting cylinder through pipe *O*. In order to keep the necessary quantity of air in the pressure tank *H* means must be provided for forcing air into it from time to time, to replenish that which will inevitably escape in one way or another. In large installations, where several pumps and possibly tanks are provided, a small air pump is installed to furnish the compressed-air supply, but in smaller plants the pump *K* is arranged so as to pump air whenever necessary. The pressure tank *H* is provided with a glass water gage, to show the hight of water in it, and also with a pressure gage. In addition, a pressure regulator is used to stop the pump when the pressure in *H* rises to the maximum, and to start it when the pressure falls below the minimum.

Fig. 248 shows a system provided with a full complement of hand valves, three of these being marked *V*, *V'* and *V"*. There are two more, one in the pump suction and one in the pump-delivery pipe *N*. When all these valves are placed in the pipe lines the inspection of the several parts of the apparatus may be done with very little trouble. If it is desired to examine or renew the cylinder packing, all that is necessary is to run the car down to the lower floor and then close valve *V*. If the main valve is to be taken apart, valves *V*, *V'* and *V"* are closed. To inspect the pressure tank *H*, valve *V"* and the one in pipe *N* are closed. If repairs or inspection of the pump are required the valves in pipes *M* and *N* are closed. Thus with all the valves shown it is not necessary to draw water from as much of the system as has been stated in previous chapters, in which it was assumed that a lesser number were used. If the discharge tank *G* is lower down than the discharge pipe *Q* valve *V'* may be dispensed with without impairing the system, and we may also add that the balanced main valve *F* can be replaced by one of the unbalanced type, such as shown in Fig. 238. The valve in the suction pipe *M* may also be discarded.

CHAPTER XLI

CONSTRUCTION AND OPERATION DETAILS OF THE HIGHEST TYPE OF PASSENGER ELEVATOR MADE BY THE STANDARD PLUNGER ELEVATOR COMPANY

For the highest type of passenger elevator the Standard Plunger Elevator Company uses the system shown diagrammatically in Fig. 249. In this arrangement it will be noticed that the discharge tank *G* is located several floors above the top of the lifting cylinder. The hight of the

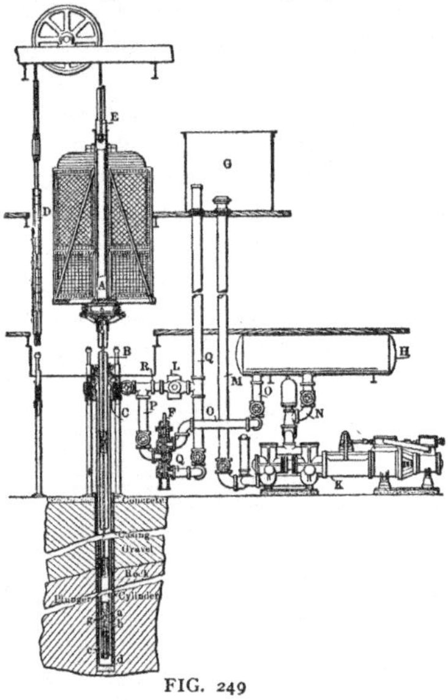

FIG. 249

discharge tank varies according to the car speed, and ranges from about 40 feet for moderate car speed to double this hight for speeds of 500 or 600 feet per minute. In addition to setting the discharge tank at an elevation, the discharge pipe *Q* is connected with the inlet pipe *P* through a branch *R* in which is inserted a check valve *L*. The object of this pipe connection is twofold; first, it prevents drawing the plunger away from the water in making stops on the upward trips, and, second, it saves a considerable quantity of pressure water, and thereby increases the efficiency of the apparatus. The valve *L* permits water to flow freely from pipe *Q* into the cylinder, but prevents water from passing through it from the cylinder to pipe *Q*. The operation of the system is as follows: Suppose the elevator is running up at full speed and that the operating valve *F* is closed quickly; then the momentum of the counterbalance *D*

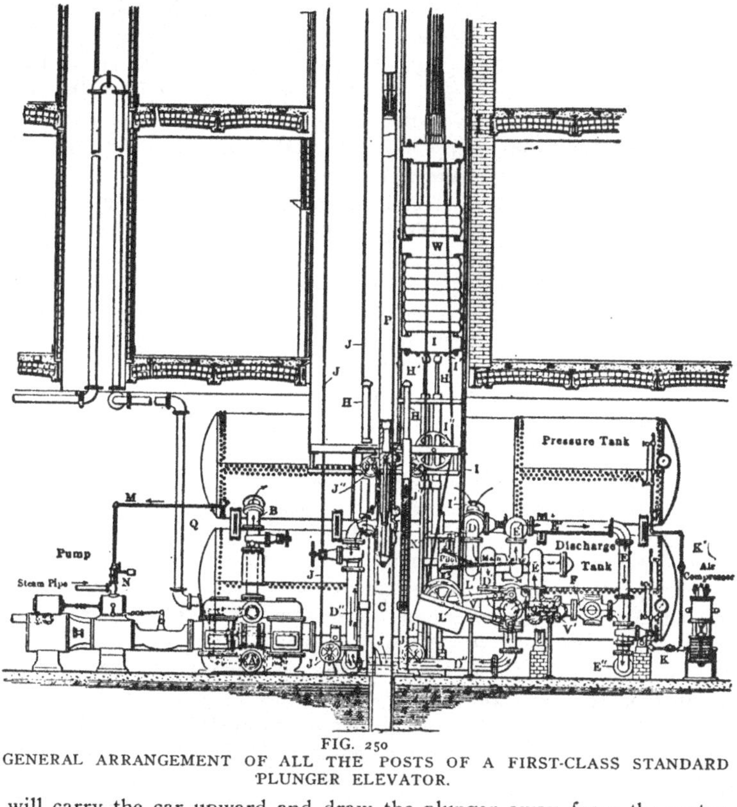

FIG. 250
GENERAL ARRANGEMENT OF ALL THE POSTS OF A FIRST-CLASS STANDARD PLUNGER ELEVATOR.

will carry the car upward and draw the plunger away from the water. This would be the effect if the pipe connection *R* were not provided, but with this connection, as soon as the plunger begins to draw away from the water, the vacuum developed, assisted by the pressure due to the elevation of the tank *G*, will cause water to run down through pipe *Q*, valve *L* and pipe *R* into the cylinder and keep the latter full. When the plunger comes to a state of rest there is no empty space under it, and as a result the car will not drop down as would be the case if water could not enter the cylinder.

To avoid drawing the plunger away from the water by too rapid a valve closure on the upward trips when the simple pipe arrangement of Fig. 248 is used, the pilot valve is adjusted so that the main valve cannot close too rapidly. With the arrangement of Fig. 249 it is immaterial how quickly the main valve is closed, providing the discharge tank is placed high enough to develop as much pressure as may be necessary to cause water to flow in through pipe *R* and follow up the plunger as fast as it moves until its motion is arrested by the greater weight of the car. All the water that is drawn into the cylinder through the pipe *R* in making stops represents energy saved, because it reduces the amount of water drawn from the pressure tank *H*.

It is not practicable in all buildings to set a discharge tank at the desired elevation, and in such cases the elevated tank *G* must be replaced by a pressure tank located in the basement. A system of this kind is shown by Fig. 250, which is far more elaborate than Fig. 249, and shows every detail of a high-class passenger-elevator system. Of the two tanks shown, the top one is the pressure and the lower one the discharge tank. The pipe *Q* leads to an inverted U consisting of two legs, as shown, the function of which is to maintain a uniform pressure in the discharge tank. This U-pipe is extended up to whatever hight may be necessary to develop the required pressure. At the bend at the top a short vent pipe is provided, which is open at the upper end, so as to prevent the inverted U from acting as a siphon and drawing the water out of the tank. The pilot-valve lever *X* is actuated by the rope *J* which runs under stationary sheaves *J'* at the bottom and over and under the two sheaves *J''* at the top of the pit at the bottom of the shaft. At the top of the building the rope *J* runs over other sheaves, as shown in Fig. 251, which represents all the apparatus at the upper end of the elevator well, and also the elevator car. The lever *L* of the top automatic stop valve *V* is actuated by rope *I*, and the lever *L'* of the down automatic stop valve *V'* is actuated by rope *I'*. The points of attachment of these ropes to the car and the way in which they are supported at the top of the elevator well are shown in Fig. 251.

The construction of the car buffers is shown in Fig. 250 at *H H*. The counterbalance buffers are of similar design, but are not generally provided with the rubber cushions shown below the spiral springs in the car buffers. If either the car or the counterbalance strikes the buffers running at a high speed, the latter are pushed down until the compression of the springs arrests the motion. The stroke of the buffers depends on the speed of the elevator, being made greater as the speed increases. The pump on the right draws water from the discharge tank through the suction pipe *A*, and delivers into the pressure tank through the pipe *B*. The air compressor forces air into the discharge tank through the pipe *K* and into the pressure tank through pipe *K'*. Each tank is provided with

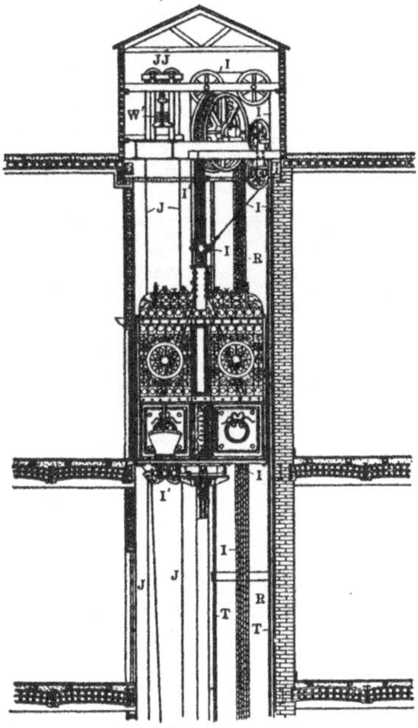

FIG. 251

gages to show the pressure and the water level. The compressor is run only occasionally, when the air supply runs low. The operation of the main pump is controlled by a pressure regulator *N* which is connected with the pressure tank by pipe *M*. This regulator controls the valve in the steam pipe and thus stops, and starts the pump whenever required by the variations of pressure in the tank.

The operation of the elevator is as follows: To start on the up trip the pilot-valve lever *X* is depressed, causing the main valve to be moved to the left; this allows water to pass out of the pressure tank through pipe *D* and the main valve to the connection *D'*, thence through the top stop valve *V* to pipe *D''*, and to the cylinder *C*, forcing the plunger *P* and the car upward. If the car is stopped on the up trip, the flow of water through this path is arrested by the closing of the main valve, and if the latter closes so rapidly that the plunger starts to rise above the water in the cylinder, then the water in the lower discharge tank flows upward through pipe *E″*, to and through check valve *F* and into the chamber of the down stop valve *V′*, as indicated by the dotted lines back of the valve; from here it runs into the cylinder until the car stops. In this drawing the check valve *F* is the same as the valve *L* of Fig. 249. When the car is running upward valve *V′* is open, so that water coming

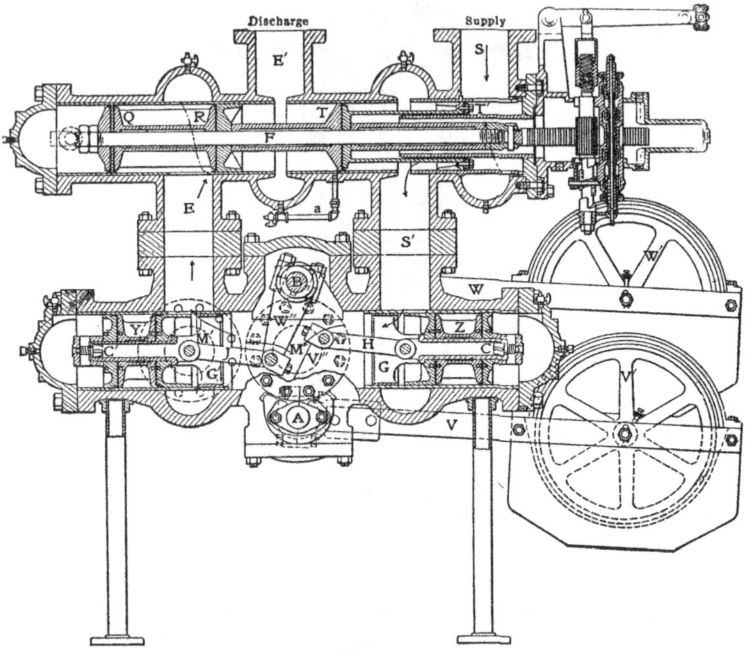

FIG. 252

from the discharge tank through check valve *F* can pass through it freely. When the car is descending the down stop valve *V′* is open until the lower floor is reached; consequently, the discharge water returning from the cylinder *C* can pass from pipe *D″* to the connection *E*, thence through the main valve to pipe *E′* and to the discharge tank through pipe *E″*.

The automatic stop valves shown in Fig. 250 are arranged slightly different from those presented in Fig. 234, this arrangement being in the position of the shafts upon which the operating levers are mounted. The main valve also is provided with a safety feature not shown on other drawings. These points of difference can be understood by inspection of Fig. 252, which is an enlarged sectional view of the main and stop valves of Fig. 250. The advantage of placing the shafts *B* and *A* one above the other is that the levers *W* and *V* can be attached directly to them, while in the construction shown by Fig. 234 one of the levers swings on the shaft of the opposite lever, and imparts movement to its own shaft through spur-gear segments. In Fig. 252 the operation of the stop valves does not appear to be practicable because it looks as if a very small

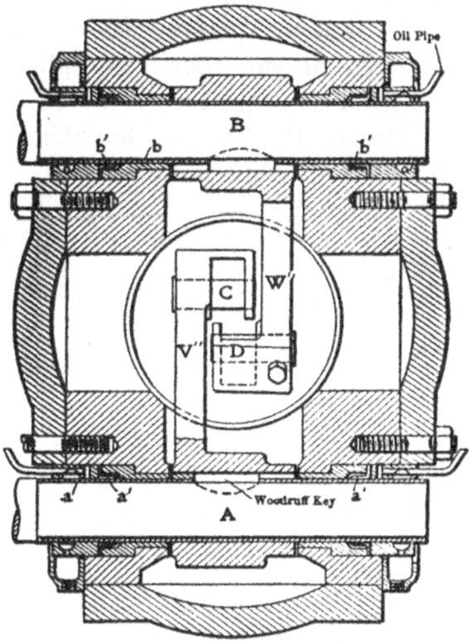

FIG. 253

movement of crank *V″* to the left would carry it into crank *W″*. This difficulty can be fully cleared up by the aid of Fig. 253, which is a vertical section at right angles to the view in Fig. 252 and passing through the centers of the shafts *A* and *B*. Looking at this drawing it will be seen that the cranks *V″* and *W″* are made so that they can swing past each other. This view also shows the way in which the bearings of the shafts *A* and *B* are made water-tight by the use of the cup packings *a′* and *b′*. The shafts are incased in brass tubing *a b* to prevent corrosion.

The safety device attached to the main valve in Fig. 250 is clearly shown in Fig. 252; it consists of the small pipe connection *a*, and its operation is as follows: Suppose the car is running upward; when it reaches the upper floor the top stop valve *Z* will close, and at the same time the main valve piston T' will move to the left, thereby locking pressure water in the space S' between the main valve and the stop valve. This pressure will force the cup packings of the stop valve Z out so as to develop possibly sufficient friction to prevent lever V and the weight of sheave V' from shifting the valve to the open position when the car starts on the down trip. When the pipe *a* is provided this cannot happen because in order to run the car down the main valve has to be moved to the left so that the piston R may be carried beyond the port and thus open communication between E and E'. As soon as the main valve moves far enough for the piston T to pass to the left of the inlet of pipe *a*, the pressure in S' drops to equality with that in E' and then the friction of the cup packings of valve Z is so reduced that the valve cannot stick. It might be said that this same result could be accomplished by putting additional weight on the lever V, but this would increase the tension on the operating rope, which is objectionable.

CHAPTER XLII

PRACTICAL INSTRUCTIONS IN THE CARE AND MANAGEMENT OF THE "STANDARD" PLUNGER ELEVATOR; ESSENTIAL FEATURES TO LOOK OUT FOR

Whenever it is desired to take out the main valve of the Standard plunger elevator, it can be removed through the back end of the valve cylinder. Before it can be drawn out, however, the rack at the end that rotates the pinion of the pilot valve must be thrown out of gear. To do this all that is necessary is to remove the hood in front of the pilot valve, into which the rack runs, and then the shoe that holds the rack and pinion in mesh can also be removed and the rack can be pushed to one side so as to clear the teeth of the pinion. When this is done the valve can be drawn out of the back end of the valve cylinder without difficulty.

To remove the automatic stop valve the cylinder head must be removed, and also the bonnet under the center. The cranks that operate the automatic stop valves are fastened to the shafts on which the operating levers are mounted by means of caps, and the screws that hold these caps can be reached when the bonnet is removed. If the cap is taken off the crank can be pushed upward and can be drawn out, together with the valve, through the end of the valve cylinder; all of which can be readily understood upon examining the valve drawing, Fig. 234. The cranks are keyed to the shafts, to prevent them from turning, and in putting the valve back care must be taken that the key is returned to position and the screws tightened up as much as they were before, so that there may be no danger of working the parts loose thereafter.

PILOT VALVE REMOVAL AND ADJUSTMENT.

The pilot valve, body and all, can be removed by taking off the end hood the same as for throwing the rack out of gear, as explained above. When this hood is removed, the bolts that hold the pilot-valve body can be reached and taken out and then the valve can be removed, together with the shaft that carries the pinion and the cams that prevent too rapid reversal of the elevator motion. A side view of all these parts is given in Fig. 254, which is a vertical section. This drawing does not show the means by which the valve body is fastened to the end casting of the main valve body; these consist of lugs that spread out on each side of the shaft *L″* at the top and bottom, opposite the bearings through which the shaft slides. A view of the valve body at right angles to Fig. 254 would show these lugs, on opposite sides of the parts *E* and *F*.

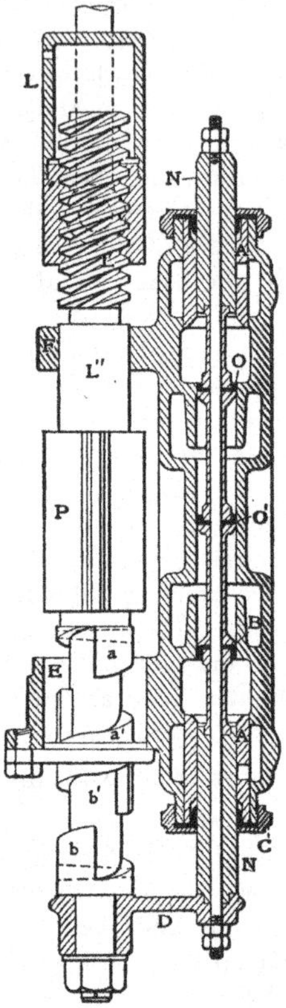

FIG. 254

To remove the valve alone, all that is necessary is to take off the connecting arm *D* and the lower cap *C*; then the valve can be drawn out through the lower end.

It can be seen that no provision is made for adjusting the position of the pilot-valve cup packings, nor for adjusting the cams *a, b, a′* and *b′*. Adjustment of the position of the cup packings would only serve to vary the lap of the valve, and such adjustment is not only not necessary but not advisable, because the manufacturers know better than any one else what the adjustment should be, and they make the valve of proper proportions. Increasing the depth of the cups will not have any effect on the lap of the valve, because they enter their seats back end first and make a joint after entering a certain distance, independent of the depth of the cup. Under certain conditions, if the edge of the cup projects beyond the end of the cylinder, water may force its way between it and the cylinder and thus leak through. This is not likely to occur, but as it may, it is wise to use cups of the proper depth, and no deeper. The cams require no adjustment, because all they are intended for is to prevent moving the lever any farther, in stopping, than it was moved in starting; and if once made of the proper dimensions to accomplish this result, they will always do so.

The only adjustment provided in the pilot valve is in the ports through the sleeves *A, A'* at the ends of the valve, and the similar adjustment on the side ports, which was fully explained in the article describing this apparatus. If in the course of time the water flowing through these port holes enlarges them so as to cause the main valve to close too rapidly in stopping, the proper adjustment can be obtained by running in the adjusting plugs a trifle. It may be found in making such changes that the car speeds up too fast in starting when the valve is partly opened in order to run at a slow speed. If this should be the case, the acceleration can be reduced by screwing in farther the plug opposite the port hole in the inner end of the sleeve *A*, and if after doing this the car does not get under headway fast enough when the valve is fully opened, the acceleration can be increased by drawing out one of the other adjusting plugs. In making these adjustments it should be remembered that a very small difference in the opening of the ports will make a decided difference in the rapidity with which the elevator will get under way; hence, the position of the plugs should be changed only a little at a time. In the type of valve shown in Fig. 234 the main valve is moved to the right to cause the elevator to start upward; it is also moved to the right to stop the elevator on downward trips. Therefore, if the flow of water through the ports of the top sleeve *A* is decreased, the effect will be to reduce the acceleration in starting on up trips, and to prolong the stopping on down trips. To stop going up and to start going down the main valve must be moved to the left; hence, if the adjusting plugs opposite the ports in the lower sleeve *A* are run in, the up stops and the downward starts will be made slower, and *vice versa*.

If the elevator is arranged so that the cylinder discharges into an open tank located on a level with the main valve, there will be no back pressure to force the water into the cylinder through the bypass connection, and the adjustment of the velocity of motion of the main valve must therefore be made so as to reduce the velocity enough to prevent jumping the plunger off the water in the cylinder when the car is brought from its maximum speed to a stop. If, however, the water in the cylinder is discharged into an elevated tank, or into a pressure tank, the valve is adjusted with reference to starting on the downward trips, so that the car may not move so rapidly as to produce an unpleasant sensation. Therefore, it will be seen that the adjustment of the plugs at the lower end of the pilot valve, opposite sleeve *A*, must be made with reference to the rapidity of stopping on the upward trips, with one method of piping, and with reference to starting on the downward trips with the other method.

The adjustment of the plugs opposite the sleeve *A* at the top of the pilot valve is made with reference to the rapidity of starting on the upward trips, and stopping on downward trips. There is little danger of starting too rapidly, because the water flowing into the cylinder has to lift the load, and it cannot very well get it under headway so rapidly as to produce an unpleasant sensation, unless the lifting capacity of the plunger is excessive, and the load in the car is light. In stopping on the downward trips, however, the reduction of speed can be so rapid as to greatly increase the tendency to buckle the plunger, hence the adjustment of the plugs at the top of the pilot valve should be made with reference to the rate of retardation of speed in stopping on the down trips, and this adjustment will be found satisfactory for the starting on upward trips.

In the valve shown in Fig. 233 the movement of the main valve is the reverse of that above explained, that is, the valve moves to the left to start on the upward trip, instead of to the right, hence the top adjusting plugs are used to do just what the bottom ones do in Fig. 234.

THE PACKINGS.

All the packings used in the valves of the Standard plunger elevators are leather cups, as can be seen by looking at the various drawings we have presented. These packings are replaced in the same manner as in the elevators of other makes previously explained, and require no further explanation here. The stuffing-box at the top of the plunger cylinder is packed either with hemp or any good soft packing, or with a specially constructed double cup leather packing. The cross-section of this packing is shown in Fig. 255. The packing is made in two parts, *A* and *B*, both of leather. These two parts are cut on one side, so that they may be slipped over the plunger from the side, and they are placed in the stuffing-box so that the joints are on opposite sides of the diameter.

To keep any hydraulic elevator in perfect running order it is necessary that all the packings be kept tight; if they are not, the car will not remain stationary when stopped at a floor, but will move gradually either up or down, according to where the leak is located. In plunger elevators, if the stuffing-box at the top of the cylinder leaks the car will settle when standing at a floor. It is an easy matter to determine whether the cylinder stuffing-box leaks or not, for if it does the water can be seen trickling over the top of the stuffing-box gland. If there is no leak at this point, then the trouble will be found in the main valve, which will permit water to flow through to the discharge pipe; hence, the packing in the piston that shuts off the discharge must be renewed.

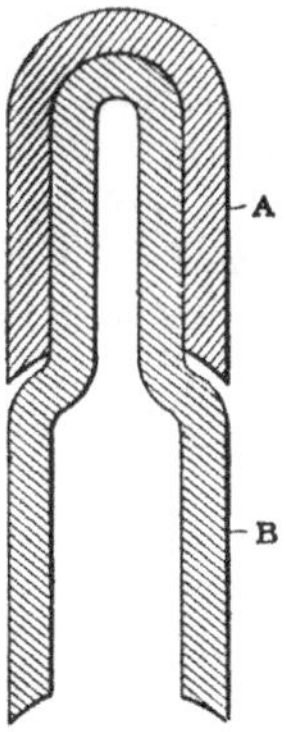

FIG. 255

If the car creeps upward after being brought to a stop it indicates that the cup packing in the valve piston that closes the supply ports is leaky. Sometimes the elevator, after being brought to a standstill, creeps up a short distance and then creeps back, and continues this alternating motion indefinitely. This indicates that the pilot valve is defective, and it is an occurrence that can take place with any type of hydraulic elevator.

In addition to keeping all the packings in good condition it is necessary that the running gear of the valves be not allowed to get out of adjustment. The rope that moves the pilot valve and those that operate the automatic stop valves must be examined frequently to see that they are in good condition and their fastenings tight, particularly as to the stop-valve ropes, because these valves are safety devices.

With the Standard plunger-elevator system in which the discharge tank is closed and a pressure is maintained therein, it is necessary that the pressure be kept up to the proper point to obtain the best results. The pressure is required to cause the water to follow up the plunger when the valve is closed suddenly in making a stop on the up trips. If the pressure is permitted to drop the plunger may be drawn away from the water in the cylinder, with the results already explained. There is no danger of getting the pressure too high, as this is limited by the hight of the inverted goose neck provided for that purpose. It is not desirable, however, to permit the pressure to rise above the proper point, because too much water will be forced out through the goose neck, and this will have to be replaced by water drawn from an outside source, which generally will be at a lower pressure; hence it will represent just so much power thrown away. It is also necessary that the supply of air in the discharge tank be well maintained; otherwise, the pressure will vary too much when water is drawn from the tank or discharged into it. Whenever the construction of the building permits, the pressure in the discharge tank is obtained by locating it at the proper elevation, as this is decidedly the best arrangement, as the pressure then cannot vary. With an elevated tank all that is necessary is to keep the water at the proper level so that the pipe running down to the cylinder may always be far enough below the surface not to draw in air.

CHAPTER XLIII

HAND-ROPE CONTROL FOR FREIGHT ELEVATORS; PUMPS AND CONNECTIONS USED WITH "SAFE LIFTERS"; LOCKING DEVICE FOR PLUNGER ELEVATORS; HAND-ROPE CONTROL

Regarding plunger elevators controlled by a simple hand rope and valve there is little information to give except in the matter of manipulating the rope. The valve proper is made substantially the same as this type of valve for other forms of hydraulic elevator, but the distance through which the hand rope is pulled to make a start or stop is slightly greater for the high-speed cars than it is with cable elevators. The reason why the hand rope has to be moved through so great a distance is not, as may be supposed, that the effort necessary to move it may be reduced, but that the valve may not be closed too rapidly by the movement of the elevator car as it approaches the upper or lower landing. In slow-running elevators the stretch of the hand rope upon which the stop balls are fastened passes through the car, and by manipulating this rope the elevator is controlled. In high-speed cars both stretches of the hand rope pass through the car and both are handled to control the movement. The advantage of this latter arrangement will be made clear by reference to Fig. 256, which is a vertical elevation of a fast-running plunger freight elevator. The stretch *B* of the hand rope is the one ordinarily used to operate the car, and this is pulled down to cause the car to ascend, and pulled up to cause the car to descend. It will be obvious, however, that if the rope has to be pulled down, say, 15 feet to make the car run upward at full speed, the operator would have a hard time doing it unless he were extremely quick in his movements; the first pull of the rope might not draw it down more than 3 or 4 feet, which would be sufficient to set the car in motion at a fair rate of speed, but not at the maximum, and the operator would have great difficulty in pulling the rope down farther because the car would be running upward. By starting the car by the aid of the stretch *C* of the hand rope the case will be very different, because this side must be pulled upward to make the car run upward; therefore, all that is necessary is to give the stretch *C* a slight upward pull, and then hold on to it until the car

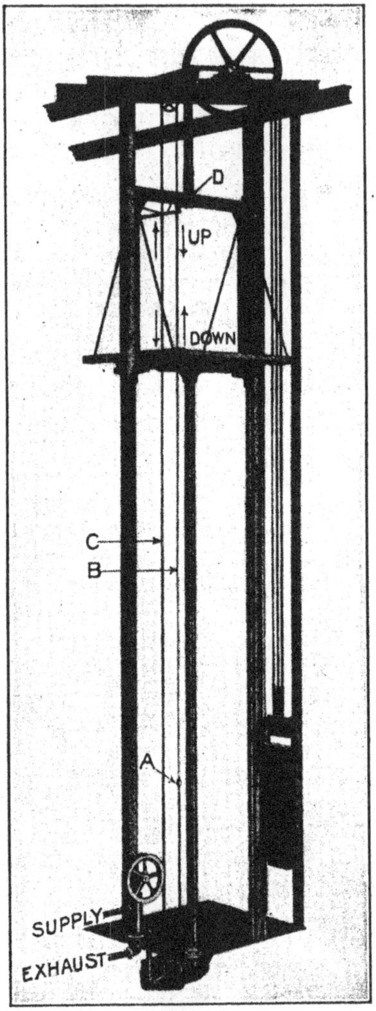

FIG. 256
FREIGHT PLUNGER ELEVATOR

attains full speed. To prevent moving the rope too far a stop is fastened on the stretch *C*, and this runs between two stationary stops set at the proper points; hence, if in starting the operator desires to run up at full speed all he has to do is to pull the stretch *C* up far enough to open the valve and then hold it until the stop on the rope strikes the stationary stop. To make a stop at any floor going downward the operation is reversed, that is, in starting, the stretch *C* is pulled down slightly and held until the desired speed is obtained, and to make a stop the stretch is grasped and held, just as in stopping on the upward trip. The stationary stops that limit the movement of the rope when the stretch *C* is held are set apart a distance equal to the combined distances through which the stop balls on the rope *B* move at the top and bottom of the well to stop the car. Thus if the top ball moves 15 feet and the bottom ball 10 feet, the stationary stops that limit the movement of the stretch *C* will be set 25 feet apart, and the stop ball on *C* will be 15 feet below the upper stationary stop when the car is standing at any floor.

"SAFE LIFTERS."

In all large buildings one of the elevators has to be designed to lift extra heavy loads, ranging from about 6000 to 10,000 or 12,000 pounds, according to the size of the building or the character of the business done by the occupants. This elevator is generally called a safe lifter, as the heaviest loads it carries are usually safes. If it were intended to carry such loads all the time it would be arranged precisely the same as the other elevators in the building except that the cylinder and the main valve would be made as much larger as might be necessary to lift the heavier load. But this elevator is only called upon occasionally to lift extra-heavy loads, and it is therefore made of the same normal lifting capacity as the other elevators, but with parts sufficiently larger than normal to give it the proper strength to carry the extra load; the increased lifting power is obtained by increasing the pressure of the water that operates it, when used to lift heavy loads. The common practice with all types of hydraulic elevator used for safe lifters is to provide a small high-pressure pump that is capable of developing the pressure required to lift the load, and this is connected directly with the lifting cylinder, so that when a heavy load is handled, all the parts of the elevator excepting the lifting cylinder and the pipes directly connecting with it are cut out of service, and are not subjected to the high pressure. The way in which the Standard plunger elevators are arranged when used as safe lifters is illustrated in Fig. 257, which shows an elevation and a plan view. In the elevation the high-pressure pump, used to lift the heavy load, is moved some distance to the right, so as to bring it out from behind the main valve and the automatic stop valves. The true position of the pump and the pipe connections between it and the lifting cylinder is shown in the plan view. The high-pressure suction pipe taps into the main discharge at the bend *D*, and the delivery pipe from the high-pressure pump connects with pipe *A* at the upper end. At the places

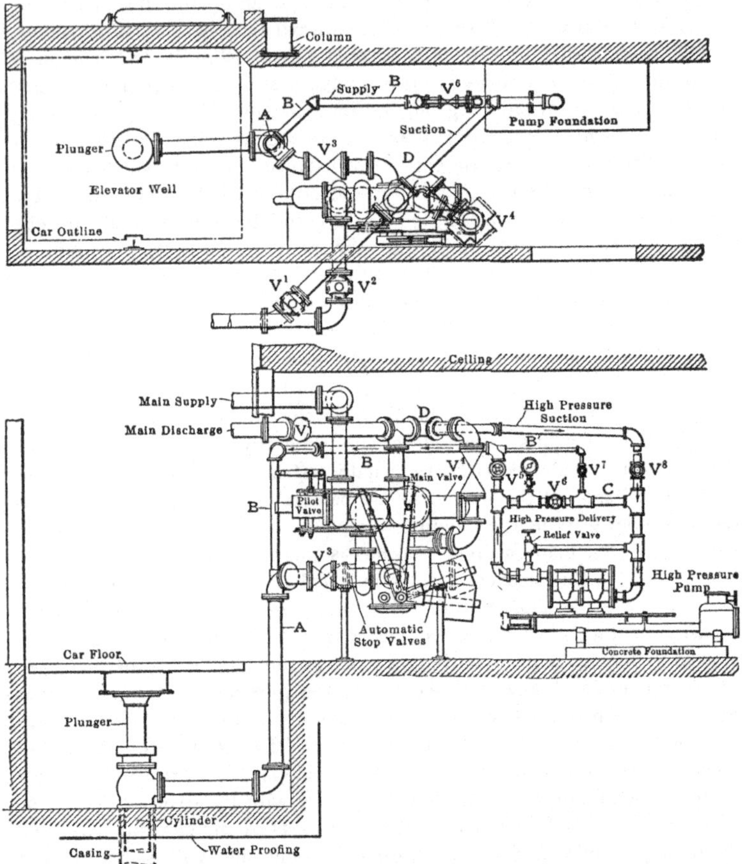

FIG. 257
PIPING PLAN FOR STANDARD PLUNGER ELEVATOR ARRANGED FOR SAFE LIFTING

marked V^1, V^2, V^3 and V^4 are located hand valves for the purpose of disconnecting the main valves from the cylinder and from the tanks. The valves V^5, V^6, V^7 and V^8 are located in the piping of the high-pressure pump, and are for the purpose of operating the elevator when used to lift extra-heavy loads. When such a load is to be lifted, valves V^2, V^3 and V^4 are closed to prevent high-pressure water from reaching the main operating valves. The high-pressure pump is started when the load is to be raised, and valves V^5 and V^8 are opened; then water is drawn into the high-pressure pump through the high-pressure suction from the main discharge pipe; and from the high-pressure delivery pipe it passes through valve V^5, and thence into pipe B through pipe A to the lifting cylinder, forcing the plunger and car upward.

As long as the pump is kept running the elevator will rise, and

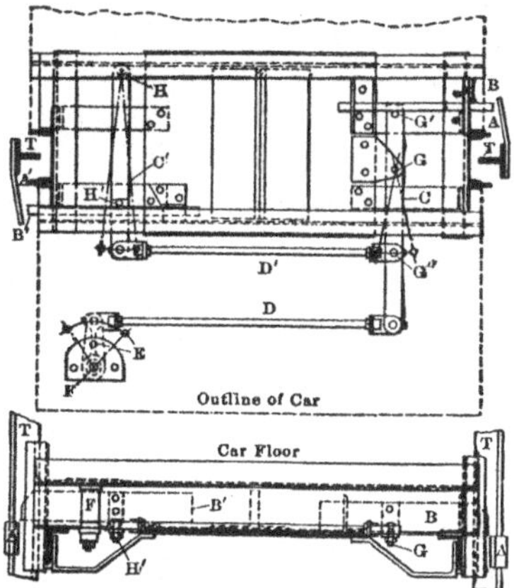

FIG. 258

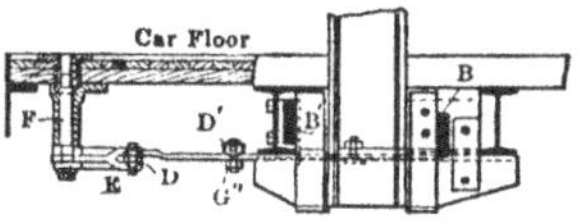

FIG. 259

when the stopping place is reached, the pump is stopped. As the movement of the car is controlled entirely by the running of the pump and the manipulation of the valves V^5, V^6, V^7 and V^8, communication of some sort must be established between the car operator and the man at the pump. This is generally done by means of electric bells or a telephone. With this method of operating the car, accurate stops at the floors of the building cannot be made at the first trial, so that the general practice is to stop the car a short distance above the floor, and then to lower it slowly to the proper position by opening the valve V^7 in the pipe B' so as to permit water to escape slowly from the cylinder. If the elevator is to be run down any considerable distance, as for example when it is lowered to the ground floor after receiving its load, the valve V^6 is opened, allowing the water to escape faster from the lifting cylinder. When the car approaches near enough to the lower floor, valve V^6 is closed and the car descends the remaining distance at a greatly reduced speed, as the water can escape only through the small pipe B' and the valve V^7. By properly manipulating this latter valve the descent of the car can be made as slow as desired. When the car is being lowered the high-pressure pump is not operated.

When an elevator is used to raise or lower a heavy load that requires some time to place upon the car, it is not advisable to depend upon the lifting cylinder to hold the car in position while the loading is going on, because during this time the car might settle enough to cause some

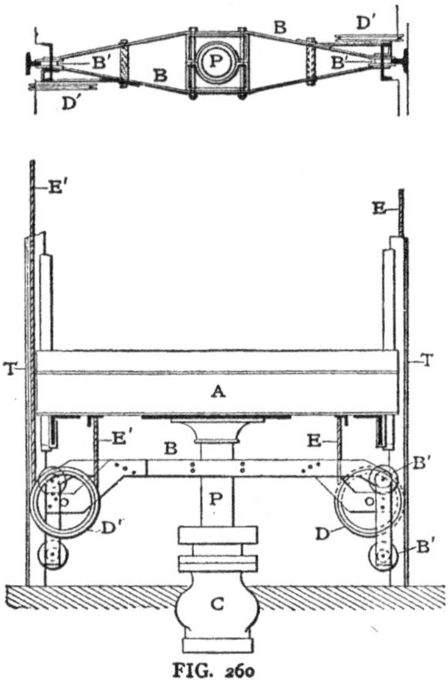

FIG. 260

trouble, if not to do actual damage. Owing to this fact it is customary to provide an elevator used as a safe lifter with a locking device at each floor that will hold the car immovable while it is being loaded. This device is thrown into action after the car has been run up a short distance above the floor, and then by opening the valve V^7, as already explained, the car is permitted to settle gradually upon the locking device. When the load is in position the first thing to do is to run the car up far enough to free the locking device, then this is drawn out of the way and the car is started for its destination.

LOCKING DEVICE FOR PLUNGER ELEVATORS.

The type of locking device used with the Standard plunger elevators is shown in Figs. 258 and 259, the former giving a plan view and a front view of the apparatus and the latter a side view. At every floor of the building strong supports A, A' are secured in the proper position upon the guide rails T, T, to hold the car level with the floor. Under the car are secured strong bars B, B' which are pushed out over the supports A, A' by means of levers connected with a vertical shaft F located in the car floor at one corner and so arranged that it can be turned by a socket wrench. On the lower end of the shaft is fastened a crank E; a rod D links the crank to a lever C, which is pivoted at G and at its end engages with a stud G' mounted on the lock bar B. A connecting bar D' is connected with the lever C by a stud G'', and through this connection any movement of C imparts motion to the lever C which is pivoted at H and moves the lock bar B' through the stud connection H'. If the shaft F is turned counter clockwise, the lock bars B, B' will be moved outward over the stationary supports A, A'. In Fig. 258 the lock bars B, B' are shown very close to the supporting pieces A, A', but when they are in their normal position they are drawn in far enough to prevent accidental striking of the stationary supports. The position of the levers C, C' is such that the shaft F can be rotated clockwise as well as in the opposite direction, and then the bars B, B' will be drawn in toward the center of the car.

When a plunger elevator is used to lift safes the compression stress on the plunger is greatly increased, as no additional counterbalance is provided to offset the weight. This extra stress is not serious in elevators of moderate rise, but when the rise is fairly great, say between 200 and 300 feet, it is necessary to provide a stiffener to reinforce the plunger and avoid liability of buckling it. The stiffener used with the Standard plunger elevators is shown in Fig. 260, which gives a side elevation and a plan view. It consists of a frame B carrying at the center a guide through which the plunger P slides, and at its ends guide wheels B', B' that run on the elevator guides T, T. The frame also carries two sheaves D, D' under which pass two ropes E, E', fastened at one end to the under side of the car and at the other end to the beams at the top of the elevator well. As the elevator runs upward, the rope ends attached to it are drawn upward, of course, and pulling the frame B upward just one-half as fast as the car moves, so that at all times the frame will be at a point midway between the bottom of the well and the car, and will brace the plunger at the central point of its exposed length.

The plungers of these elevators are made as nearly water-tight as practicable, but they are liable to be leaky sometimes. If a plunger leaks, the effect will be that the load to be raised will be increased by whatever the water in the plunger may weigh. In extreme cases, in very high buildings, the accumulation of water in the plunger may be sufficient to prevent the elevator from lifting its maximum load. If the plunger leaks, it is not an easy matter to make it tight, but it is a very simple thing to remove the water, and this should be done. The do it is to drill a hole about 1/4 inch in diameter in the lowe pipe, just above the end casting, say 2 feet above the lower pipe, and draining the water out. After the water is out th be plugged up. This is easily done by tapping the hole and a brass plug, which should be filed off flush with the plunger smooth.

INDEX

INDEX

Accidents, 105.
Accumulators, construction of, 205.
  for high pressure machines, 190.
  on Otis elevators, 205.
  reserve, 231.
Air, chamber, auxiliary, 228.
  check valve on outlet feed of, 229.
  effect in the lifting cylinder, 90.
Alinement of cylinders, 116.
Anchoring of cylinders, 114, 168.
Arrangements, diagram of, for high pressure elevators, 230.
Automatic stop-valve, 56.
  with accumulators, 228.
  adjustment of, 155.
  cleaning and inspection of, 179, 221.
  construction of, Otis type, 273.
  on Crane elevators, 159.
  danger in changing adjustment of, 224.
  derangement of, 156.
  electrical control of, 75.
  non-interfering, 75.
  on Morse and Williams, 176.
  operation of, 136, 177.
  on Otis high pressure machines, 212, 291.
  on plunger elevators, 240, 252.
  on "Standard" plunger elevators, 304.
  for stopping cars at top and bottom landings, 100.
  troubles of, 104.

Buffer, rubber, for car and for counterbalances, 242.
  construction of on "Standard" elevators, 302.
  on travelling sheaves, 138.

Cables, electric, care of, 128.
Check-valve, in outlet from air chamber, 229.
Control, 52.
  by magnetically operated valve, 66, 68, 70, 73.
  by pilot valve, 52.
Controller, floor, 76, 124.
Counter balancing, 11.
  independent, 11.
  for the lifting ropes, 25.
  in plunger elevators, 19, 265.
Cup packing, for high pressure elevators, 188.
Cushion, spring, chamber, 232.
Cylinder, alinement of, 116.
  anchoring of, 114.
  care of, 163.
  construction of, 139, 240; for high pressure machines, 194.
  Crane, cylinder, packing of, 159.
  friction in, 121.
  lubrication of, 121.

Differential valves, 98.
Double piston rods, 33.
Double power systems, 30, 79.
  Otis double power elevator, 79.
Drain pipes, inspection of, 127.
Drip-pans, 173.

Electrical, device for controlling the pump, 210.
  features of hydraulic elevators, 123.
  care of, parts, 210.
Equal pressure, device to obtain, 9.

Floor controller, non-interfering, 76, 124.
Freight elevator, simple type, the Otis, 257.

Gear ratio of elevator machines, how to determine, 6.
Gravity system, 30.
Guide shoes, 64.

Hand-rope control for plunger elevators, 313.
Hand-wheel used in place of pilot valve in Morse and Williams elevator, 184, 202.
Heavy service elevators, 232.
  operation of, 233.

High pressure elevators, 14.
  hydraulic elevators, 187.
  use of accumulators with, 205.
Horizontal cylinder machines, 21.
  compared with vertical, 23.
  description of "pushing" and "pulling" type, 131.
    operation of "pulling" type, 142.
    operation of "pushing" type, 133.
    operation of Whittier, 142.
    operation of Morse and Williams, 147.

Lifting-cylinder, design of, 292.
Locking device for plunger elevators, 318.
Low pressure system, 17.
Lower casting, design of, for plunger, 295.
Lubricating, the cylinder, 121; of a horizontal machine, 182.
  sheaves in horizontal machine, 170.
  sheaves, overhead, 173.
  slide on which crossheads run, 183.

Magnetic control of valve, 66, 68, 70, 202.
Magnets, care of, 126.
  for high pressure elevators, 204.
  removing, 128.
Main valves and pilot-control valves, arrangement of, 254.
  operation of, 135.
Morse and Williams, 147, 154, 176.

One-to-one gearing for elevators, 3.
Operating valve, construction of, 261.
  main, with Otis plunger machines, 244
Otis elevators, 30.
  double power vertical, 79.
  automatic stop-valves on, 271.
  ball and socket bearing for sheaves on, 172.
  main operating valve on, 244.
  governor on, 38.
  plunger type, 280.
  running rope system on, 85.
  simple type of, for carrying freight, using hand or rope control, 257.
  travelling sheave frames on, 113.
  vertical, 209.

Packing, for accumulator, 225.
  for cylinder on high pressure machines, 226.
  for piston, 93.
  for main and cylinder pistons of crane elevator, 159.
  for Morse and Williams piston, 150, 181.
  for plunger on high pressure machines, 226, 268.
  how to renew packing for plunger on, 267.
  for "Standard" piston, 307.
  for stuffing boxes, 96.
  for stop-valves, 101.
  for valves, 96.
  for valves, high pressure, 225.
  for Whittier piston, 169.
Passenger elevator, construction of, 238.
  "Standard", 300.
  management of, 307.
  operation of, 302.
Pilot valve, 46, 47.
  adjustment of, for speed regulation, 129, 246, 286, 307.
  auxiliary, to ease stopping of plunger, 245, 248.
  control of, 52.
  location of, how to decide upon, 129.
  magnetic, 68.
  operation of, 135, 286.
  on high pressure vertical machines, 201.
  secondary, 70.
  removal of, for adjustment, on "Standard elevators, 307; on Whittier elevators, 144.
Piping connections, 297.
Pistons, 43.
  air as a cause of "settling" of, 91.
  on Morse and Williams elevators, 150.
  packing of, 92.
  packing of, Crane type, 159.
  weighting, the reason for, 120.
Plunger elevators, 18.
  construction of plungers for, 238, 240, 242.
  high pressure, 194.
  described, 236.
  diameters of plunger, usual, on, 265.
  hand-rope control for, 313.

Plunger elevators, locking device for, 318.
management of, 307.
operation of, 302.
pilot valves on, 245, 248; the adjustment and removal of, on, 307; the packing of, on, 310.
plungers for, 26; design of, 294.
design of lower casting for plungers on, 295, 297.
design of valves for, 269.
how to remove and set valves on, 270.
weights of plungers for, 267.
Plungers, construction of, for lifts, 319.
how to disconnect, 275.
packing, to renew the, of, 267.
Pressure, arrangement to obtain equal, 9.
high, systems, 14.
low, systems, 17.
regulator, 30.
tank discharge, 232.
tank system, 30.
Pump, 231.
adjustment between pump and stop-balls, 228.
control by electrical devices, 210.
Pushing and pulling horizontal cylinder machine, 21.

Rack and pinion Valves, 263.
Ropes, care of, 264.
effect of stretching, 106.
how to shorten, 107.
in the "pulling" type of horizontal elevators, 143.
inspection of, need of continual, 106.
in Otis running rope system, 85.
replacing of, 111.
shackles for, 109.
standing or running systems, 47.
stretching of, 90, 106.
stretching of, in Whittier machine, 167.
unequal stretching of, 66.

Safe-lifters, 315, 319.
Safety, plank 37.
device, 60.
Otis, governor, 38, 60.
Settling, causes of, 90, 269.
Shackles for elevator ropes, 109.
Sheave frames, travelling, 112.
supports for, how to strengthen, 116.
Sheaves, carrier rollers for rope sheaves, 179.
construction of, in high pressure machines, 196.
for horizontal machines, 171; lubrication of, 170.
overhead, 171; lubrication of, 173.
Otis ball and socket bearing for, 172.
in "pulling" type horizontal elevators, 143.
traveling, construction of, 138.
Shoes, guide, 64.
replacing worn, on Otis plunger elevators, 274.
Simple balanced valve, 8.
Speed, controlling, of elevators by valve movement, 260.
high speed valves, 263.
regulating by pilot valves, 129.
regulators, 52, 58, 217.
Standard plunger elevator, 279, 284, 291, 292, 300, 315.
Spring-cushion chamber, 232.
Stop, automatic, at top and bottom, 20, 35, 100.
balls, 20, 100.
motion gearing on Whittier machine, 167.
valves, 101, 155.
valves, automatic with accumulators, 228.
Strainers, the importance of, 174.
Stuffing-box, casting, 268.
replacing, how to determine whether it needs, 269.

Two-to-one geared machines, 4.
Three-to-one geared machines, 5.

Upright plunger elevators, 6.

Valves, automatic stop, 56.
with accumulators, 228.
on plunger elevators, 240, 252.
on Otis elevators, 271.
comparison between, on vertical and horizontal machines, 23.
construction and operation of, 261; on Whittier machines, 144.

Valves, on Morse and Williams machines, 149, 151, 153, 184.
circulating and non-circulating, 83.
differential, 98.
freight elevators, on, 291.
high-speed, 263.
leak in, caused by settling, 91.
magnet, control of, 66, 68.
main, operation of, 135.
modification in, controlled by magnet, 202.
on Otis machine, 244; latest design, 249.
Valves,
operation of pilot, 135; on plunger elevators, 287
Valves—*Continued.*
packing of high pressure, 225.
pilot, and main, control, arrangements for, 254; on high pressure machines, 198.
piston, on plunger elevators, 253.
rack and pinion, 263.
removal and setting of, 270.
relief pipe for discharge, on, 173.
simple balanced, 8.
stop, 100.
speed regulating, 58.
Water-balance elevator, 3.
Whittier Machine Co.'s "pulling" type elevator, 142, 147, 165, 215; stop valve mechanism, 166, 215.

Spotted something that isn't quite right?

Let us know and we'll fix it for future reprints.
And as a thank you for your diligence, we'll
send you a free Sothis book of your choice.

publisher@sothis.press

This page intentionally left blank